COASTAL HEROES

Miles O. Hayes

With Contributions from Stewart Farrell, Tim Kana,
and Duncan FitzGerald

July 2011

PANDION BOOKS
a division of Research Planning, Inc.
P.O. Box 328
Columbia, South Carolina USA 29202

Email: permissions@researchplanning.com

A CIP catalog record for this book has been applied for from the Library of Congress.

ISBN 978-0-9816618-2-7

All photographs are by the author unless otherwise indicated.

Illustration and design by Joseph M. Holmes.

SUSTAINABLE FORESTRY INITIATIVE
Certified Fiber Sourcing
www.sfiprogram.org

The paper used in this book meets the minimum requirements of the American National Standard for Information Sciences – Permanence of Paper for Printed Library Materials, ANSI/NISO Z39.48-1992

Printed in Lillington, North Carolina USA by Edward Brothers, Inc.

Cover Illustration: COASTAL HEROES Victoria Byrne (on the back), John Gifford, and Walt Handy running a beach profile on Plum Island, Massachusetts circa 1969.

DEDICATION

During January/February 2011, while I was putting the finishing touches on the rough draft of this book, came the sad news that two of the more influential people in my professional career, John Robinson and Maylo Murday, had died. This book is dedicated to their memory.

John, who died of a brain tumor in February 2011 while in his late 60s (see accompanying photograph), was the most influential single person in getting us into the oil spill business. John was an engineer who got his start with NASA. He was always an unusually creative and innovative manager. Not only that, he had such an engaging management style that the people who worked for him were totally devoted to his vision. *Vision*, that's the word that best fits John Robinson and his contribution to the oil spill response business. He had the vision to start and nourish NOAA HAZMAT, the most experienced and effective scientific spill response unit in the country and, I assume, in the world. He was also responsible for the creation and management of the NOAA *Mt. Mitchell* cruise to the Arabian Gulf to determine the impacts of the Gulf War Spill of 1991. This cruise, and the associated studies on land, in which both my business partner (and wife) Jacqui Michel and I participated, was one of a kind.

We obtained our first contract with John in 1978, while I was still at the University of South Carolina. I remember the almost daily phone conversations I had with John in the first year or two as the program was getting organized. Our contract with NOAA HAZMAT, which was eventually transferred to our private science technology company, Research Planning, Inc. (RPI), continues even until this day.

As John's former associates were commiserating on our loss, an e-mail came to us from Brad Benggio, a NOAA Scientific Support Coordinator, who will appear more than once in the chapters that follow, which said: "Hopefully, we can all remember John and what he brought to HAZMAT, and the individuals he inspired. I hope our next leader has some of John's abilities to lead and inspire us as we head into the next era with vision and purpose."

I first met Maylo Murday, a native of the Republic of Mauritius, in Lomé, Togo in 1979. He eventually became a main partner in our adventures in West Africa, which are described in Chapter 13 – WEST AFRICA. In fact, in reviewing my write-up of that chapter, I found that I had used the word Maylo twenty four times. He was the office manager of our company RPI/Nigeria in Port Harcourt, Nigeria for several years. Later, he joined our staff in Columbia, South Carolina, and earned his Ph.D. at the University of South Carolina.

On a personal level, I found Maylo, a follower of the Baha'i religion, to be one of the few truly religious people that I could have a meaningful conversation with. He told me that Baha'is believe in the unity of all humankind and, therefore, they reject prejudice–racial, political, or otherwise–and stress ethical teachings, such as world peace, education, and sexual equality. If, for some bizarre reason, I ever felt the need for a religion, this one would be my choice.

Recently, Tim Kana and Jacqui were beginning to organize a gathering of my ex-students and associates at the University of South Carolina in October 2011. In trying to locate Maylo, Jacqui contacted his ex-wife Jeanne, who responded, "I have the sad duty to tell you that Maylo passed away in September of 2002. He always felt a great debt of gratitude to Miles and you, and remembered his time at the University of South Carolina and RPI fondly. He spoke of you both often."

He was a very special friend to me during those times.

TABLE OF CONTENTS

LIST OF FIGURES

LIST OF COLOR PLATES

(midsection)

FOREWORD: WHO AND WHY

Our greatest glory is not in never failing,
but in getting up every time we do.

CONFUCIUS

When I was a child, I spoke as a child, I understood as a child, I thought as a child: but when I became a man, I put away childish things.

1 CORINTHIANS 13:11

It ain't the parts of the Bible that I can't understand that bother me, it's the parts that I do understand.

MARK TWAIN

The major purpose of this book is to present the significant aspects of how coastlines evolve, stressing some original ideas regarding the origin and morphology of the coastlines of the world that my students, co-workers, and I have made over the years. Our chosen profession is coastal geomorphology, or, some prefer to say, coastal geology. Also, with most of the ideas or projects presented in the different chapters, side stories are told to present the history of their development, as well as an introduction to the reader of the diverse and unforgettable people–scientists, students, and otherwise–involved. I have been lucky enough to experience a scientific career that has lasted over 50 years, involving field projects on all the major continents except Australia. I also have conducted studies near the Magnetic North Pole and the South Pole, and along the entire coastline of Alaska. In addition to those areas, most of the shoreline of the Arabian Peninsula, the coast of West Africa, and many other areas (in 42 countries and still counting) have been investigated.

A pertinent question relative to this book may be why does one become a scientist in the first place, particularly in a relatively esoteric field like coastal geomorphology? Neither of my parents attended high school, with the closest person to a scientist in my entire family, or close friends and acquaintances in the small community of Oakley in the mountains of North Carolina where I grew up, that I knew of being my older brother, Edwin. He obtained an electrical engineering degree at Indiana Tech under the GI Bill as a result of his years in the U.S. Air Force during World War II.

Most geologists that I now know were drawn to the field by a love of nature and the outdoors, the same calling I had. My father was an ardent naturalist in his own way; he knew every bird song, the names of all the trees, and so on. But my first love was fishing, which was engendered early during fishing trips with my Grampaw Worley and Uncle Luster.

Once when I was about twelve or thirteen years old, a textile-mill hand named Hoojer Smith and I were on a two-mile walk along a railroad track on our way to fish the Swannanoa River. As we plodded along the crossties, Hoojer asked, "Hey Murse (one of my many nicknames, short for Muscles), what you gonna do when you graduate from high school? Darlene (his ethereal wife) says your Granny is always braggin' about what a good student you are."

I answered, "Well, Granny and her sister, my Aunt Marthy, have it in their heads that I'm going to make a preacher. But, except for Brother Hamby, I've never met any preacher that I could even stand to be around, much less model my life after. I'm going to college. Going to be a scientist. Decided that yesterday, while I was collecting some books out of a big pile of discards from the school library that they threw out behind the annex."

"What books?"

"Old books. One by a geologist named Lyell written in the 1800s. You should see the pictures and sketches of fossils in that book. But the one I liked best was called *Physiography of North America*, or something like that[1]. It had lots of stuff in it on rivers. I want to study rivers."

"Hell, that's about all you do now, ain't it? That and play baseball. I see you and Marvin Jones and some of your other pals heading for the river about three or four days out of the week."

Make a preacher? That is a fairly common occurrence in the Bible belt that surrounds Asheville, North Carolina, the town near where I grew up in a family of born-again Southern Baptists.

Without much further discussion of this matter, I will just point out that, despite my aversion to preachers in general, my best friend, Adrian House, and I were "born again" together ourselves at age 15 or so, becoming full-fledged members of the Oakley Baptist Church. By the way, we were not "prayed through" by the preacher. That was done by Tommy Butler, our Sunday School teacher. During the years Adrian and I were in high school, the Oakley Baptist Church became our home away from home, that and the ball field and the rivers and the mountains.

About three years after Adrian and I found God, the summer after I finished high school, I was invited to go to the Oakley Methodist Church by the man who organized and ran the Christian Youth Fellowship of Oakley (boys only). The youngsters in that assembly all attended the Methodist and Reed Memorial Baptist Churches, with none being from my "lower classed" Oakley Baptist Church. Therefore, I was normally not a part of that group but, because I had been valedictorian of my class of 36 stalwarts, the leader wanted to show the assembled that just because you were a boy growing up in Oakley, you didn't necessarily have to go around acting like a dumbass. I was the first male valedictorian in the school in fifteen years. Actually, I don't think I deserved the honor *per se*, I think it was a tie. That is, if the recognition was based on grades. Wonder of wonders, there was another boy in my class named Ian McCleod, who made good grades too. Unlike me, Ian acted like a true intellectual, not bathing too often, building his own TV sets, and generally making others look bad if they challenged his thinking capacity. Because I was a member of the baseball and football teams, I could bathe regularly, after each workout, even in the wintertime. At this point, I probably should mention that in my early years, I lived in a four-room house without indoor plumbing, which meant the old Saturday night bath in a galvanized washtub.

Anyway, while getting these excellent showers in the locker room at school, I worked very hard at being humble, a practice I abandoned as soon as I received my Ph.D. (explained in more detail later). Thus, I'm pretty sure that the Oakley School faculty decided to give the valedictorian award to me alone and not include Ian, because they feared what he would say in the valedictory speech at the graduation ceremony. To no one's surprise, Ian moved away and became a very successful engineer in Ohio. In other words, he was not "too dumb to make it in no northern town," as was implied about the boys from the South in Randy Newman's great song, *Rednecks*. In the ACKNOWLEDGMENTS, I give a brief introduction of a couple of my great high school teachers.

Okay, so I was named valedictorian of the class by the teachers, but what about my 35 fellow students, who voted on certain honors for the graduating seniors? Clearly, I was not in the running for either BEST LOOKING or MR. PERSONALITY and, in fact, not even for MOST ATHLETIC, because ours was a basketball school, and I never played one minute of basketball in my entire life. Also, several of my classmates were three-letter men (football, basketball, and baseball). Of course, my buddy Ian was voted MOST INTELLECTUAL. What was left? How about MOST LIKELY TO SUCCEED? Guess I had to win something.

I left the North Carolina mountains and attended Berea College in Kentucky on a work scholarship. At first, I majored in agriculture, taking courses in animal husbandry and agronomy. In my third semester, however, I enrolled in an introductory geology course, which brought back memories of the books by Lyell and others that I appropriated from the abandoned pile behind the Oakley School annex. I was hooked and took a major in geology, which became my passion for the next three years, that and playing baseball, where I was a four-year letterman (and starter) on the baseball team that was coached by the one and only Monarchy Wyatt.

The biggest shock of being at Berea College was the realization that I wasn't necessarily the smartest kid there. Almost everybody else was also the valedictorian of their small high school classes back home in the mountains. It was tough at first, but I learned to study harder and by the time I was a junior, I was doing okay.

From Berea, I went to Washington University in St. Louis for a Master's degree in geology, and then to the University of Texas, where I was awarded a Ph.D. degree in Geology and Marine Science in 1965. After that, I was a professor of Geology at the University of Massachusetts (1964-1972) in Amherst, Massachusetts, and from the fall of 1972 until 1983, I was a member of the faculty of the Geology Department at the University of South Carolina. Presently, I am semi-retired, but I am also Chairman of the Board at Research Planning, Inc. (RPI) in Columbia, South Carolina (more about this company to follow).

So, I became a scientist all right, but what happened to the "born again" Christian who left the North Carolina mountains? I guess a lot of scientists evolve, as I have, when it comes to matters of faith. Faith and science are ancient enemies, I guess you could say, so maybe that's why I abandoned the monotheistic religions somewhere along the way.[2] After being a self-proclaimed agnostic for over thirty years, I spent some time studying Buddhism, but The Way appears to be a tough road for me to follow. Be that as it may be, I now read Thomas Paine, Christopher Hitchens, Michel Onfray, Michael Parenti, Sam Harris, Daniel Dennett, Richard Dawkins, and puzzle a little bit over the works of the great Spinoza, Hesse, and Nietzsche. I have also read all of the "Holy Bible," parts of it many times, as well as a good bit of the Quran and the Book of Mormon, but that's another story.

A few years ago, I wrote a book entitled *Black Tides*[3] that was published by the University of Texas Press. In that book, I focused on the work that my colleagues and I carried out in the field of oil spill response, starting with a study of an oil spill in 1975 caused by the wreck of the tanker *Metula* in the Strait of Magellan, Chile, and concluding with work on the "mother of all spills" along the coasts of Kuwait and Saudi Arabia during the first Gulf War in 1991.

In this book, in addition to some in-depth discussion of how the coast was made, I make a special effort to recall the natural spectacles that one is bound to experience during pursuits such as camping on the ice in Antarctica and in the wadis of the Oman desert, crashing an airplane in the frigid waters of Alaska, mapping the shoreline of the Gaza Strip in Palestine, and so on. Also, I have made a conscious effort to introduce to the reader some of the unusual characters I've been lucky enough to meet over the years, namely great scientists, mountain-climbing bums, crazy bush pilots, recluses that live alone in the woods, and even a few academicians of note.

Chapter 1 is a discussion of my dissertation study of the geomorphological impacts of Hurricane *Carla* on the South Texas coast in 1961. After covering activities carried out in several different areas around the world in Chapters 2-17, three of the last four chapters discuss projects on the Arabian Peninsula: 1) A shoreline erosion survey on the coast of Oman (Chapter 18. THE SOUTHERN CROSS); 2) A mapping project on the coast of the Gaza Strip (Chapter 20. THE ENVIRONMENTAL SENSITIVITY INDEX (ESI)); and 3) the last chapter in this book (Chapter 21. NIGHTMARE IN JUBAIL CITY), which focuses on the greatest of all my field projects, a survey of the impacts of the world's largest oil spill on the Gulf shoreline of Saudi Arabia, which was carried out in 2002-2003.

Every man is a creature of the age in which he lives, and few are able to raise themselves above the ideas of the time.

FRANCOIS MARIE AROUET (VOLTAIRE) in ESSAI SUR LES MOEURS ET L'ESPRIT DES NATIONS, 1756.

We are not here concerned with hopes or fears, only with the truth as far as our reason permits us to discover it.

CHARLES DARWIN

These quotes are important in the context of this book, because they represent why the study of evolution, history, philosophy, and even coastal geomorphology are important because, if we do this (rise above prevalent ideas), it allows us to escape the prejudices of our time. Hopefully, I did this, in my own small way. In fact, so small as to be miniscule compared to the intellectual giants such as Voltaire, Spinoza, Nietzche, Jefferson, Darwin, and Einstein. Okay, so don't laugh yet because, obviously, I still think this a narrative worth telling, so read on.

(MOH on 4 July 2009–Looking out the window at the beautiful wooded peaks of the Appalachian Mountains that surround our mountain home in western North Carolina).

ACKNOWLEDGEMENTS

First and foremost, I wish to acknowledge my wife (and business partner) Jacqui Michel (JM) for suggesting that I write this book in this particular format (i.e., two sections for each chapter – The Science and The Story), and for her careful editing of the original manuscript, which has benefited greatly as a result of her help. Joe Holmes, RPI's cartographer and illustrator, has done an outstanding job putting all of the illustrations in their final form. Wendy Early, also at RPI, is acknowledged for her editorial and production efforts on the finished product. Tim Kana, Stewart Farrell, and Duncan FitzGerald are also acknowledged for contributing stories for the book.

This text was also aided greatly as a result of reviews by Stewart Farrell, Al Hine, Tim Kana, Orrin Pilkey, and Walter J. Sexton. All of their suggestions are much appreciated. However, I should add that I take full responsibility for any of the scientific discussions or personal accounts that are given here.

The primary theme of this book concerns the efforts of the graduate and undergraduate students and other research associates who endured my rants and raves in my research groups and programs over the years. The graduate students who received their degrees at University of Massachusetts (UMass) are listed in Appendix A. These graduate students, as well as several of the undergraduate students at UMass, are also listed and discussed at some length in Chapter 2 – ON THE ICE, Chapter 3 – THE FIELD TRIP, and Chapter 4 – *SUZANNE,* EARTHQUAKES, AND TSUNAMIS.

More specifically, in addition to the UMass graduate students listed in Appendix A, there were a number of undergraduate students who participated in the program. Victoria Byrne, Walt Handy, and John Gifford, who are pictured running a beach profile on Plum Island, Massachusetts (circa 1969) on the cover of this book, were in this group, which also included Robert Bozeman, Karl Geller, Linda Sandanato, Philip Durgin, and several others. There were three undergraduates on the coastal-zone field team surveying the south coast of Alaska in the summer of 1970 (discussed in Chapter 4). They were Bob Henry, Frank J. Raffaldi, Jr., and Paul Hague, an ex-Navy, more senior undergraduate student, who became my lifetime friend, and about whom you will hear several times more in the following pages. Woody Hobbs, who did an M.S. thesis, was also on that field team. John Gifford and Bob Henry wrote senior honors theses, both of which were well done. At UMass there was only one "repeat offender" (i.e., someone who did both their M.S. thesis and Ph.D. dissertation in my program), Stewart C. Farrell, who went on to become, and remains, a Professor of Marine Science at Stockton College on the New Jersey shore. Also, he is presently the Executive Director of the Richard Stockton College of New Jersey Coastal Research Center.

The graduate students who received their degrees at the University of South Carolina (USC-East) are also listed in Appendix A. These graduate students, as well as several research scientists and technicians in the South Carolina programs, are discussed at some length in Chapter 5 – KIAWAH ISLAND, Chapter 6 – BRIDGES OVER TROUBLED WATERS, Chapter 7 – THE MEN THAT DON'T FIT IN, Chapter 10 – PRICE INLET, and Chapter 14 – THE ROCK RECORD.

Perhaps special mention is in order for those graduate students who earned their M.S. theses based on work carried out at UMass (or did a significant amount of research there), and then made the trip to South Carolina where they completed their Ph.D. dissertations at USC. This august group includes Jon Boothroyd, Al Hine, Dennis Hubbard, Ray Kaczorowski, and Ed Owens.

There were also four "repeat offenders" who received both degrees at USC–Tim Kana, Larry Ward, Dennis Hubbard (based partly on work carried out while at UMass), and Chris Ruby.

In addition to the graduate students at USC listed in Appendix A, there were a number of undergraduate students who participated in the program, who should be mentioned here, including Mike Waddell (who also worked with us as he completed his graduate studies), Leita Jean Hulmes, Steve Wilson, Mark Cable, Jane Zenger, and Fearless Frank Lesesne.

Special mention should also be made of four of the research scientists in the Coastal Research Division (CRD) at USC–Dag Nummedal, Mike Stephen, Pulak Ray, and Gary Zarillo. They worked at CRD for several years, not only assisting in maintaining the research production, but also playing a major role in the educational development of the graduate students. Larry Ward and Tim Kana also served in this capacity for a period of time.

Fellow scientists who worked specifically on the coastal research programs at RPI, whose long-term efforts to keep the company moving are greatly appreciated, include Jacqui Michel, Tim Kana, Erich Gundlach, Dan Domeracki, Walter J. Sexton, John Barwis, Chuck Getter, Chris Ruby, Pete Reinhart, Chris Reel, John Hodge, Geoff Scott, Maylo Murday, Linos Cotsapas, Todd Montello, Craig Shipp, Bob Stein, Colin Plank, Tom Freeman, Scott Zengel, Ken Finkelstein, Steve Covell, Zach Nixon, David Betenbaugh, Mark Jordana, Rick Kennan, Jason McKinlay, Travis Darden, and Dallon Weathers.

In this book, I talk rather extensively about the field seminars we conducted on the South Carolina coast for oil-and-gas geologists during the thirty-year period between 1976 and 2005. Chapter 14 – THE ROCK RECORD is devoted to this topic. Almost all of the graduate students in the program at USC-East worked on this course from time to time, with Tim Kana and Walter J. Sexton playing primary roles as instructors and logistics providers for many years. Others who deserve special mention include Ailene Duc, Mike Bise (master trench supervisor), Joe Holmes, Tim Eckard, Ray Levey, C.Y. McCants, Chris Reel, Helen Mary Johnson, Chris Reel, Doug Imperato, Bo Tye, Craig Shipp, Pete Reinhart, Chris Ruby, Jim Saad, Dallon Weathers, Tom Freeman, and Norm Dodson.

Although not directly involved with coastal research, I was very lucky indeed to have had a number of brilliant and creative associates and partners in the RPI companies devoted to research (as service companies) in the oil-and-gas industry (RPI/Colorado, RPI/Canada, RPI/Pacific, and RPI/Texas). These include John Horne, Dan Domeracki, Tom Ryer, Skip Rhodes, Tor Nilson, Alan J. Scott, Dave Noe, Clarence Campbell, and Hugh Monroe.

Next, I want to touch briefly on my personal educational background, for which I am indebted to a large number of outstanding teachers. It is kind of sad for me to witness the rapid rate of deterioration of the public schools in the United States, particularly in the state of South Carolina where I now live, because I and the erstwhile 35 other members of my graduation class were extremely lucky to graduate in the late spring of 1952 from the excellent Oakley High School, located on the outskirts of the town of Asheville, North Carolina. Things were different then. For example, my high school teachers were the pillars of the community. Many of them were rich too, owning apple orchards, a lake house, and so on.

In the FORWARD, I mentioned how the discovery of some ancient geology books being discarded from the school library initiated my first interest in the subject of geology, which eventually became my life's profession. Beyond that, however, were the great teachers in the Oakley School. Four years of English classes under Mrs. Blankenship provided a foundation of learning for which I have been forever grateful. Then there was Mr. Wells, ex-Navy captain in World War II, who taught history and chemistry. I will never forget when in our ninth grade history class he read to us the entire text of the Robert Service poem, *The Cremation of Sam McGee*, one of the many ways he tried to broaden our interests beyond the limited horizons of the mountains.

Later, after I had left for college, Mr. Wells became a pioneer in the industry of raising Christmas trees, a wide spread business throughout the North Carolina mountains at this time. I had several other very good teachers in high school, but Mrs. Blankenship and Mr. Wells are the two that I remember most.

I also had great teachers at Berea College. At Washington University in St. Louis, I studied under the best teacher I ever had, James C. Brice, who gave me a solid foundation in sedimentology and geomorphology that has stood the test of time. In Chapter 1 – THE HURRICANE, I mentioned the three widely heralded scientists that I had the privilege to work under, Robert L. Folk, the renowned sedimentologist, W. Armstrong Price, the father of modern coastal geomorphology, and H.T. Odum, the ingenious marine ecologist. The four years plus that I was in the Geology Department at the University of Texas I felt that I had the free rein to pursue any new idea that came along. Always with the encouragement and cooperative spirit of Professors Alan J. Scott and Earle F. McBride, I had the freedom to learn and explore new research directions. Office mates Murray Felsher and Peter B. Andrews, and other fellow graduate students, including my long-time friend Richard A. Davis, Jr., were equally excited to be where we were, and it was a great experience. We were lucky indeed to have come along at just the right time.

1 THE HURRICANE

When I went to UMass and was one of the few professors there (in Geology) who was not a graduate of an Ivy League University, I was quite proud to say that I was a graduate of the University of Texas.

MILES O. HAYES

The Science

Imagine, if you will, that you are a serious, intellectually curious 26-year-old graduate student with an M.S. degree, who finds himself at a major University working under one of the world's greatest scientists in his field which, in this case, happens to be a combination of coastal science and sedimentology. You have worked for over a year and a half on your Ph.D. research, which included spending a considerable amount of time in the field, in this situation on the coast of Texas. Numerous sediment cores and samples have been taken along the shoreline, and sampling cruises on the continental shelf aboard the University's vessels have been completed as well. Everything is proceeding according to plan. Then, suddenly, your dissertation area is struck by one of the largest hurricanes of the century, which basically completely reorganizes the geomorphology of your study area, destroys much of the samples and data you have collected, and generally wreaks havoc along the coast. Your research up to that time is almost completely wiped out. This is what happened to me in September 1961, and it was the best piece of good luck that I ever experienced in my entire professional career. How so? That story follows.

In late August of 1959 my bride of a little over a year, Barbara Anne Hyder, and I packed our belongings in our 1951 Ford and drove to Austin, Texas, where I had gained provisional admission at the University of Texas (UT), with hopes of beginning work on a Ph.D. degree in Geology. I had just concluded the fieldwork on my M.S. thesis at Washington University (WU) in St. Louis on some sandstones in western Missouri, about which I eventually published a paper in the AAPG Bulletin (Hayes, 1963). While at WU, I was lucky enough to study under the best teacher I ever had, James C. Brice, whose teaching established a foundation in geomorphology and sedimentation that would be of great service to me throughout the rest of my scientific career.

The initial day I walked into the Geology Department in Austin, the first person

I met with was a graduate student named Tom Freeman, who showed me around the Department and gave me some friendly advice. After graduating with his Ph.D., Tom went on to become a well-known carbonate petrologist and Professor at the University of Missouri. In the fall of 1980, his son, Thomas, became one of my graduate students (for an M.S. degree) at the University of South Carolina. As noted in the discussion in the Chapter 18 – NIGHTMARE IN JUBAIL CITY, Tom's son, whom I always called simply "Freeman," played a prominent role in our follow-up studies in 2002-03 of the Gulf War oil spill on the Saudi Arabian Coast.

Once at UT, my plan was to work under a brilliant, up-and-coming sedimentologist named Robert L. Folk, but first, I had to take a general geology entrance exam administered to all the incoming graduate students. Folk said my score was "highest of all the imports," so I guess that was one of the reasons he took an interest in me, finding me a job as a student researcher at the Defense Research Laboratory (DRL) on the UT campus. This appointment was very helpful, considering that I had no other visible means of support at that time, except for Barbara Anne's new job as a medical technologist at a local hospital. As time passed, I also eventually received the usual oil-company fellowships that UT grad students were prone to land, and served as Folk's teaching assistant in a couple of his graduate courses.

Altogether, I spent five years at UT, splitting my time between Austin and Port Aransas on the coast (see map in Figure 1). During much of that period, I also continued to work part time, up to thirty hours per week, as a research scientist at the Defense Research Laboratory (DRL) in Austin. At DRL, I worked on a "classified" anti-submarine warfare project, in which I studied SECRET data and other information on the shorelines of the whole world. This project was probably the most significant learning experience of my life, as far as future research on coastal geomorphology was concerned. Through this work, I was able to see global trends in shoreline and continental shelf morphology and sedimentation that would guide my thinking in all future research projects (Hayes, 1964c; 1967c). Especially noteworthy was the observation of the effect of tides on the morphology of depositional coasts, such as is illustrated in Figure 2A. I later summarized these observations in a field trip manual called *Terrigenous Clastic Depositional Environments* (Hayes, 1976), in which I noted that: a) River deltas and barrier islands are best developed on microtidal shorelines (tidal range <6 feet); b) Salt marshes are most abundant in macrotidal shorelines (tidal range >12 feet); and c) Tidal inlets and tidal deltas are more abundant on mesotidal shorelines (tidal range 6-12 feet). These shoreline types are illustrated in Figure 2B.

[NOTE: Throughout this book, I have used both English and metric units, sometimes interchangeably. For the most part, I used the system that we most frequently used at the time when the described event occurred. Or, as Tim Kana (pers. comm.) suggested, perhaps this merely reflects the schizophrenic practices we so often employ in our business.]

At the same time, I began a study of the coastal environments of the south Texas coast under Folk's supervision. The advent of Hurricane *Carla* in September 1961, at which time my study on the coast was underway, provided for me the opportunity to observe and record the impacts of a major hurricane on a coastal plain shoreline, the first major study of its kind, which was aided significantly by the abundance of detail on that particular shoreline that was available to me. My

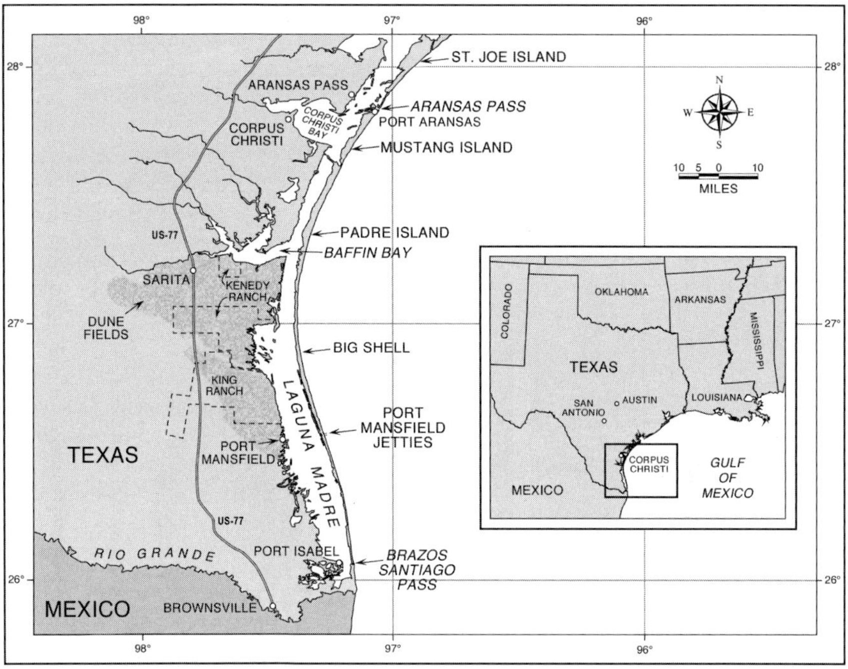

FIGURE 1. The south coast of Texas.

results were first published in a field trip guidebook in 1964 (Hayes, 1964a) and in my dissertation in 1965. Two later, more broadly distributed, publications were published in 1967. I can now immodestly claim at this late date that this study has been frequently referred to in the literature as the "classic study" on this topic. Some of the major findings of that study follow:

- A detailed account of the dynamic processes associated with hurricanes, emphasizing storm surges and wave action. The global causes of such storms and their common, historic paths over the Texas coast were also covered in detail;
- Overall assessment of the geomorphological impacts of the storm, such as the cutting of 16 new deep channels through Padre Island, with detailed observations on the washover fans thus created, as well as the history of the channels' closings (see examples in Figure 3);
- Erosion of the beachfront and the history of its recovery;
- Erosion of the coastal dunes and their patterns of recovery; and
- Overall sediment distribution patterns created by the storm.

Because I had taken sediment cores on the inner continental shelf in the months preceding the hurricane, I was able to discover the impact of the storm on the shelf sediments by repeating a significant number of the cores. I discovered that a widespread and moderately thick sand layer was deposited on top of the

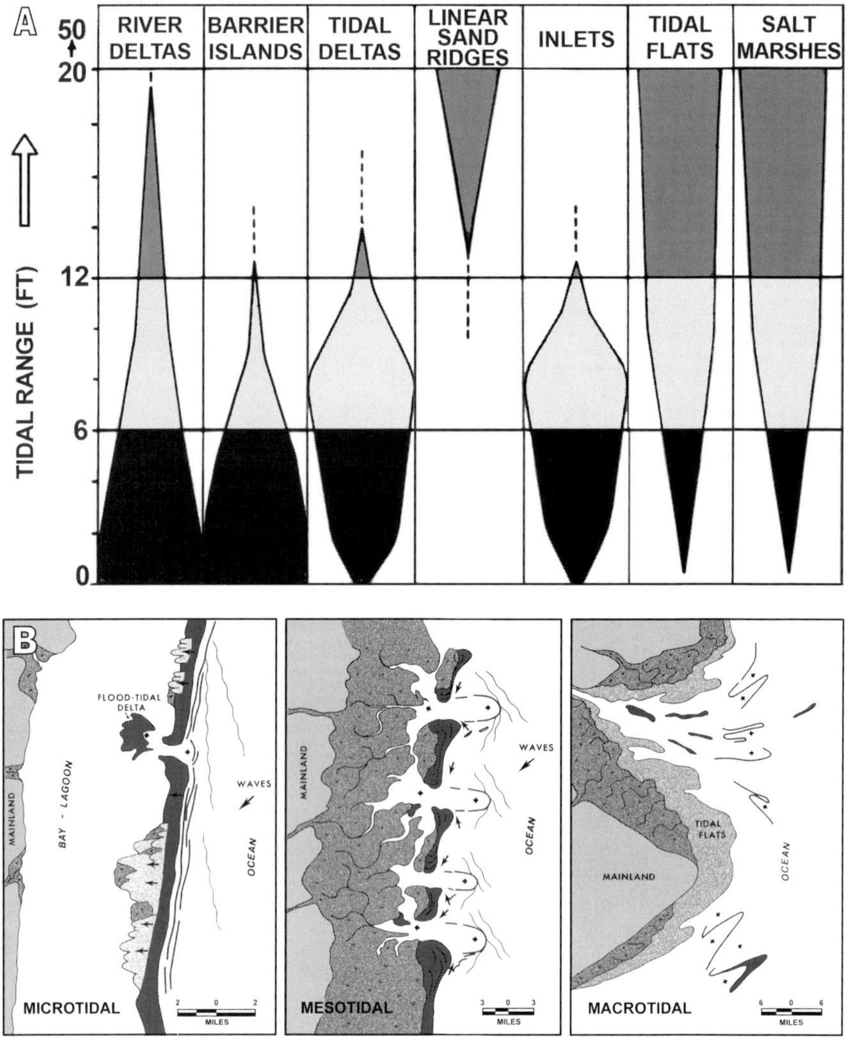

FIGURE 2. Tides and shoreline morphology. (A) General distribution of shoreline types on depositional coasts related to tidal range. Based on a study of the world's shoreline carried out in the early 1960s while I was an employee of the Defense Research Laboratory at the University of Texas. From Hayes (1976). (B) Generalized models of the three basic types of depositional coasts (excluding major deltas). (Left) Microtidal (tidal range less than 6 feet, usually wave-dominated); (Center) Mesotidal (tidal range = 6-12 feet, usually mixed-energy); and (Right) Macrotidal (tidal range greater than 12 feet, usually tide-dominated). From Hayes and Michel (2008).

previously muddy bottom by currents directed offshore as a result of the receding storm surge, which I concluded was deposited by a bottom-hugging density current (illustrated in Figures 4 and 5). This discovery led to numerous papers citing similar types of sediment layers, called *tempestites* by some authors, in the rock record, most notably in the Cretaceous rocks of Alberta, Canada. Also,

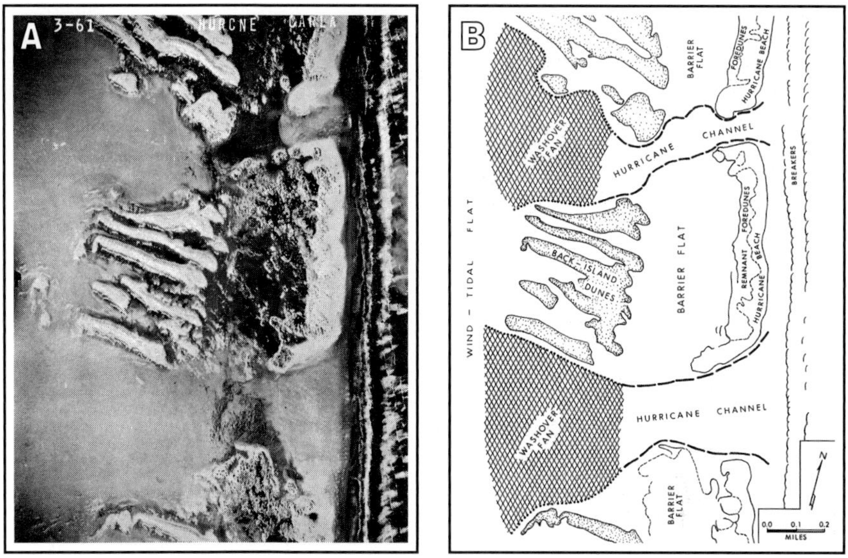

FIGURE 3. Hurricane channels and washover fans activated on Padre Island, Texas by Hurricane *Carla*, September 1961. (A) Vertical aerial photograph of a portion of Padre Island, located about three miles south of the Port Mansfield jetties (Figure 1), illustrating the profound effect of Hurricane *Carla* on the island's morphology. Photograph was taken on 3 October 1961, three weeks after *Carla* crossed the coast (courtesy of the Galveston District, U.S. Army Corps of Engineers). (B) Tracing of photograph in A. Two large hurricane channels and washover fans are shown. Other morphological features shown are hurricane beach, remnant foredunes, barrier flat, and back-island dunes. Note the long, E-W trending seif dunes in the back-island dune fields. From Hayes (1965).

some later workers challenged the notion that storm-surge ebb currents were responsible for the new layers, relating the cause to normal storm-wave-generated processes. However, studies of the shelf sediments off the South Carolina coast after Hurricane *Hugo* in 1989 by Gayes (1991) showed that the ebb of the storm surge played a major role in the occurrence of the post-*Hugo* scour channels and sediment layers on the shelf. In short, I was right all along. So there![4]

The original theme of this dissertation was centered on doing a large number of grain size analyses of the sediments in the area, which I did. One of the results of this activity is given in Figure 6, which shows that the sand on the dunes of the backbeach of Padre Island is bimodal, with a coarse mode being derived from the Rio Grande to the south, and the finer mode from other rivers to the north. These results not only showed the futility of using grain size analyses of sand to determine environment of deposition in the rock record,[5] but it also clearly showed that there is a convergence of longshore sediment transport directions in the middle of Padre Island.

The Story

(October 1959 - Corpus Christi Bay, Texas)

The lights of the crowded van faded abruptly into the fog bank in front of us as we bounced along in the dark on the narrow, one-way rickety wooden

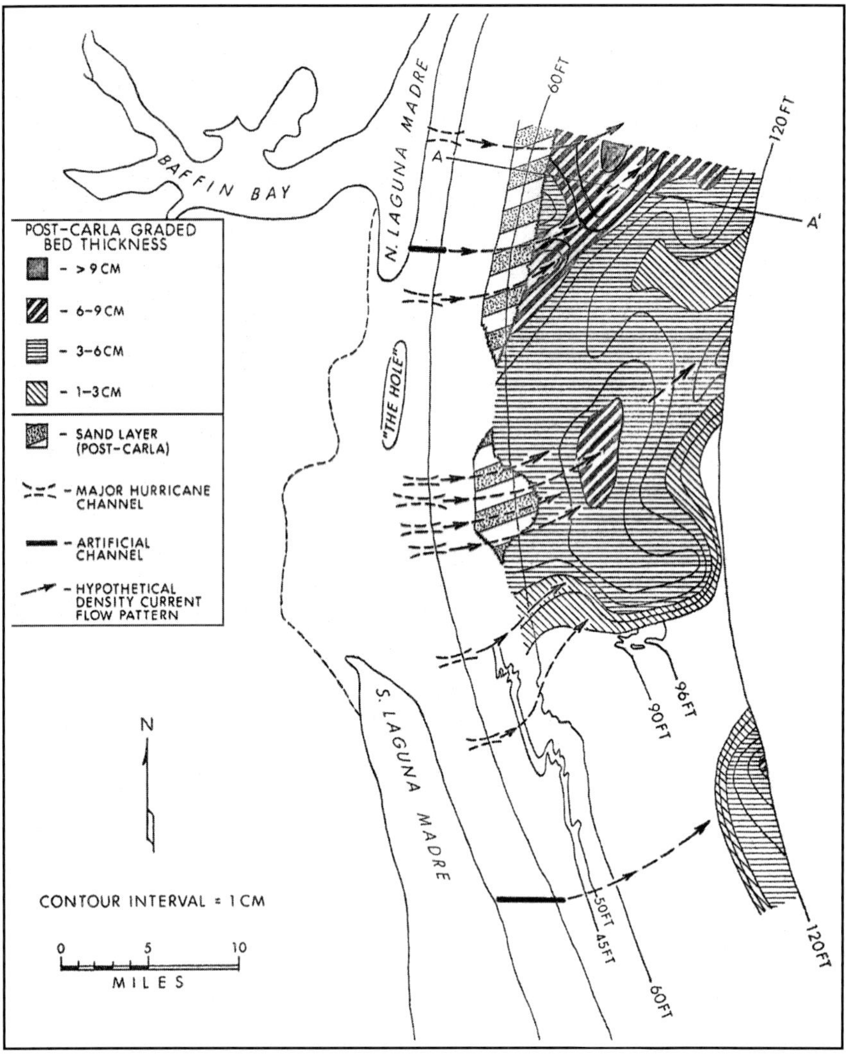

FIGURE 4. Sediments deposited in the inner neritic zone (60-120 feet depths) off central Padre Island, Texas during and after Hurricane *Carla*, 1961. The distribution of the sand layer deposited over nearshore mud (48 to 60 feet depths) during and shortly after the storm is shown. An isopach map of the post-*Carla* graded bed over the homogeneous mud of the inner neritic zone is also shown. Note the correlation of the graded bed and sand sheet with the location of major hurricane and artificial channels through the island. This factor, as well as other evidence (e.g., graded bed, slope correlation, etc.) implies a density current origin for the post-*Carla* sand layer and graded bed rather than an origin due to bottom wave action. The role of storm-surge ebb currents is also demonstrated. From Hayes (1965, 1967a).

causeway. The night was eerie and scary, as if we were driving offshore into the pitch-black bowels of the Gulf of Mexico. No lights could be seen ahead of us, only the white railing on either side of the narrow span disappearing into the fog.

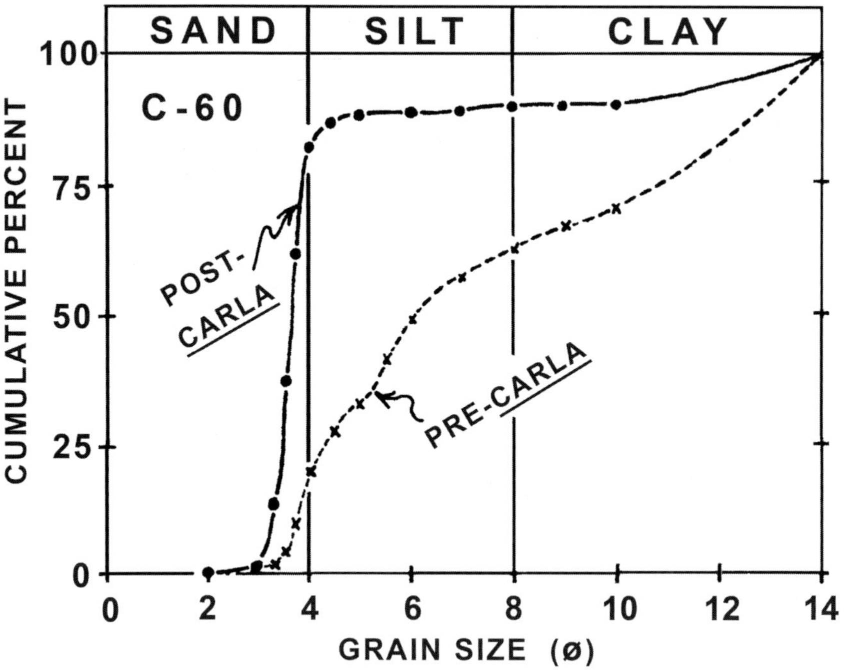

FIGURE 5. Hurricane *Carla*, 1961 sand layer and graded bed. These two curves are cumulative grain-size distribution curves for bottom sediment samples collected at the same depth (60 feet) on the same transect (directly off the mouth of Baffin Bay; Figure 1). The pre-*Carla* sample was collected on 27 July 1961, and the post-*Carla* sample was collected on 27 June 1962. The post-*Carla* deposit was a relatively clean sand layer (6-9 centimeters thick) overlying homogeneous pre-*Carla* mud. From Hayes (1965, 1967a).

The windows were down, and we could hear the waves lapping against the skinny pilings beneath us. As a new Ph.D. candidate at UT, this was my first field trip to the Texas coast, where I intended to carry out the research for my dissertation starting the following summer. We left Austin around 7 p.m. on a Friday night for a weekend field trip as part of Professor Al Scott's paleoecology class. It was almost midnight when we drove onto the wooden causeway that spanned Corpus Christi Bay. Our destination was our sleeping quarters at the UT Marine Institute at Port Aransas across the bay on Mustang Island (Figure 1). As I looked back at it, the van following us bounced rhythmically. I could see also the same concern and trepidation that I was feeling in the eyes of the other rookies in the back seats of our van.

We had been on the causeway for what seemed to be an eternity when I asked Walt Siler, who was driving, "My God, Walt, how much further is it?"

"Hell, it gets further every time, seems like. And we still haven't crossed over on the damn ferry, assuming that numbnut ferry captain isn't asleep on the other side of the channel." Walt had learned that rough language while flying airplanes all over the world for the U.S. Air Force. He had retired from the service and, like most of the rest of us in the van, was pursuing his Ph.D. in geology. He had several

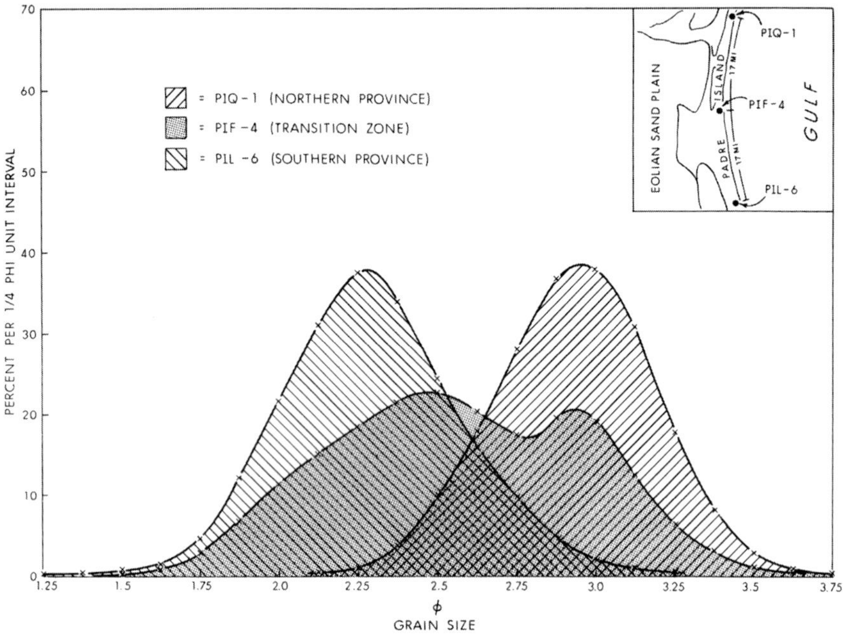

FIGURE 6. Frequency distribution curves for three dune sand samples from Padre Island, Texas. Locations of the samples are given on the inset map. Note near normality of the "coarse" southern end member, originally derived from the Rio Grande, as well as for the "fine" northern end member, originally derived from rivers to the north (Hayes, 1964b). The strong bimodality of sample PIF-4 is a result of the two modes being mixed in the middle of Padre Island due to converging longshore sediment transport directions in that area (from the north and from the south). From Hayes (1965).

years on all of us, including Professor Al Scott, who was only one year older than me. Therefore, he was more or less the *de facto* leader of this expedition, because it was Professor Scott's first year at UT as well.

"Rest easy, boys," Walt continued, "you are in for a wonderful treat at 0700 in the morning. Al says that the great HTO is going to give us a tour of the *Instituto* and then Herman Moore, their *muy antigua* boat captain, is going to take us for a spin in their world-class, heh, heh, research vessel, *The Ciencia*. Oh boy, who else besides H.T. Odum would give their piddly-assed 30-foot fishing boat a handle like C.I.E.N.C.I.A. Science, ha! Wait till you see HTO. A damn red-haired giant, he's at least six-four, shaped a little bit like an upside down pear. So full of energy that sparks seem to fly off him. Never sleeps. Nobody could keep up with that guy. He says he's developing an energy flow system for the whole universe, or something like that. Hell, its way over my head. I'm just a practical, middle-aged fellow that used to fly airplanes and wants to learn how to interpret rocks."

Somebody in the back said, "He's from the famous Odum family of ecologists, and he's published some of the most original papers in the field of marine ecology."

"Well, he might be a great scientist, all right, but his management style leaves a little bit to be desired," Walt replied. "He's so beloved by his staff and students that their nickname for him is 'red on the head like a dog's dick.' Everybody except that ass-kisser Ph.D. student of his, can't remember his name, you'll know him when you see him. His technical staff is in its third cycle," he continued. "That is, everybody on the staff has quit, or been fired, twice. All the members of the present scientific staff have been there less than a year. Yep, you fellas are in for a real treat at 0700 in the morning. That is, if we ever get there."

Next morning, as promised, the great HTO showed us the grounds and facilities of the UT Marine Institute, which consisted of three or four modest-sized wooden buildings built on stilts, and a small boat basin and boathouse tucked into the side of the Aransas Pass waterway. It was comfortable and isolated and reeked of the smell of the ocean, a scent that invigorated me during all the time I spent at the coast. Joining us on the tour was HTO's newly appointed administrative assistant, John Thompson, a lean, tall, blue-eyed and suntanned lieutenant-in-waiting. Following HTO's orders, John had posted signs all over the place telling the uninitiated new students how to behave on campus. The one that sticks out most in my mind was placed in the men's room in the main conference center. It read - DO NOT STAND ON COMMODE! Hey, I'm not making this stuff up. Only the great HTO could dream up a sign like that.

Next came the memorable ride on *The Ciencia* with its redoubtable captain, Herman Moore. Much later, I asked Herman, a retired mill-hand in the local carbon black plant, how he came to be captain of *The Ciencia* and the larger *Lorene* that would soon replace it. He said, "Well, Dr. Odum was looking for a boat captain one day and I was cutting grass in the yard up there by the boat basin, and he looked over at me and said how about you Herman? I said sure and here I am. Gosh, I never captained no boats before, I put in twenty five years over there in that plant where you can see the smoke rising fixing broke down furnaces and such."

It wasn't long after that conversation, while on my first sampling cruise out to blue water with Captain Herman on the *Lorene,* that I learned he couldn't read a compass. Therefore, I instantly proclaimed myself the chief navigator for the cruise.

I asked, "Captain, how the hell do you find your way back to Port Aransas when you don't have someone like me on board?" We were well out of sight of land.

"I just look for the smoke from the carbon black plant and head for it. Always takes me right back to the jetties."

When he took a group of unseasoned students out for their first sampling cruise, Captain Herman's favorite trick was to cut the engines at the first station and let the boat waft around parallel to the swell and assume its maximum roll. Then he would open up a smelly can of sardines and circulate among the group offering them some. After finishing off the sardines, he would light up one of his favorite cigars and go over and offer to help the students with their sampling. Then he would smile and return to the wheel while they as a group dashed to the rail and said good-bye to their breakfast.

Living and working at the coast at Port Aransas and surrounding environs was considerably different from the mountains of western North Carolina, to say the least, as well as from the hills of eastern Kentucky that surrounded Berea

College, and from midtown St. Louis, where Washington University is located. The countryside adjoining the shoreline inland for a hundred miles or more is flat, endlessly flat, with broad cotton fields and grassy prairies dotted with cattle feeding off the issue of the rich, black soil. But each time in the two hours or so that it took me to drive across that monotonous landscape, I would always look ahead to making the trip from the end of the road and down the beach to the middle of Padre Island, the longest barrier island in the whole world, where I would wade in the surf for hours, feeling the waves breaking against my legs and chest and then climb to the top of the dunes behind the beach and look out to sea past the three lines of breaking waves in the emerald green water and then turn north, south, east and west to view the hemisphere of the horizon in this new world *where the land meets the sea* that would become my focal point for so many years into the future.

During my tenure as a graduate student at UT, I studied under two other geniuses of note, besides the great HTO that is, namely Professor Robert L. Folk of the Geology Department at UT and Dr. W. Armstrong Price, an independent consultant who operated out of Corpus Christi. Folk, a smallish, wiry black-haired bundle of energy in his mid thirties with a flair for the dramatic in his teaching and research, was not only my dissertation supervisor, but also invited me to work as his field assistant for two summers along the Mexican coast. Part of my job during the Mexican work was to keep a close eye on him, because he could not swim a lick, having the world's highest specific gravity among human beings. In other words, once in the water, he sunk like a rock. In the summer of 1960, we were part of a research project on Alacran Reef off the coast of the Yucatan Peninsula. One day while doing some sampling of the reef platform on the exposed part of the reef, Folk, who was wearing a life jacket at the time, slipped into one of the outlet channels and was swept out to sea by a strong ebb current that rushed him toward Cuba or Panama or beyond. I jumped into the channel and swam out to pull him out of the main current and back onto the reef front.

That summer, I did a brief study of an island on the backside of the reef that disappeared every winter (*Isla Desaparecida*), and Folk published a paper on the grain size and composition of the sediments on the reef. In the acknowledgments of his paper, which became something of a classic in its field, he gave special credit to "M.O. Hayes - lifesaver."

Armstrong Price, in contrast, was a tall, laid-back southern gentleman who bore a strong resemblance to the actor Gary Cooper. A graduate of The John Hopkins University, one of his fondest memories was competing with the great Indian athlete Jim Thorpe in a college lacrosse game. Though in his late-sixties or early-seventies, he was in such great shape that I had trouble keeping up with him during our walking tour of the Padre Island beaches, and I was in my mid-twenties at the time! During a brief tenure as a professor in the School of Oceanography at Texas A&M University, he wrote several creative, seminal manuscripts on the topic of coastal landform development, which have served as a catalyst for much of my thinking on that subject, particularly with regard to the concepts of tide-dominated and wave-dominated coasts (Hayes, 2005b, c).

When I was working on my dissertation, Dr. Price had a small consulting business located in Corpus Christi that catered mostly to the oil-and-gas industry, but he still actively pursued independent coastal research on the side. Therefore, he took an immediate interest in my study and served as an outside member on my dissertation committee.

I last saw Dr. Price at a scientific meeting, where he gave a talk that outlined his final thought-provoking paper on coastal geomorphology. He was ninety-nine years old at the time. Some of us boys took the old guy out for a nice dinner after his talk. As we dropped him off at his hotel room, his parting words were, "Thanks fellows, I enjoyed partying with you tonight."

As I walked back to my room after the dinner, I thought, "Yes, old Armstrong Price surely has had his act together all right. Playing lacrosse with Jim Thorpe. Working out his brilliant ideas about waves and tides shaping the coast while running his comfortable consulting business in beautiful Corpus Christi. Three adoring wives to look after him over the years. Presenting a paper when he was 99 years old. Yep, he sure has."

When I arrived in Austin, Professor Folk was heavily engrossed in the study of grain-size characteristics of sediments collected from a number of sedimentary environments, having himself invented some graphical statistical techniques for measuring grain-size parameters. I proposed to continue that type of work in the coastal zone of south Texas for my dissertation, which entailed sampling sediments from as far offshore as 120 feet water depths and onto the mainland into the dune fields of Kenedy County (Figure 1). On the other hand, Dr. Price was primarily a coastal geomorphologist, which means that he interpreted the processes shaping landforms in coastal areas, such as the wave characteristics that molded the beaches, wind effects on dune forms, and so on. Over time, I dutifully collected the sediment samples needed to carry out my proposed project, but as the work progressed, my interest focused almost entirely on the dynamics of coastal processes and their impact on coastal landforms, in particular those generated by Hurricane *Carla*. In other words, I had decided to become a coastal geomorphologist, *a la* Dr. Price, hoping to perhaps enlarge upon that field.

This interval while I was in Texas, from 1959-1964, was a time during which several notable incidents occurred that I will never forget, such as when President Kennedy was shot in Dallas (discussed further later). I also remember a time during one of the innumerable four-hour-long drives from Port Aransas to Austin when I joined legions of others, including former President Jimmy Carter, by observing a formation of UFOs. It was late, around midnight, when I pulled off the road to stretch my legs and arouse myself from the numbing drive. The road I was driving along ran almost due north from Corpus Christi. I stopped at the crest of a hill a few miles short of where the road joined the major thoroughfare between Austin and San Antonio (Figure 1). At that intersection, I normally turned to the northeast toward Austin. There was no other traffic on the road I was traveling, but I could see the numerous headlights of the heavy traffic moving along the major highway ahead of me. As I looked north, I observed a number of lights moving in single file two to four thousand feet above the cars on the main highway. At times, those lights would abruptly change course, mostly moving vertically a few hundred feet, and then follow along over the traffic for another mile or two. I watched this strange phenomenon for fifteen or twenty minutes, during which time only one or two cars passed by me, and then the mysterious lights all abruptly disappeared, as if turned off in mid flight.

A rash of UFO sightings were reported at that time, but the only person I told about seeing them was my wife, Barbara Anne, who responded, "You were just getting tired from the long drive. Boy, you surely have a major sunburn this time! I'll get some lotion."

[NOTE: I don't recall ever seeing an explanation of this particular episode of UFO sightings. Goldsmith and Owen (2002) reviewed reports of UFOs that have been traced to meteors, weather balloons, early morning rising of the planet Venus, and a discarded Soviet rocket booster. Overall, they were skeptical that this planet has ever been visited by extraterrestrials, stating that "even if a million civilizations exist in the Milky Way, and even if one percent of these civilizations are devoted to interstellar exploration on a grand scale, there would still be 30 million stars to explore to find each civilization." They also intimate that the earth may not be a place special enough to warrant such a search. Oh well, another fantasy shot to hell.]

And on another night around midnight, I was driving home from the office at school listening to the car radio. As I pulled into the parking area beside the flimsy "married housing" World War II barracks where Barbara Anne and I lived, a ratchety voice of a person I pictured to be an ancient hillbilly musician came over the radio singing about answers blowing in the wind. Of course it was Bob Dylan, whom I had not yet heard of. I will never forget that moment when I sat transfixed in the car glued to every word. For once, I was thinking about something else besides coastal geomorphology.

During the early part of the summer of 1961, I began sampling the bottom sediments on the inner continental shelf in the study area I had chosen for my dissertation work, a rectangular expanse encompassing the central portion of Padre Island and extending offshore to the 120 feet depth contour, as well as inland for about 50 miles. Working aboard the *Lorene* with Captain Herman and several field assistants supplied by the Marine Institute, we collected bottom samples along a number of transects oriented perpendicular to the coast. Inshore, where the bottom was sandy, we used a metal grab sampler that dug into the sand and, further offshore, we used a Phleger corer, which penetrated into the muddy bottom sediments and collected a two-inch diameter plastic tube of sediments a couple of feet long *in situ* (i.e., as a relatively undisturbed vertical layer inside the tube).

This fieldwork was interrupted for a while in mid-summer that year when Barbara Anne and I joined Professor Folk and his wife Marge on a field excursion to a place called Isla Mujeres off the east coast of the Yucatan Peninsula in Mexico. That trip began with a long bus ride from Austin south to the border, along the coast of Mexico and across the Yucatan Peninsula, which entailed numerous ferry rides for the bus (minus the passengers who came on the next crossing), dodging palm trees in the dirt road, and so on. In the field, I was the grab sampler we used to collect bottom sediments in the channel between the island and the mainland. That is, I dove off the boat we had rented and collected the bottom samples by hand. Before we left the dock, the gentleman we rented the boat from looked me up and down and asked, "The boy is *muy practico*?" Well, anyway, we survived, even if I did think I might drown a couple of times as the man with the world's highest specific gravity was poised above in the boat waiting to save me.

One time we pulled the boat up on a mainland beach and went exploring in the semi-arid jungle behind the beach. As we worked our way through the heavy underbrush, we suddenly came upon the remains of a Mayan ruin. Located a long ways from any highway or human habitation, it was in a state of considerable disrepair.

As we poked around in the rubble, Folk lifted a piece of rock and shouted, "Look! It's CHAC's hand!

"What?"

"You know, the rain god, CHAC. This is his hand. Let's keep it!"

We debated at some length the ramifications of keeping (i.e., stealing) the hand, with me also questioning its authenticity. But Folk insisted it was the real thing. A man with unbounded intellectual curiosity and enthusiasm, he had studied the Mayan culture, and he was, in fact, fairly fluent in the Mayan language. All I knew about the Mayans was that they were mighty handsome, especially the women, and that, although only six-feet tall, I towered above them, feeling like a giant, as I walked along the crowded streets of Merida and the other towns. When we left for the long bus ride home, CHAC's hand was nestled snugly in the bag containing the sediment samples we had collected.

Toward the end of the summer, I returned to Port Aransas to continue the offshore sampling with Captain Herman and a field assistant named Wayne.

One day in September, as we were nearing completion of the offshore work far out to sea at the end of one of the transects, Captain Herman walked out on the deck and said, "Hey Hayes, guess what?"

"You have engine trouble again? Out of sardines? You're lost?"

"No, smart ass, there's a hurricane coming. A big one, called Hurricane *Carla*. We had better head on in."

"Aye, aye, Captain, anything you say!"

In the few hours it took to get back to Port Aransas, we listened to the radio for weather forecasts and, by the time we made port, the weather service was predicting landfall in Miami.

"It's still going to get pretty rough out in this part of the Gulf for a few days, why don't you go on back to Austin and visit with that pretty wife of yours while this thing blows over," Herman suggested, as we passed between the jetties.

"Good idea. Hey Captain, mind if we leave this bunch of cores and samples we collected on this cruise on the boat? If I make a push for it now, I may make Austin before nine or ten o'clock."

"Sure. I'm not going out again until you get back. But, listen, I won't take them off here for you."

"Fine. See you in a few days, Captain." I waved good-bye to him and Wayne as I dashed by the small boathouse on the way to my car.

All the cores and samples taken during previous offshore cruises were stored neatly in the boathouse beside the boat basin, where Herman was gently bringing in the *Lorene* to tie it up.

I drove back to Austin, speeding along as judiciously as possible, had a happy reunion, and laid around the apartment resting, not paying any attention to the news. A day and a half later, I turned the TV on and learned that *Carla* was making a bee-line for the Texas coast and was scheduled to cross somewhere near Port Lavaca, which is located about forty miles north of Port Aransas.

When *Carla* made landfall in early September 1961, it generated a storm surge of 24 feet at Port Lavaca and elevated water levels all the way from the Rio Grande to the Mississippi River. The entire shoreline of Texas was subject to hurricane-force winds, which exceeded 200 mph in places. It was one of the largest hurricanes to ever cross the Texas coast.

Folk bought some land in the hills outside of Austin, and he asked Gus Cotera, another one of his graduate students, and myself to help him build a primitive structure on the land that would be his escape hideaway where he could spend the weekend, work on his papers undisturbed, and so on. We completed the rickety structure and he was happy with it. That is, until *Carla*. The high winds generated by *Carla* blew the "hideaway" into smithereens. Folk blamed the incident on the "wrath of CHAC," because we had stolen his hand from the jungle in Yucatan. Folk kept the hand on his desk at school, and from then on, he always had it sandwiched between two bags of sugar, hoping to appease the angry god.

As the storm blew in, I wondered nervously what the "wrath of CHAC" would do to my dissertation area and, more importantly, to the collection of samples and cores in the boathouse and on the *Lorene* that we had labored for months to collect.

Before the storm had completely passed, I drove to the coast. Listening to the news reports as I drove along, I expected to find that Mustang Island had been planed off completely by the hurricane and was nothing but barren sand flat. At Aransas Pass, I learned that the causeway was washed out and that the only way to the island was by boat. The wind was still up some and the waters in Corpus Christi Bay were quite rough. And it was still raining hard. The only way I could find to get to the island was to hire a drunken fisherman, who took me across the bay in his 18-foot skiff. We had a memorable, extremely rough and wet ride, but I was determined to get there as soon as possible.

Well, the main buildings at the Institute were pretty beaten up, but still there, and they appeared to be restorable. But, guess what? Not a trace was left of the boathouse where my cores and samples had been stored! It had been almost completely inundated during the storm. One piece of good news was that Captain Herman had taken the *Lorene* across the bay to Corpus Christi before the storm, so the boat, and the samples and cores we had left on board, were still intact.

HTO and I walked down the path together and stood by the side of the boat basin at the former site of the boathouse. He said, "You know what Miles, I bet quite a few of your cores are still at the bottom here in the basin. If we were to swim down there, we might find some of them."

Next day, the great HTO and yours truly spent hours clawing our way to the bottom of the boat basin feeling for cores. The water was so muddy that you could not see an inch through it and the basin was filled with all kinds of miscellaneous debris. Once, while I was hanging on the side of the basin gasping for air, HTO's red head popped out of the water and he surfaced like a huge whale holding a bent core triumphantly in his hand.

"Got one!" he exulted.

I could hear some derisive comments from the crowd that had gathered to watch this exhibition, but I had no choice but to go on. My pride was at stake, even if I did cringe at the thought of grabbing a big snake each time I grappled around in the debris. A large number of huge rattlesnakes had been washed off St. Joe Island onto Mustang Island, and people were killing them by the dozens in the debris piles located all along the waterfront. We did find some of the cores, but the bulk of them were missing, probably having been washed out into the main channel.

This may be hard to believe, but in some ways, I was glad the hurricane demolished my dissertation project. I thought the sediment study had the potential to be somewhat boring and not lead to many original contributions. But, now, what about the geomorphic impacts on the coast of one of the largest hurricanes of all time? I had some baseline data, especially with the cores and samples that were saved on the *Lorene*. By resurveying the offshore area and comparing the new core data with that taken before the storm, I would have the first detailed evidence ever collected on the effect of hurricanes on the bottom sediments on the inner continental shelf. Therefore, I began at once to document the effects of the storm on the coastal geomorphology of my study area, and to plan for a resurvey of the shelf. That work eventually became the major focus of my dissertation project.

Except for some fishing notes taken while in high school, I had not kept a journal at any time before *Carla*. However, with the decision being made to do the hurricane study, plus my confident feeling that I was doing something that might be truly noteworthy and historic, I started keeping a detailed, personal journal of my activities, particularly of the fieldwork. Once I started that habit, I have practiced it off and on ever since. Much of the remainder of this chapter, as well as the rest of this book, are based on the notes in my journals.

We started the resurvey of the offshore area in early May 1962 by moving the *Lorene* south to a small harbor called Port Mansfield, located on the landward side of Laguna Madre opposite a jettied channel cut through Padre Island (Figure 1). That channel marked the southern limit of my study area. It was an easy trip to the Gulf from that location; however, our first few days of effort were fraught with difficulties, namely equipment problems and high winds that were a bear to work in, considering the unstable, near unseaworthiness of the *Lorene*. John Thompson made the three-hour drive from Port Aransas to go out with us on the first cruise and became woefully seasick, as we struggled with sampling equipment that didn't work. On May 12th, I wrote the following in my journal: "We went out today at approximately 5:45 a.m. Sea was pretty rough with an odd, very high swell. For some reason, I had a terrible spell of sea sickness, and threw up between every station (25-30 times, I guess)."

For most of the month of May, heavy winds blew out of the southeast. Meanwhile, HTO and Thompson were worried that we were keeping the *Lorene* in Port Mansfield too long. They noted that there had been five beautiful days in a row in Port Aransas. Why were we not going out?

We kept trying. I wrote the following about the trip we took on 31 May: "Left for the Gulf at 6:00 a.m. Met with high swell in the channel. Since Gulf was so rough, we decided to drop HTO's drift bottles only–too rough to collect bottom samples. On the way out, we were tossed about pretty bad. Sometimes when 'the swells ran out from under us' (Captain Herman quote), the bottom would smack pretty hard and crack and pop. Sometimes several inches of water would wash across the deck. Wayne and I both held up well as far as seasickness goes. I felt fine the whole trip." Then I made a list of proper practices to employ to avoid seasickness. We finally gave up for a while as the winds continued to blow, and we went back to Port Aransas to work on Padre Island and in Laguna Madre, leaving the *Lorene* in Port Mansfield.

Most of the summer of 1962 was a scramble for money to keep the research going and the weather delays were nerve wracking. HTO would promise money for a time and then change his mind, back and forth. Once I went to Austin to plead my case for more financial support with Folk. His first comment was, "You should be running up and down the beach getting ideas while you are sitting around waiting for the weather to change." He also lectured me on how he did his dissertation work at Penn State by riding the public bus to his study area every day, which resulted in an original treatise on the classification of limestones. I told him it would take an inventive technique beyond my capabilities to ride a bus out to 120 feet depths on the continental shelf. He concluded our conversation by saying, "Oh well, guess I will write Odum a letter some day and ask him who shot Santa Claus."

In the midst of all this, on Tuesday, May 29, I made a quick trip to Houston to visit the research laboratory of the Humble Oil Company (now known as Exxon). The boss of that facility was a famous geologist named Harold N. Fisk, who had done original work on both the Mississippi River Valley and the Mississippi Delta and published a number of classic papers, which are still in use today. The Humble and Shell research laboratories in Houston had the reputation of being the ultimate in geological research on sedimentology, including work on modern shorelines of the type I was doing for my dissertation. I was especially anxious to meet Dr. Fisk and have him critique my results. Also, inasmuch as I was strapped for financial support, I thought I might be able to secure a little help from Humble.

My notes from that trip observed that the Humble Research Center was "very impressive from the outside, but looked like Sing Sing Prison on the inside. Each 'researcher' has his own private cell to sulk in–approximately 50 cells per corridor." I was told there were forty geologists working in that office.

I went first to Fisk's office, who was confused as to why I was there and foisted me off to his underlings as quickly as possible. They showed me some of their techniques for working with sediments, and seemed about as puzzled as the old man was as to why I had come.

Then, I kind of invited myself back into Fisk's office, whose attitude had changed to "brusque and brow-beating." At that point, the young graduate student and the grand old man were about to get into a yelling match which, no doubt, the young graduate student would have lost but, for some reason, he seemed to back off a bit, and we engaged in an intensive two-hour conversation. I later wrote, "Actually, I think he was very nice to me in his own way. Probably treats everybody like that." I learned later from friends who had worked in that facility that my assessment was not far off.

Some direct quotes from old Harold:

"Humble Oil Company is not in the habit of doing other people's work for them."

"It's purely academic (a line of research I was suggesting). WON'T PROVE A DAMN THING!"

We eventually discussed my study area, where I knew his group had done some research. He turned around behind his desk, unlocked a display case, and pulled out a massive, hard-backed book of well over a thousand pages and waved it at me, saying "This is a summary of our work on your south Texas study area, mainly on the stratigraphy." I reached over to take a look at the wondrous document, and he pulled it away teasingly, immediately placing the weighty volume back in the display case from where he had removed it.

"This information is classified and confidential," he told me after securing the lock. He said their group had done over 3,000 borings in that study.

"Wow," I said, "an analysis of the grain size and structures in those cores would add a lot to my project. Think I could get some samples?"

"Nah, I don't think so, I believe we threw all of that stuff away."

And so it went, and I departed more-or-less empty handed. I wrote the following concluding statement in my notes. "Can't say that I learned a single thing that I didn't already know. No new ideas or techniques being used as far as I can see." Looks as if I was in the Age of Arrogance even before I received my Ph.D.[6]

But I had something of a point that has stuck with me through all these years. The operatives in the big companies with their seemingly unlimited budgets sometimes thought, especially at that time, that they had a corner on the market of originality and creativity of the research in our field. I was unconvinced, especially after that visit to Houston. You see, I had studied the piles of maps in the cluttered office of W. Armstrong Price in Corpus Christi, and had listened "at his feet" as he explained them. With the possible exception of Fisk himself, Dr. Price had not a single peer in that whole Humble operation when it came to originality and creativity in understanding how the coast was made.

Like a Labrador retriever that had crawled out of the water and up into the boat, I shook and shook myself free of the austerity, hypocritical trappings, and cynicism of the mighty Humble Research Center, and headed down the coast to Port Aransas to go back into the field and find things out for myself.

Professor Folk had two quotes posted on the front door of his office:

1) "Believe not what the ancients and sages of the past have told you, believe only that which you have tested for yourself and found worthy." Gautama Buddha

2) "Poor indeed is the student who does not exceed his master." Confucius

These were excellent words by which a young researcher should live.

Dr. Price had worked with Fisk on the project that the report he waved at me in his office was based on. He explained to me later, after my visit to see Fisk, that there had been a dispute as to who owned the land in the central portion of Laguna Madre, which covers and uncovers with water at the whim of the southeasterly and northerly winds. This was an area thought at that time to have large oil-and-gas reserves, and many companies were intensely interested in obtaining rights to drill there. Humble had the exploration rights on the King Ranch, and possibly the Kenedy Ranch as well, which covered a large area landward of the lagoon (Figure 1). Humble's lawyers said the land area the company already had under lease should extend "into the dry lagoon," because they had already secured relatively cheap leases from the ranch owners. Another company, I believe it was Sun, disputed this claim and said that the contested area was part of the lagoon, which belonged to the state of Texas. If so, that land could be leased by them from the state. Many millions of dollars were at stake in this dispute, and Humble hired Fisk, Price, and others to the tune of about one million dollars, which could buy a lot research in those days, to bring some science to bear on this problem. Hence the 3,000 borings and Dr. Price's maps of the lagoon and dune fields, etc. According to Dr. Price, the court decided to throw out all of the geological evidence and "take the word of the man on the street" with regard to what was lagoon and what was "dry land." Humble lost! When I heard this story, I didn't know whether to laugh or to cry. No wonder Fisk threw those cores away. In any event, Dr. Price didn't think the court's decision was very funny.

Combating high winds, a sick and contrary boat captain, and a lack of funds, by mid-June we were back in Port Mansfield trying to finish the sampling out on the shelf. On June 20, we were caught offshore in a heavy squall and thunderstorm, the worst weather of the whole trip. Next day, Captain Herman conveniently found that the water pump was leaking. It was pretty rough out there! This led to a long Chinese fire drill with John Thompson bringing a mechanic down from Port Aransas, the mechanic putting on the wrong part, Thompson and I driving to Harlingen to buy the wrong part, John and I getting into several long squabbles about the lack of logistical support I was receiving, and so on. I had a plan to finish as quickly as possible by adding an additional boat captain for this last leg of the work and anchoring offshore overnight, for which Herman suggested that we would need a second anchor. I put in a request for John to bring one down when he came, and proceeded to ask him where it was when he arrived.

His response - "Miles, I can't shit you an anchor!"

"Okay," I said, "we'll do it without a second anchor because, heck, we only need two more days to finish this."

John took the Captain aside and discussed my plan. Herman told him it couldn't be done.

After Thompson came back and told me Herman's response, I said, "Oh come on John, you know the old Captain, he's just being a little difficult because we had that bad weather the other day."

"Compared to you, he's being angelic!" was his quick response.

We finally set out to finish up the offshore sampling at 8:00 a.m. on June 26 with a backup captain, Captain Nutt, and my trusty field assistant, a high-school football player named Reese Brown, on board. As the work progressed, I saw a chance for finishing early if we worked straight through the night using a spotlight attached to the top of the cabin that lit the stern deck. Reese and I wore life jackets as we worked and conditions were generally good, the winds calm for a change. Captain Herman was sick most of the time, so Nutt did an iron-man performance from 1:00 p.m. until midnight. We finished the last transect (CC) at 3:35 the next morning. With the last core secured in the cabin, Reese and I danced for joy, because we didn't have to heave that heavy sampler on board any more.

The triumphal return of the *Lorene* to its homeport took place at 8:00 a.m. the following morning, as Herman nosed it into the small, now restored boat basin. HTO was overjoyed to see us, and he and I had a long discussion of the offshore reefs we had discovered, the water samples collected, and so on. Herman, Reese, Nutt, and I were the heroes of the Institute for the next couple of hours (the first COASTAL HEROES - ?).

About noon that same day, Reese, Nutt, and I left Port Aransas for the long drive to Port Mansfield to pick up our carry-all and deliver Nutt to his car. Reese and I arrived back at Port Aransas at 9:30 p.m., dead tired.

Despite the fact that I had decided the main focus of my dissertation would be the changes wrought by Hurricane *Carla,* I still wanted to finish sampling the sediments of all the coastal environments as I had originally envisioned in order to carry out the initial goal of the study, as well as to make observations of hurricane effects along the way. Working with small boats supplied by the Institute, Reese and I completed most of the scheduled sampling of the sediments in Laguna Madre in the month of July. It was hot, sometimes over 100°F., so we welcomed the opportunity to dive into the lagoonal waters for the samples where appropriate.

The sparsely populated dune plain located landward of Laguna Madre and extending for tens of miles inland was another, as yet, unsampled environment. Most of the land in that area was the property of two large ranches, the King Ranch and the Kenedy Ranch (Figure 1). A major highway, U.S. route #77, ran for 48 miles through this area with only one crossroads, called Sarita, and no gas stations along the way.

Toward the end of the month of July, I sampled the beach and dune sediments on the King Ranch, while accompanied by a couple of game wardens. I had never seen so much wildlife–quail, doves, hundreds of deer, wild turkeys, and thousands of ducks in the waters of Laguna Madre. The countryside was essentially unblemished and superbly beautiful in all of its many aspects–the grasslands on the black soil prairie, the oddly wild chin oak country, and the lush green of the new plants and flowers growing on the recently dried lake beds.

The wardens told me that the "Mexicans," who did most of the work on the ranches, could count on long-term employment, but they mostly spoke Spanish and didn't vote.

"They can't hold court down in Kenedy County, because they don't have enough people to form a jury," they laughed. "There are only about a dozen votes in the whole county and everybody holds some kind of political office."

On July 30, I drove to Kenedy County to see one Mr. Francis French, the point man for the Kenedy Ranch, in order to obtain permission to collect sediment samples on that ranch. His house, the first one on the left when you reached the small hamlet of Sarita, resembled a gambling hall inside, with a square gaming table in the middle of his living room and empty liquor bottles all around in the windows. He was probably about 55 years old with a slight build, graying hair, and bulging, blood-shot eyes. All the time I was talking with him, he sat at the gaming table in a sleeveless undershirt keeping three separate tops spinning on the tabletop. A deputy sheriff slumped in a chair in one corner, and a fellow who was introduced to me as some kind of a foreman on the ranch leaned back against the wall in another corner.

"How many more times are they going to do this?" Mr. French asked. "Some other guy from the University was here a while back taking soil samples, and then there was that ass-hole from the Humble Company and old Armstrong Price. Hell, Armstrong already has a sample of every square foot of the ranch."

I would have to jump through a few hoops, but he was receptive to my request, seeming to be glad to have someone new to talk to, even if he was considerably preoccupied with keeping the three tops spinning. He gave me a long discourse on the few citizens of the county, including all of the "Mexican" ladies, who "made a living on their back."

"You can go anywhere you want to, just don't drive on the road by the big house and disturb old lady Kenedy" were his parting words.

The graduate students at the Institute occasionally gathered on one of the vacant lots on the island for a game of softball. At one of the games, I talked with one of the male students by the name of Hadley, a tall gangly fellow with red hair who hailed from Oklahoma. He wasn't much of a softball player, but somehow I liked his style. In many ways, the nature of his rapport with the other students, and some of his personal traits reminded me persuasively of Hoojer Smith, a mill hand in the mountains of North Carolina, whom I used to fish with when I was a teenager (he was introduced in the FOREWORD). So, naturally, we became fairly good friends. We had many talks about our place in the world in general, and what we each were aspiring to do at the coast. He said I should read *Poor No More*, by Robert Ruark (1950), a book about the evolution of a poor boy from North Carolina, hinting that we were both a lot like that. I read it and strongly disagreed with Hadley. Heck, I didn't care anything about money, except that I wished I had enough to buy a four-wheel-drive vehicle so I could finish my sampling in Kenedy County, and down the beach on Padre Island.

Hadley wasn't exactly a hippie, but he was definitely different from the usual gang of ass-kissers that hung around HTO. Back in Austin, he invited Barbara Anne and me over to his pad for dinner with him and his live-in girl friend. Barbara Anne was a strict adherent to the Southern Baptist philosophy, and neither of us used alcohol, not even a beer now and then. When we arrived, Hadley had the lights down low with incense burning and Bob Dylan singing in the background. He offered us a variety of alcoholic drinks, which we politely refused. His girlfriend, I forget her name, was tall with black hair and had a phenomenal body; very friendly and quite amused by Hadley's effort to tempt the Southern Baptists. The evening was a bust. Barbara Anne refused to see them again, because they were living in sin, among other reasons. Thus, Hadley and I did not see each other as much as I would have liked, because when I talked with him, he also made me think about something else besides coastal geomorphology.

After a long hassle trying to find a four-wheel-drive vehicle with which to complete the fieldwork, I paid a few hundred dollars for an old, beat-up World War II jeep that was said to have seen service during the Battle of the Bulge. It constantly broke down on me, necessitating long walks on the Kenedy Ranch, along the beach on Padre Island, and elsewhere. In my journal notes of 7 April 1963, I wrote:

"On the way back up the island this morning, both the gas pedal and the brakes went out. I used a string for a hand throttle. It was quite a feat to steer, change gears, and man the hand throttle. It was also a little hard to get used to having no brakes, but I guess they're not really necessary on the beach."

The beach on Padre Island contains abundant shell material, particularly a zone exactly in the middle of the island called Big Shell (Figure 1). The shell material is quite soft, therefore a four-wheel-drive vehicle was absolutely essential to travel up and down the island.

I loved it down there on the beach, that effort being my favorite part of the study. Working by myself, most days I didn't see a single other soul, and the waves at high tide had usually obliterated the tracks of the fishermen who had been there the weekend before. This part of Padre Island, one of the most remote and pristine shorelines in the lower forty-eight states, is now a National Seashore, part of the National Park system.

In early September of 1963, Hurricane *Cindy*, a much smaller storm than *Carla*, crossed the Texas coast and waves generated by that storm reworked the beach on Padre Island. My jeep was no longer usable by that time, so I talked John Thompson into borrowing a Scout from the state wildlife department for us to make one more run down the island to carry out observations on the impacts of *Cindy* on the beach and finish up the last of the sampling in the dunes (see grain size curves for the dune sands given in Figure 6). The drive down the island on the 18th of September was a rough one, as we had to cross over a new cuspate berm that the post-*Cindy* waves had built on the seaward side of the hurricane beach left by *Carla*. We camped that night on the backbeach near the Port Mansfield jetties.

By the light of a flashlight, I wrote the following in my journal:

"It's really a lovely night. I've never seen so many stars–so close. The Milky Way is the sharpest I've seen. Jupiter is only a few degrees off the Gulf–no moon at all."

Nothing beats camping on the beach on a clear night under the stars miles and miles away from the nearest man-made lights.

Next day, John and I sampled three transects across the dunes, walking about 15 miles over the soft sand in the process. It was one of the toughest days of the entire project. When we lay our sleeping bags out on the sand on the backbeach that night, the sky was crystal clear again, but about 2:00 a.m., it started to rain. As I lay there half asleep with the rain pouring down, puddles of water all around me and the roar of the large waves generated by the storm in the background, I half-dreamed that the tide was rising up over the beach and that we were being washed out to sea. I jumped up, scared out of my wits. After calming down a little, John and I spent the rest of the night cramped in the front seats of the Scout. That was one of the hardest rains I had ever seen up to that time.

Two more days of sampling with John, and the fieldwork was over for me at long last.

Back at the University, I shared an office in a corner of the ground level of the Geology Department with two of Folk's other graduate students, a New Yorker named Murray Felsher, who was then and remains a good friend,[7] and Peter B. Andrews, a New Zealander, whom I haven't heard from since he went back home. We were a tight-knit group, and I have to give credit to Peter B. for his insistence that I focus my efforts on the changes *Carla* had made on the offshore bottom sediments, which is probably the part of my study that has received the most notoriety over the years. Also joining us in that crew was Dr. Rikii Shoji, a middle-aged professor from Tohoku University in Japan.

Dr. Shoji once told me, "Mr. Phase, Doctor Wholk's rimestone crassification is very famous in Yahpon." Thus, Mr. Phase became one of the many nicknames I have had over the years.

Some time around 1:00 p.m. on 22 November 1963, I was walking down the hall toward the office to meet up with Felsher and Peter B. I had made arrangements with Barbara Anne for us to go see President Kennedy's motorcade, which was scheduled to pass along the highway a few hundred feet down the hill from our office later on in the day. I was thinking about asking them to join us. As I walked into the office, they were huddled by the radio listening to a description of the shooting in Dallas. Many, but not all, of the graduate students in the Department were enamored by Kennedy's charisma and inspirational leadership. Count Felsher and me among them. We were both speechless, looking at each other in disbelief. We felt totally disenfranchised to be in Texas at that juncture, feeling sadness, disgust, and frustration all at the same time.

After about half an hour of listening in, Peter B. said, "I'll see you fellows tomorrow. This is not a good place for me to be right now." Then he left.

I hung around the office until 4:30 p.m., at which time I made notes in my journal about how quiet the campus was and that the stillness, unlike any I had seen there before, was overwhelming.

Nonetheless, life went on for us, and I was bearing down on wrapping up the lab work and a good bit of the writing on my project toward the end of the spring semester in 1964. Late one afternoon that spring, I believe it was, while walking across the campus, I noted HTO slowly heading toward me along the same path. He had a sad, dejected look about him, sort of like he had just struck out in the bottom of the ninth with the bases loaded, or something.

"Hi there Dr. Odum, what brings you to the hill country?"

He kind of looked away, almost not stopping.

"Oh, hello Miles well, they flunked Chen today," he blurted out.

Then I remembered that one of his Asian Ph.D. protégés, a hard working, unassuming student who quietly went about his business at the Institute, was taking his qualifying exam on this day. I could tell that HTO was heartbroken, nearly on the verge of tears. I had always liked HTO a lot and, to be truthful, was totally in awe of his original thinking. But to see him so broken up over the misfortune of a mere graduate student revealed a side of him I never knew existed.

After a short visit, during which I said a few kind words about his student, I walked away fuming, empathizing with him.

"There is not a one of those academic pimps on that committee worthy of carrying HTO's note pads," I thought. "And now, they've flunked one of his best students, because of their petty jealousies of a genius the likes of which they could never be."

So it goes when working out of an isolated field station. I guess Chen could have choked. I don't know for sure. But this was a pretty good object lesson in professor/student relationships that I would take with me to the University of Massachusetts in a few months, where I would become a lecturer in the Geology Department.

Soon after that, HTO left Port Aransas for Puerto Rico to blaze more new ground in the field of ecology and eventually settled in at the University of Florida. The last time I heard about him, in January 2001, I was told that he was still going into the office almost every day.

2 ON THE ICE

The greatest obstacle to the progress of science and to the undertaking of new tasks and provinces therein, is found in this - that men despair and think things impossible.

FRANCIS BACON

The Science

In September of 1964, I started a teaching job in the Geology Department of the University of Massachusetts (UMass), which was located in the town of Amherst in the west-central part of the state. My primary goal was to start a coastal research program examining the shoreline of New England, which has tides up to around 10 feet. This was part of the continuum of my long-range plans to study the effect of tides on the world's shoreline, starting with the small tides of the Texas Coast (microtidal), moving to the intermediate tides of the New England coast (mesotidal), and eventually moving on to coastlines with tides over 12 feet (macrotidal), which turned out to be in Alaska, Chile, France, and elsewhere. When I first arrived at UMass, I was a little short of research funds, so the Department Chairman, H.T.U. Smith, asked if I would like to work on a research grant he had to do studies in Antarctica, and I accepted his offer. The program we worked under was called the U.S. Antarctic Research Program (USARP), sponsored by the National Science Foundation (NSF).

Therefore, in the winter of 1965/66, I led a three-man geological expedition in Antarctica to examine a sequence of the Beacon Sandstone of Devonian to Triassic in age (400 to 250 million years old) that is located in the Dry Valleys area (Figure 7). These rocks are very similar to other units of similar age in South Africa and India, containing huge petrified trees that grew in extensive forests, resulting in the accumulation of thick coal beds, among other artifacts. The correlation of these very similar sandstones around the southern hemisphere was supporting evidence geologists used to determine the makeup of the original "southern" supercontinent of Gondwana, which had earlier collided with the "northern" supercontinent Laurasia, to form a single land mass called Pangea. This single land mass, Pangea, was starting to break up into separate components by the end of the Permian period (around 250 million years ago), after which the newly created fragments of Pangea, including the ones that originally made up Gondwana, except for the continent of Antarctica itself, which apparently stayed put, eventually migrated to

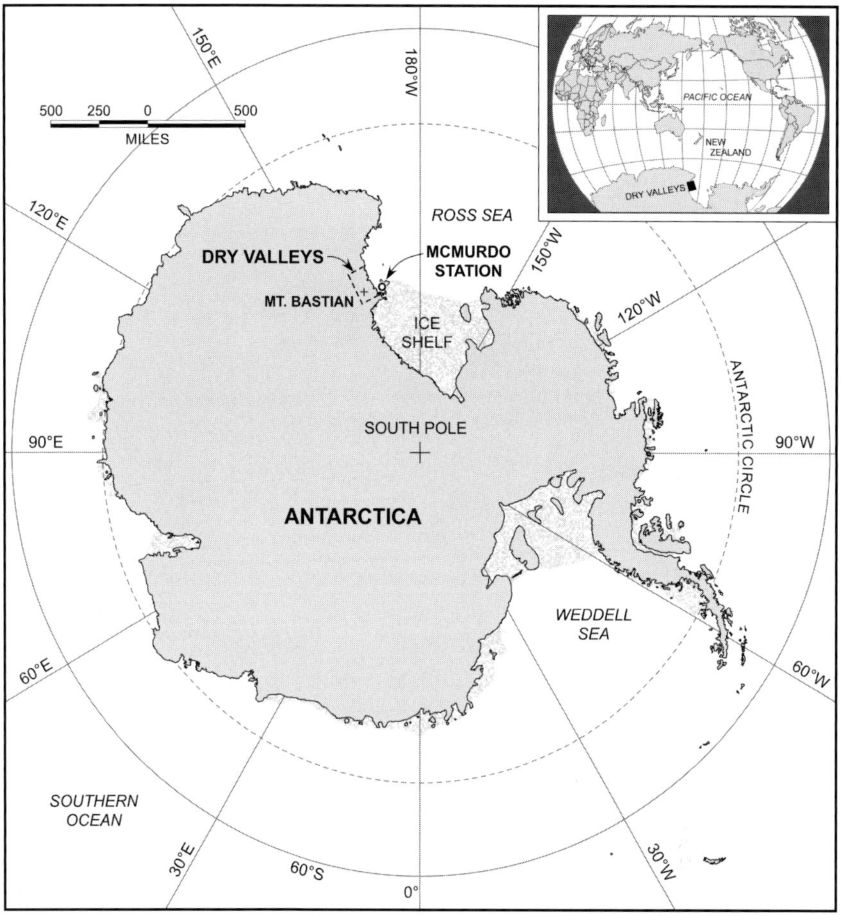

FIGURE 7. Map of Antarctica showing the field areas visited in the winter of 1965/66.

their present positions.

Our field team also found evidence of a major glaciation episode during the Permian period (Matz and Hayes, 1966; Pinet, Matz, and Hayes, 1967; Pinet, Matz, and Hayes, 1971). The abstract of this third paper, published in the Journal of Sedimentary Research, reads as follows:

"A thin pebbly mudstone unit which lies on a Devonian-Permian disconformity at Mount Feather and Mount Fleming in the Dry Valleys of Antarctica is interpreted as a glacial-fluvial deposit, and is correlated with Upper Paleozoic tillites farther to the south. These glacial beds are composed of two lithologies: mudstone (till) and channel sandstones (water-laid sands). This marks the northernmost known exposure of Paleozoic tillites in Victoria Land."

We also made readings of paleocurrent directions, postulating that some of the sandstones were deposited as sandy point bars in river channels (see example in Figure 8A) and other environments (e.g., rippled tidal flats, as is illustrated

FIGURE 8. Outcrops of the Beacon Sandstone in the Dry Valleys of Antarctica (Figure 7). (A) Dave Matz measuring crossbed dip angles and directions in some preserved cuspate megaripples on the surface of a sandy point par. The current that deposited the megaripples was flowing from the lower right to the upper left. (B) Preserved ripples on the surface of a sandstone layer. Both photographs were taken in the winter of 1965/66.

in Figure 8B). As noted earlier, this work provided one more piece of evidence regarding correlation of the Beacon Sandstone equivalent units among the original pieces of the ancient continent of Gondwana. These findings obviously support the concept of continental drift, or the moving apart of tectonic plates (African, Indian, and Australian plates in this case).

Seeing those petrified trees and coal beds made us think originally that the continent of Antarctica had migrated to its present position from warmer climes. However, according to New World Encyclopedia (2010), *Antarctica has been near or at the South Pole since the formation of Pangaea (since about 280 million years ago).* Therefore, all of the other fragments of Pangaea moved away from Antarctica, leaving it at the extreme southern end of the earth. Also, the southern parts of South America and Africa, as well as the peninsula we now call India, began to move far away from each other as well as away from the present continent of Antarctica.

Relatively recent research (e.g., M. Pagani, pers. comm.) reveals that long periods of geological time have existed during which the major ice fields presently found on earth did not exist, including the one in Antarctica. Among other interesting ramifications of these findings is the fact that if all this ice were melted, sea level would be about 220 feet above its present level. Also, presumably then, with the ice gone and under conditions of more favorable temperatures, huge forests could conceivably have grown in Antarctica, despite the fact that it is located so far to the south. This is a fascinating topic about which I hope to do some future research.[8]

The Story

(21 December 1965; 11:15 p.m. - Field camp on the flank of Mt. Fleming, Antarctica)

We were spending a third night in our small Scott tents (see Figure 9) at an elevation of about 6,000 feet on the side of Mt. Fleming in the Trans-Antarctic Mountains on the continent of Antarctica (Figure 7), a place all the hardy souls who worked there commonly referred to as "On The Ice." We were working hard and there was little time to spare, so even though it was past 11:00 p.m., I sat down and started to catch up on my field journal. I was working with two fellow scientists from UMass, Dave Matz, who was starting a Masters thesis under my supervision on the depositional history of a sedimentary sequence known as the Beacon Sandstone that was discussed briefly earlier, and his field assistant, Larry McGee, an undergraduate student noted for his mountaineering skills.

Thumbing back through my notes before starting to write this, I noted my descriptions in the plane while flying into our jumping off point, Christchurch, New Zealand, which included viewing the snow-capped volcano called Mt. Egmont on the North Island and the phenomenal braided streams of the Canterbury Plain on the South Island. It is interesting, at this stage of my life, to read how astounding to me the first sightings of features such as these had been. Starting with the beaches of Oahu, where I spent a day driving around the island, almost everything I saw during that trip was a new learning experience for me.

A certain manager of the NSF projects in Antarctica, who was stationed in Christchurch, New Zealand, didn't seem to think my project was necessarily the

FIGURE 9. Our Scott tents in one of the Dry Valleys in Victoria Land, Antarctica (Figure7) in the winter of 1965/66.

most important one going on down there, and so we had a bit of a run-in right away. This man, who I referred to as Freddie USARP in my notes, was turning down my requests right and left. Didn't he understand that I had broken new ground with my brilliant study of Hurricane *Carla*, and that more great science awaited my team in Antarctica? He was busy badly screwing up my carefully thought out work schedule, the kind of thing that bothers me no end even to this day.

I wrote the following in my journal the night after that run-in, which was two days before our plane was scheduled to depart for The Ice. "Probably the thing that bothers me most is the NSF and its handlers. It's the intellectual pimps like–(I'll leave the names out, but you can probably guess that Freddie USARP's was among them)–that truly bother me. They parade like peacocks in a world of false importance, carefully guarding their nest egg. Probably the thing that really bothers me is that they are hood-winked by the old-time operators. A young man must crawl for them, unproductive oafs that they are. God deliver me from this. If this is science, then where will it all end? Has-beens and blow-hards control the purse strings, that's a fact. It is terribly discouraging to know that in order to carry out a decent research project, I have to crawl through the slime also."

This ranting came from an individual who was just entering the Age of Arrogance which, almost always, happens as soon as one gets a Ph.D.,[9] and I had only had mine for a few months at that time. As later chapters in this book will disclose, I learned right away that my kind of research could not exist without money, thus, before long, I could mix it up fairly well with the has-beens and blow-hards.

Meanwhile, back on The Ice, I started listing in my journal some random thoughts about what I had learned during my first three days in the field. Strangely, for a young man from North Carolina, I took the cold pretty well. Our clothing was excellent, and my feet had not been cold once, even though the temperature never rose above freezing during any of our field excursions. I was amazed at how quickly things froze. In fact, we only cooked one meal a day, in the evening, because of the time it took to gather the ice and snow to make water, thaw out the food, which was supplied by the Navy facility at McMurdo Station–an amazing potpourri of cuisine ranging from rainbow trout to rib-eye steaks–and to cook the thawed out food. For the other two meals, we just stuffed our pockets with candy bars. During the four weeks we were there, I lost 15 pounds, and I wasn't overweight to start with.

I had a little trouble sleeping at night at first, because it never got dark, and the light penetrated the yellow canvas covers of my tent (Figure 9). That was a bit of an adjustment, but the three most trying times of the day, in order of severity, were: 1) Getting out of my sleeping bag in the morning in the sub-freezing temperatures; 2) Using our makeshift head, which consisted of a wooden commode seat balanced between two rocks; and 3) Climbing into the cold sleeping bag at night.

I made a list of the things I had been "overwhelmed by" up to that point:

a) The beauty of the place. We were in the Dry Valleys, where the precipitation is possibly the lowest on the planet; therefore, the rocks in the mountain sides were not completely covered with snow and ice, with their variant colors partially exposed to view. The panoramic views from the ridge tops were truly spectacular.

b) We were camping in one of the most isolated spots on earth, many tens of miles from McMurdo Station and several hundred miles from the South Pole.

c) There was no evidence of life anywhere–no birds, no plants, nothing. Only the fossilized Permian tree limbs and trunks and the tracks and trails of Devonian invertebrates in the rocks (see Figure 10).

d) Every day I walked in areas and saw sights that had never been traversed or observed by another human, except my two associates and me.

During this first excursion, we stayed in the field for nine days. It was a rather rigorous existence, and I hadn't worked that hard since John Thompson and I sampled the dune sand on Padre island (discussed in Chapter 1 - THE HURRICANE). The sandstone layers were more-or-less horizontal, thus we measured their thickness and described and sampled them by climbing up the side of the mountains, a process called measuring a stratigraphic section. It took two or three days to describe some of the sections. Often, we would leave camp early in the morning and climb up the steep slopes for hundreds of feet to where we had left off the day before (see Figure 9). When we reached the start points, we were commonly drenched with sweat under our long underwear. After a few minutes working at one level, the sweat froze into icicles along our spines.

As we climbed up the sections, we collected rock samples, which we carried in backpacks. After a few hours of collecting, these packs sometimes weighed 50 to 60 pounds. When the day's work was finished, we looked for an ice-covered scree slope, which were fan-like accumulations of debris located down slope from gaps in the sides of the mountains. Typically, snow blown through these gaps was

FIGURE 10. Bedding plane of a sandstone layer, a slump block, in the Beacon Sandstone in the Dry Valleys of Antarctica. The circular features are preserved vertical burrows of invertebrate organisms.

later frozen into an icy cover over the scree fans, which extended from the gap to the bottom of the valley in slopes as steep as 30-40 degrees. Carrying the heavy rock samples, we would slide down the slippery ice slopes on our backs using our ice picks as brakes, reaching the valley floor, where we were camped, in 15 to 20 seconds. However, we were very careful to choose the ice fans that were not too steep for this stunt.

On Christmas Day, 1965, we were camped at a place called Mt. Bastion (Figure 7). There had been a light snow Christmas Eve night, and the ground was covered with snow the next morning. Imagine that, a white Christmas in Antarctica. We had set that special day aside to describe the upper part of the Mt. Bastion section. This turned out to be 1,025 feet of sandstone, which Matz and I finished describing at 10 p.m. in the midst of a cloud that enshrouded the mountaintop. Thus, Christmas turned out to be one of our longest and most productive days, because we were so near the top of the section and none of us wanted to climb Mt. Bastion again. This was our third day on that outcrop. In general, our spirits were high all day, although the new snow made the climbing quite tricky. We sang songs and worked at a brisk pace. I think McGee was amazed at my repertoire of country songs. We finished our evening meal of steaks at around midnight, followed by a round of poetry from one of McGee's books.

We had scheduled a pick-up by one of the Navy helicopters at around nine in the morning on the 28th, thus we feverishly broke camp that morning and packed all the gear neatly for transport back to McMurdo Station. Right on time, we saw the chopper coming toward us about three valleys down the mountain chain; but, strangely, they started circling round and round inside that valley and not the one we were in. We fired flares and cursed and yelled and fired off more flares. However, they soon turned around and headed back to McMurdo without us. We called the Station right away, a call that had to be relayed through the South Pole Station, though McMurdo was at least 150-200 miles closer. We were depressed by the thought that we would have to set the camp up again, which took a long time. Luckily, the logistics office at McMurdo received our message and we were picked up around 3:30 p.m. that same day. The pilots never admitted that they were lost on the first try, not realizing that we had watched them circling in the valley below us that morning.

Back at McMurdo, the shower was outstanding. Ten days of grim were enough.

On the 31st, Matz and I were dropped off at a place called Allan Nunatak, a massive and isolated island of bedrock sticking up through the flat surface of the Polar Ice Cap. McGee had taken sick and couldn't make the trip. Once on the ground, we had trouble setting up our tents in the howling wind.

New Years Day, 1966, we started describing the best rock exposure at Allan Nunatak, which was on the south side of the mountain. It was the coldest day we had worked so far, and the wind, blowing straight at us from the South Pole, was blasting against the steep slope we were climbing. If you faced the wind for long, your nose and lips would turn numb, causing us some concern about frostbite. The lower part of that section was pretty dull, so it turned out to be a fairly grim New Years Day.

We made a significant error in our estimate of the thickness of the section we wanted to measure at that location, having only allowed ourselves two full days to do it. We scheduled a helicopter pick-up on the 3rd so we could move to another site, thus we were forced to complete the section on the 2nd. This we did, completing it at 1:45 a.m. the next morning. We finally made it back to camp and to bed at 3:00 a.m. You may be wondering how we were able to carry out these observations in the middle of the night. Don't forget, the sun never set while we were on The Ice.

I wrote the following in my journal about this lengthy exercise. "However, this was a good day. The wind finally died down that night and we found the upper part of the section to be very interesting." In those rocks, we discovered a magnificent petrified tree 60 feet long, only a few hundred miles from the South Pole!

After that, we were back and forth between the field and the Navy Base, enduring a whiteout or two in the process. On the 6th, I had a little disagreement with the USARP logistics officer on the scene, Ken Moulton. Having such encounters with logistics officers, which began with my relationship with John Thompson at the Port Aransas marine lab (discussed in Chapter 1 – THE HURRICANE), was to become a life-long habit of mine. Anyway, that afternoon Moulton had turned down my request for a stop-off at Mt. Fleming during a recon flight to the famous Beacon Valley that he had scheduled for us on the following day. I wanted to re-examine a critical part of the section there.

Moulton said, "Why that's impossible! I just can't conceivably justify such a side trip to the Navy. After all, it is their helicopter you'll be using."

That night I wrote a long tirade in my journal about useless logistics coordinators that I won't repeat here. But I will include the write-up I did on the trip the following morning.

"This was our day for a VX-6 recon flight to Beacon Valley. In place of the two geologists I had requested, Moulton substituted a Russian geologist named Lev Klimov, a metamorphic petrologist (in other words, someone who wouldn't know much about the sandstones we were studying), and some oaf from the Navy, whose major contribution was to comment on how cold he was at each stop.

"We were gone 5 hours: 1.5 hours to fly back and forth, 0.5 hours for our recon of the Beacon Valley, and 3 hours for the pilot and co-pilot to look for ash trays and ventifacts."

And so it went. Moulton eventually became more cooperative regarding my requests, and we continued to tally hundreds of feet of measured section, in between bouts with bad weather.

On January 11, we went back to the Beacon Valley by helicopter and described a part of the section at Maya Mountain. This is a spectacular area, looking a whole lot like Glacier National Park, without the trees, of course. Klimov, a very nice gentlemen in his 50s, went with us again, and we had some good talks. He was apparently very well known in his field, and he gave me a copy of a major textbook he had written (in Russian).

It had snowed again the night before the visit to Beacon Valley, five or six inches, which is a lot for that area, so climbing was difficult. So was the geology, inasmuch as parts of the section were completely covered with snow. However, we easily found the base of the Permian section, our goal for the day. On the way down the mountain at the end of the day to meet the helicopter, we slipped and fell many times, and during one of my plunges, I lost my prized watch that Barbara Anne had given me for a graduation present.

Speaking of falls, one of the sections we were describing during an earlier trip was partly covered by one of the ice fans. This fan was exceptionally steep, but we decided we needed to get to the other side of it, where we could complete measuring the section up that side of the mountain. We were working our way across the steep slope by chipping footholds in the ice with our ice picks. Matz was taking the lead when one of the footholds broke loose and he slipped onto his back, headed down the steep slope in what could have been a fatal slide. In a flash, he followed the procedure we had learned in our training sessions before the expedition, by digging the sharp end of his pick into the ice, instantly stopping his slide. McGee and I had to do some fancy chopping of footholds to reach him, but we managed to retrace our steps back to safety. I often wondered if I could have reacted that quickly under the same circumstances. That was the closest we came to having a serious injury during the trip.

When we returned to the base the night we did the research in the Beacon Valley, Moulton informed us that a crack was developing in the runway, which was on a thick sheet of ice over the open water, and that we should plan to leave for Christchurch the next day.

Weather conditions were dubious the following morning, but it was decided later that the flight would leave around 4:00 p.m. This led to a mad scramble of packing by me and McGee. Matz had resigned himself to staying behind. It was a cold day, one of the coldest of the summer, as we gathered at the strip for departure. This was a scene of the greatest flurry of activity I had seen on The Ice.

Earlier during the day, I began to feel poorly, so I obtained the usual outlay of cold pills from the dispensary. I was also developing some urinary problems that caused me a good deal of concern.

However, as we boarded the plane, the covey of helicopters, noisy Nodwell tracked vehicles, bright sun, chilling wind, and general enthusiasm of the crowd raised my exhilaration to the point of forgetting about my physical problems. We were not long underway, however, until I realized that the flight was going to be a painful one. Turned out to be the most miserable airplane ride I had taken up to that time.

At first, I spent some time talking to a fellow geologist, Larry Frakes, about his work around the southern hemisphere, and how our study might complement his, but, as time passed, I became more and more ill. During that nine-hour trip, I made 7 painful journeys back to the head, each one more excruciating than the last. I was beginning to get a few looks and whispers from the other passengers, becoming a little embarrassed at my predicament. My malady resembled diarrhea of the bladder and, in between trips to the head, I was intermittently feverish and suffering chills, always nauseated.

Eventually, in the half-stupor between sleep, the fever, and the buzzing eardrums, I began to hallucinate about dying. Kind of wishing I would, as a matter of fact.

After arriving in Christchurch at about 1:00 a.m., I reported to the Naval Dispensary and was told that I had a bladder infection. The doc gave me some pills, put me on a high-liquid diet, and ordered me not to "live it up in Christchurch." Little chance of that!

McGee and I checked into a hotel, and I was deathly ill all the next day, suffering the same symptoms that I had on the plane, plus the nauseating side effects of the highly potent pills. I had never been that sick before.

Next day was a little better, but I stayed in bed. The bathroom interval remained about every 1.5 hour, and was still very painful. Along towards the afternoon, I began having visions and half-dreams, still thinking this might be it. Later, the makings of a novel passed through my mind. Whole chapters would flash by as I lay there in my stupor. I couldn't force myself to get up, but thought "How I wish I could record these words and phrases, so perfectly formed." The segments of the book I had in mind, titled *Child of The King*, kept going through my mind over and over. That is, until the next morning, when I arose from the dead and walked about the streets of Christchurch, making ready to catch the plane for home around midnight. I was feeling so good that McGee and I went over and made peace with Freddie USARP.

And on the flight between Pago Pago and Honolulu, McGee and I were watching a most remarkable sunset. At one time, there was a lower tier of evenly spaced, coal gray, puffy cumulous clouds that looked like a herd of grazing sheep. I smiled because that reminded me of the New Zealand I had been so happy to survive. The sunset colors showed only on the upper tier of clouds. We could observe the sea below as a checkered pattern. The sunset lasted for a long time, climaxing with a thin line of the deepest crimson along the horizon, the richest color I had ever seen in a sunset. Probably the most beautiful scenes that I had seen on the trip had been from the air. I remembered viewing the sun shining on the giant icebergs as we neared the continent of Antarctica. Close rivals had been the beaches along the Oahu coast and the Beacon Valley.

Then, breathing a sigh of relief at having survived the cold of Antarctica and the painful and scary urinary infection, I pulled my famous orange camera bag out of the overhead rack, took out a book, and started reading another of my favorite novels. Over the years, I read all of William Faulkner, and then all of Kurt Vonnegut, Jr., and then all of Ernest Hemingway and Jean-Paul Sartre, and parts of Saul Bellow and John Steinbeck and Robert Penn Warren and Vladimir Nabokov and many others, thinking, "Ah, if only I could write like that." But I had no story to tell–yet.

3 THE FIELD TRIP

When I walked into the Inn,
I was struck by a New England recollection,
a hollowness, a nostalgia, a feeling of old
things, trim, in place, and preserved, set in a
splash of color before the onslaught
of the cold, icy northern winter.

BILLY WHISKERS on 7 October 1978 at Black Point Inn, Scarborough, Maine

The Science

After that stint on The Ice in Antarctica in the winter of 1965/66, I never worked on rocks again in much detail for many years, concentrating most of my program's research efforts, which involved a large number of graduate students, on modern coastal processes and sedimentation.

My first academic year at UMass was 1964/65 and my last one was 1971/72. While I was there, the graduate students listed in Appendix A completed theses and dissertations under my supervision. As mentioned in the previous chapter, Dave Matz and Paul Pinet did M.S. theses on the Beacon Sandstone in Antarctica. Also, Jim Wessel did a M.S. thesis on the Triassic rocks in western Massachusetts, in which some phenomenal preserved antidunes and ball-and-pillow structures were discovered in a unit called the Mt. Toby Conglomerate (Hand, Wessel, and Hayes, 1969; see also Figure 11). All of the other graduate student studies were on coastal geomorphology and sedimentation on the New England coast. Fayez Anan, a fellow graduate student at the University of Texas, who joined our team at UMass, did a Ph.D. study of the coastal sediments of the Massachusetts and New Hampshire shore that involved the invention of a settling tube for measuring the grain size of sand. Victor Goldsmith's Ph.D. project was on the evolution of barrier islands on the Cape Cod shoreline, and Al Hine's M.S. thesis was on tidal inlet and beach geomorphology also at the Cape. Stewart Farrell did both M.S. and Ph.D. studies on the coast of Maine. Allan D. Hartwell and Joan DaBoll did M.S. theses on the estuaries on the north shore of Massachusetts. Woody Hobbs, a member of the 1970 field team in Alaska discussed in the next chapter, studied a segment of the coastline of Cape Cod Bay, and Peter Rosen analyzed the cuspate spits on Nantucket Island, both for their M.S. theses. Three-dimensional stratigraphic

FIGURE 11. Primary sedimentary structures in the Mt. Toby Conglomerate of Triassic age in an outcrop located near Amherst, Massachusetts. (A) Close-up view of a preserved antidune on an alluvial fan. Water creating the antidune was flowing from left to right. Photograph taken circa 1969. (B) Ball-and-pillow deformation structures in the same unit as pictured in A. Photograph taken circa 1976. JM for scale.

work was completed on the Holocene sediments in the marshes landward of Plum Island in a Ph.D. study by C. Larry McCormick, and Skip Rhodes did a seismic and coring study of the Merrimack Embayment for his M.S. thesis (see Story Two).

All of this work on the Massachusetts and New Hampshire coasts led to a field trip on 9-11 May 1969, with an accompanying 462-page field trip guidebook that we called the *White Elephant* (Hayes, 1969). Ten graduate students contributed to this guidebook as authors and co-authors of 15 separate articles, as well as of the descriptions of the 29 field locations visited during the three days we were in the field. The location of the area covered during the field trip is shown in Figure 12.

Stew Farrell recently reminded me that putting that "White Elephant" together within a limited time frame was no mean task. Several of the key players had just recently returned from a long summer field project in Alaska that will be discussed in Chapter 4 – ALASKA: *SUZANNE,* EARTHQUAKES, AND TSUNAMIS. Some of the activities involved in creating the new guidebook included Jon Boothroyd doing the chart-pack drawings, Marsha Reiber doing the typing, and Stew, Al Hine, and Joan DaBoll "fixing the edits." Skip Rhodes single-handedly cranked out the photos in the dark room. The whole crew worked way into the night for many days, getting the book ready to publish just in the nick of time.

Without leaving the *White Elephant,* I should point out the paper on the gravel beach at Rye Beach, New Hampshire by Barry Timson (Timson, 1969), which was also a stop on the field trip. This was a beginning of a long series of encounters with gravel beaches, culminating with detailed studies of them at the *Exxon Valdez* oil spill site (Hayes, Michel, and Betenbaugh, 2009; discussed in Chapter 17 – GRAVEL BEACHES AND FAN DELTAS).

Perhaps the most memorable of the studies, that I personally worked on anyway, were those dealing with the sand beach cycle, primary sedimentary structures of sand beaches, tidal inlet morphology and processes, and intertidal bedforms.

The Sand Beach Cycle

After a year or two measuring beach profiles on the New England coast, it became painfully obvious that something was amiss with regard to the generally held theories about beach cycles. The concept of the cyclical change of beaches from a flat, erosional profile in winter to a wide, depositional berm in summer had been well ingrained in both the popular and scientific literature for many years following research carried out in the 1940s. The concept originated from detailed studies on the coast of Southern California (e.g., Shepard, 1950; Bascom, 1954). The idea was that, generally speaking, erosional storm waves are more common in the winter, and flatter, depositional swell waves are more common in the summer on the California coast, hence the terms *summer* and *winter beaches*. Beach profiles measured at Scripps Pier in La Jolla in the 1940s commonly showed wide beaches in the late summer and fall, and narrow beaches in the winter and spring.

However, this is an idea that has not withstood the test of time for the New England coast and, as I learned later, neither has it for most other shorelines of the world. At about the time of the publication of the "White Elephant" in 1969, Cy Galvin and I (Galvin and Hayes, 1969) discussed the striking contrast between beach cycles on the U.S. east coast and those presented in the literature for the west

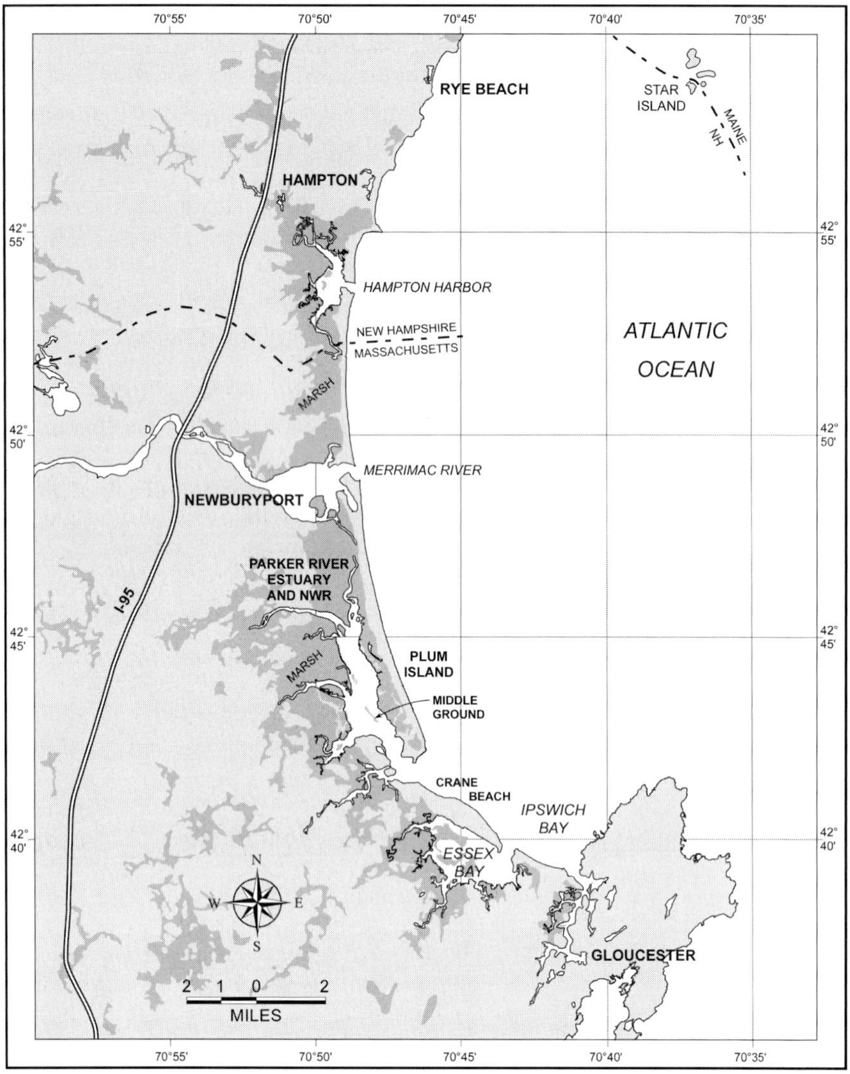

FIGURE 12. The Massachusetts/New Hampshire field area discussed in this chapter.

coast. Our conclusion was that eroded winter profiles and strongly depositional summer profiles are not a dominant theme on northern Atlantic beaches, particularly in the New England area. The beach cycle in that area is controlled by the passage of individual storms, principally extratropical cyclones that are called "nor'easters" in New England (discussed further below). This also appears to reflect, at least to some degree, less severe wave climate and relatively smaller seasonal changes in wave climate on the northern Atlantic coast of the U.S. Mean wave heights and periods are about 50 centimeters and 7 seconds along the northern Atlantic coast, considerably smaller values than those for the California Coast (75 centimeters and 12 seconds). These dissimilarities in wave climate result, to some extent, from differences in their locations on western and eastern edges of oceans with predominantly westerly winds. The winds blow predominantly onshore on the California coast and offshore on the East Coast.

In 2010, my wife and business partner Jacqui Michel (hereafter referred to as JM) and I published a book titled, *A Coast to Explore: Coastal Geology and Ecology of the Central California Coast,* in which we discussed the winter/summer beach concept. With regard to the comment made earlier about the West Coast being a windward shore, Ed Clifton (pers. comm.) stated that "Local winds are less a factor than the fact that the largest storms (typhoons) develop on the western side of the ocean, and that every large weather system in the western Pacific (and the Gulf of Alaska as well) generates long period swell which arrives on the U.S. west coast." We finally concluded that the largest waves that occur along the Central California coast are generated in one of two ways: 1) When major *El Niño* storms cross the coast; or 2) When some of the largest swell from especially huge offshore storms reach the coast.

In the 2002 Coastal Engineering Manual, published by the U.S. Army Corps of Engineers, the following statement appears: "The seasonal onshore-offshore exchange of beach and nearshore sediments ... is now known to apply best to swell-dominated coasts, such as the U.S. west coast, where wave climate changes seasonally." Better late than never.

Therefore, the general consensus today is that the terms *winter beach* and *summer beach* should not be used with reference to erosional and depositional cycles on beaches, although, as we found when doing research for our book on the Central California coast, many published discussions of the beaches in California still use these terms, or else refer to "normal seasonal changes," which presumably means more-or-less the same thing.

Paul Komar, working on the Oregon coast, suggested the terms "storm profile" (flat beach) and "swell profile" (well-developed berm). I think this is a good alternative to calling them winter or summer beaches.

With regard to our work on the beach cycle in New England, during six years of research between 1964 and 1969, we found that northeasterly storms play the dominant role in the generation of cycles of erosion and deposition of the sand beaches in that area (Hayes and Boothroyd, 1969). Observations at 40 beach profile stations surveyed every two weeks over that six-year period revealed the following stages of low-tide morphology relative to storm occurrences (examples of these types of profiles are illustrated in Figure 13):

1) <u>Early post storm</u> (up to 3 or 4 days after the storm) – Beach is flat to concave upward, and the beach surface is generally smooth and composed of medium- to fine-grained sand. Severest storms leave erosional scarps along the dunes.

2) <u>Early accretional or constructional</u> (usually 2 days to 6 weeks after the storm) – Small berms at the high-tide line and intertidal bars and troughs are quick to form.

3) <u>Late accretional or maturity</u> (6 weeks or more after the storm) – Landward-migrating intertidal bars weld onto the backbeach to form broad, convex berms.

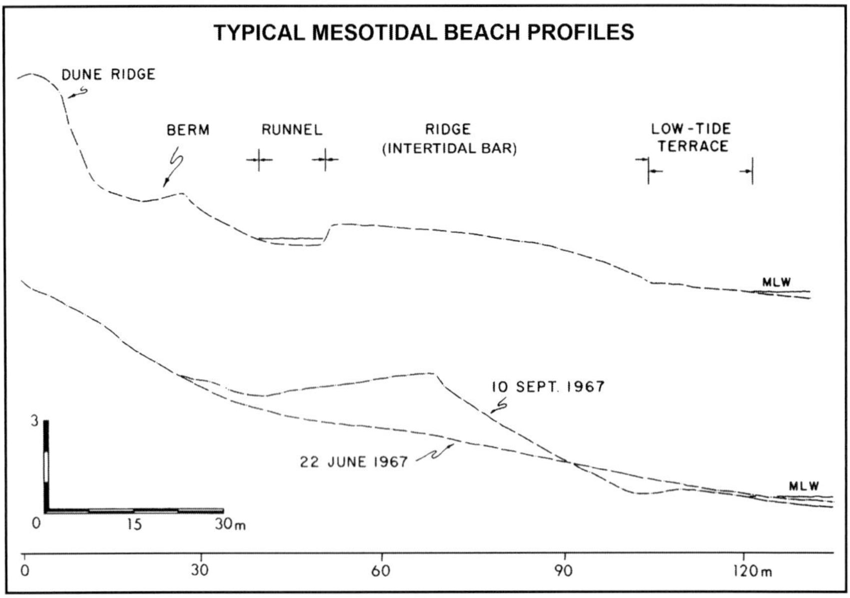

FIGURE 13. Examples of the three most common types of beach profiles occurring on the northern New England coast (tidal range = 2.7 meters, mesotidal). (Upper) Typical constructional profile, the most frequently occurring profile type. Crane Beach on 22 August 1967. (Lower) Two profiles on Plum Island. Late-accretional mature profile on 10 Sept. 1967. Same station after a brief summer storm on 22 June 1967. From Hayes and Boothroyd (1969).

Primary Sedimentary Structures of Sand Beaches

During my tenure at UMass, my students and research associates dug hundreds of trenches across the beaches of the New England area, one of which is illustrated in Figures 14, 15, and 16. One of the key discoveries of this work was that the landward dipping crossbeds created by landward migrating intertidal bars are commonly preserved in the middle to lower portions of depositional berms. The photograph of the middle of a big berm on Plum Island that was dug on 23 August 1967 (Figure 16) shows abundant landward dipping beds in the core of the berm. The photograph in Figure 16A was reproduced in several publications, including at least one book of note (*Waves on Beaches,* 1972; R.E. Meyer, Editor).

Seaward dipping, low-angle sand layers in the trenches reflect the deposition of upper flow regime flat beds on the surface of the beachface. Flat beds that dip slightly landward on the berm tops, which commonly contained antidune crossbedding, were also present in many of the trenches. Except for the bedding created by the migrating slip faces of the intertidal bars, the structures of beach sediments were generally dominated by upper flow regime deposits. The physical sedimentary structures that occur on sand beaches were discussed in some detail in Hayes (1972).

Although geology has long been considered to be primarily a field science, the greatest breakthrough in the understanding of the smaller-scale physical

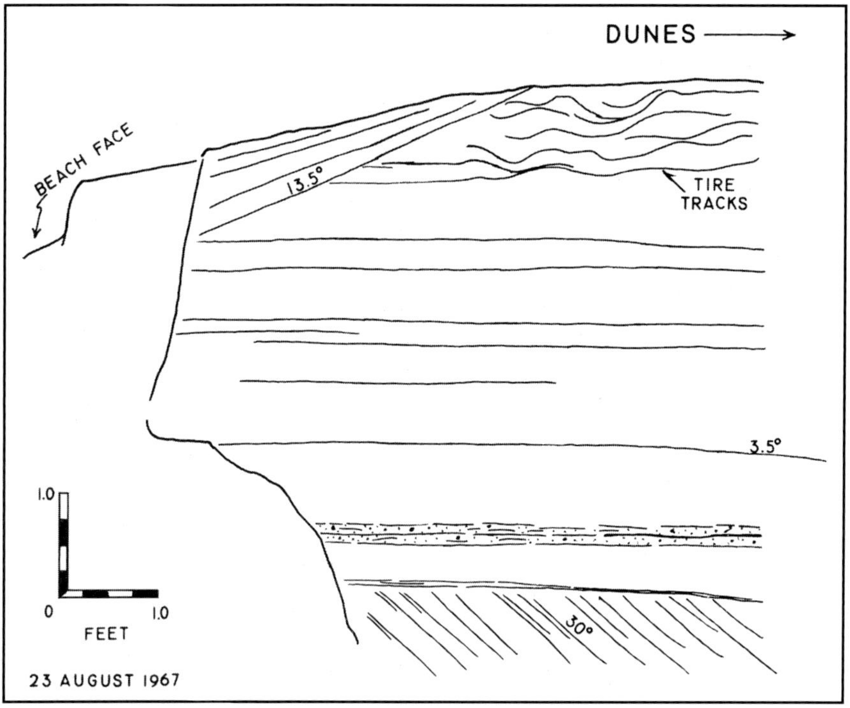

FIGURE 14. Internal structure of seaward edge of a welded berm at Camp Sea Haven, Plum Island on 23 August 1967 (Figure 12). The three major types of bedding are: (a) Low-angle seaward-dipping beds formed by onlapping of sand layers over an erosional surface that dips seaward at an angle of 13.5 degrees; (b) Horizontal bedding, which occupies most of the central part of the trench, formed by vertical upbuilding of a large intertidal bar as it migrated landward. These beds are plane beds in the flow-regime scheme; hence, they were formed under conditions of relatively shallow, very high velocity flow; and (c) Landward-oriented, high-angle (30 degree) crossbeds at the base of the trench. These crossbeds were formed by landward migration of the intertidal bar on which the horizontal beds higher up were deposited in its initial stages of development. From Hayes, Anan, and Bozeman (1969).

sedimentary features that occur on beaches and tidal flats was the result of laboratory flume studies carried out by hydraulic engineers (e.g., Simons and Richardson, 1962). In flowing water, sand grains start to move at velocities of about 10 inches/second. These workers demonstrated that the sediment bed of a laboratory flume (box-like trough with a sand bottom and a mechanism to generate water flow across the sand bed) could be made to pass through the following sequence of features (called **bedforms**) by simply increasing the velocity of the water flowing across the sand bed or decreasing the depth of the water:

flat bed → small ripples → large ripples (megaripples)
→ washed-out megaripples → flat bed → antidunes

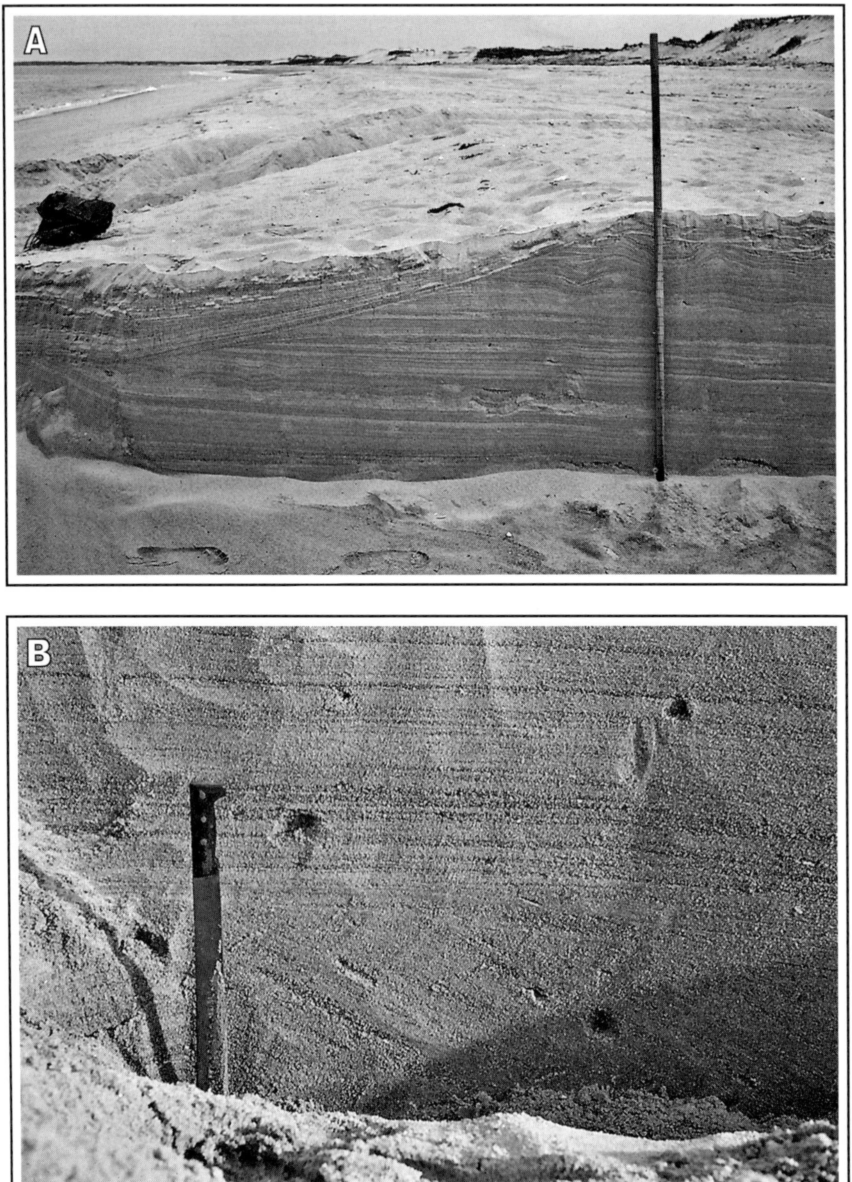

FIGURE 15. Photographs showing the beach structures illustrated in Figure 14. (A) Upper portion of the trench. (B) Lower part of the trench. Note the high-angle crossbeds at the base of the trench that dip landward at an angle of approximately 30 degrees. From Hayes, Anan, and Bozeman (1969).

FIGURE 16. Internal structures of the large, late-stage welded accretionary berm at Camp Sea Haven, Plum Island on 23 August 1967, the seaward portion of which is illustrated in Figures 14 and 15. (A) General view looking south, showing tire tracks on top of the berm. The COASTAL HERO in this picture is Fayez Anan. (B) Close-up of the middle of the trench shown in A. Note the landward and upward growth of the crossbedding formed by the advancing slip faces of landward-migrating intertidal bars. From Hayes, Anan, and Bozeman (1969).

While doing this work, they came up with an empirical relationship that dictates which of these different bedforms are active at any given time. It is called the **Froude number**, which equals:

$$\textit{Froude number} = \frac{V}{\sqrt{gD}}$$

V = velocity
g = gravitational constant
D = water depth

Therefore, as the water velocity increases, the Froude number also increases, and as depth increases, the Froude number decreases.

The first part of the sequence up to the washed-out megaripples occurs when the value of the Froude number is <1.0. This part of the sequence is called the **lower flow regime**. By increasing the Froude number to values >1.0, by either increasing the velocity of flow or decreasing the water depth (see formula), the sand bed goes into the **upper flow regime**, where either flat beds or antidunes are formed.

The lower flow regime bedforms are asymmetrical, with the sand grains moving in discreet steps by rolling or bouncing up the gently sloping upcurrent side of the bedform and sliding down the steeply dipping surface on the downcurrent side (called the slip face). Thus, these forms move along in the direction the current is flowing, sometimes several feet during a single tidal cycle.

The most common bedform type under upper flow regime conditions is the **flat bed**, over which the flow of the sand grains is more-or-less continuous, streaming in long streaks across the sediment surface. This is the most common type of sand surface occurring on sand beaches, where water depths are typically small and wave-generated water flow is generally rapid. Thus the Froude number is typically >1.0 (see formula again). Under conditions where the velocity is more rapid than normal, for example, when the water flows down a subtle slope (e.g., over the crest of an intertidal bar or down a more steeply sloping high beachface), a feature called an **antidune** is formed. Under this condition, the sand surface develops a series of sinusoidal forms oriented perpendicular to the direction of water movement. A relatively thin layer of water flows in phase across the top of these features, which are linear and parallel with a typical spacing of about 35-50 centimeters (on intertidal portions of sand beaches). The water flowing down the slope on the downcurrent side of the sinusoidal antidune picks up sand grains from that relatively steep slope only to drop them out again as the water flows up the slope on the upcurrent side of the next antidune in line, which creates bedding that dips upcurrent (opposite to the direction in which the water is flowing). Actually, a more accurate definition of the shapes of antidunes is that of a trochoid (Hand, Wessel, and Hayes, 1969), a word created by Gilles de Roberval for the curve described by a fixed point as a circle rolls along a straight line (Wikipedia Encyclopedia).

Intertidal Bedforms

With the moderately large tides, strong tidal currents, and abundant sand on the northern New England coast, it was hard to ignore the abundant bedforms that decorate the surfaces of the tidal sand flats, tidal deltas, and tidal point bars in the area. In the Glossary of Terms at the close of the "White Elephant," written by Anan, Farrell, Hayes, and Greer (1969), we defined bedforms as follows (with some slight modifications of the original):

The sequences of bedforms that develop under conditions of increasing flow regime are:

1) RIPPLE: Asymmetric bedform formed by unidirectional flow. Wavelength less than two feet.
2) LUNATE-LINGUOID RIPPLE: Asymmetric bedform with concave scour pits located on their downcurrent ends.
3) SAND WAVE: Wavelength greater than twenty feet. [NOTE: We still were not sure which of the two major bedform types, megaripples or sand waves, were formed under higher flow regime conditions. We also puzzled over the effect of water depth (sand waves are more common in deeper water)].
4) MEGARIPPLE: Wavelength between two and twenty feet.
5) LINEAR MEGARIPPLE: Megaripple with straight crests.
6) SCOUR MEGARIPPLE (also called CUSPATE MEGARIPPLE): Megaripple with undulatory to cuspate crests and well-developed scour pits downcurrent of the slip faces.
7) PLANE BED: Flat bedding surface upon which parallel laminations of sand are deposited, formed under upper flow regime conditions (Froude number >1.0).
8) ANTIDUNE: Symmetrical, trochoid-shaped bedform also formed under upper flow regime conditions that is in phase with the water surface. Produce low-angle, bidirectional crossbedding. [NOTE: I now think this last statement is wrong. The antidune beds mostly dip upcurrent (Hayes and Kana, 1976; Barwis and Hayes, 1985). See also the antidune pictured in Figure 11A.]

In any event, these studies of intertidal and subtidal bedforms were more or less championed by Boothroyd (1969) and later elaborated in more detail and improved upon by Boothroyd and Hubbard (1975).

In her paper on the Parker River Estuary in the "White Elephant," Joan DaBoll (1969) produced the map of the sand waves on the ill-defined flood-tidal delta in the estuary entrance shown in Figure 17. In describing this particularly complex intertidal flat, she said: "In the flood bifurcations off the main channel, flood-oriented sand waves have formed with maximum wave lengths of 225 feet and heights of 8 feet. These … (like those sand waves submerged in the main channels) … are also largest near the mouth of the estuary and decrease in size up the channel. On the flats, flood-oriented sand waves dominate, governed by two factors. The first factor is the pattern of flood-current asymmetry and maximum

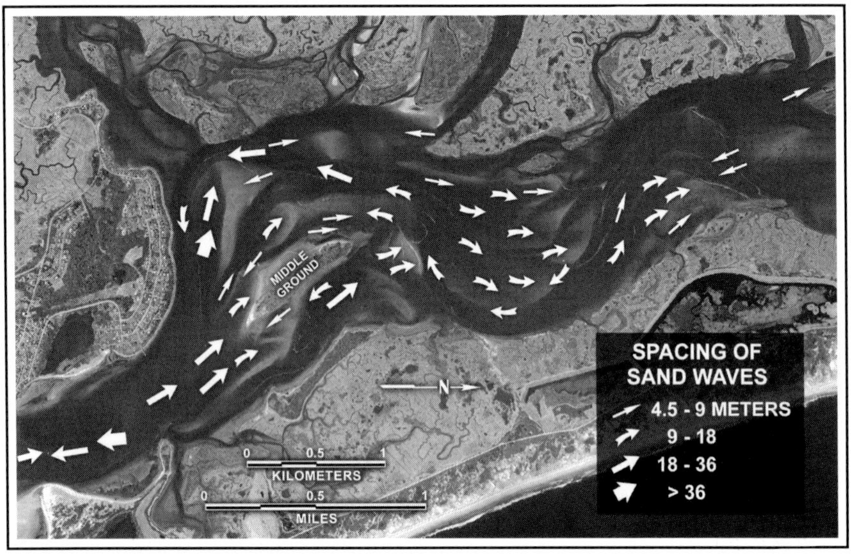

FIGURE 17. Lower portion of Parker River Estuary, Massachusetts. Note the dominance of flood-oriented features. The major sand body at the middle left of the diagram is an ill-defined flood-tidal delta called Middle Ground. It is about one mile from the throat of the tidal inlet (Figure 12). After Daboll (1969) and Boothroyd and Hubbard (1975). In this case, the original bedform orientations and sizes shown by these two sources are approximately superimposed on an orthophoto acquired in April 2008, courtesy of the office of Geographic Information (MassGIS), Commonwealth of Massachusetts Information Technology Division. An oblique aerial view of these sand waves is given in Plate 1.

velocity over the flats, and the second factor is the shape of the flats. Most of these flats rise gradually to a well-defined high ridge on the up-stream side." An oblique aerial view of these sand waves is given in Plate 1.

Tidal Inlet Morphology and Processes

We spent a lot of field time in the tidal inlets at Essex Bay, Parker River Estuary, Merrimack River Estuary, and Hampton River Estuary (Figure 12). Also, in both his M.S. and Ph.D. research, Stew Farrell studied the dynamics of two tidal inlets in Saco Bay, Maine, namely the Saco River and Scarborough Estuary entrances. Essex Bay Inlet, which is illustrated by the vertical image in Figure 18, was particularly interesting because it contains ebb- and flood-tidal deltas of near equal size, a fairly unusual occurrence, as will be discussed in Chapter 10 – PRICE INLET. However, rare or not, both tidal deltas were well exposed in the Essex Bay Inlet and relatively easy to study. Ultimately, these tidal deltas became our original representatives for creating general models of these two types of tidal deltas, which have now become well ingrained in the coastal geomorphology literature.

As shown by the general model of ebb-tidal deltas given in Figure 19, the morphological components of ebb-tidal deltas include a main ebb channel (with slightly stronger ebb currents than flood currents), flanked by linear sand bars on both sides, and a terminal lobe at the seaward end. The main ebb channel is bordered by a platform of sand (swash platform) dominated by swash bars (intertidal bars built up by wave action that migrate very slowly landward). The swash platform is separated from adjacent barrier beaches by marginal flood channels (which are dominated by flood currents).

JM and I noted in our Central California book (Hayes and Michel, 2010) that the largest ebb-tidal delta in the world, as far as we know, is situated just outside the entrance to San Francisco Bay. The general configuration of the ebb-tidal delta off the Golden Gate Bridge is strikingly similar to the general model given in Figure 19. However, this is not a big surprise, because the ebb-tidal model is quite consistent around the world, especially in areas with moderate-sized tides (mesotidal; tidal range = 2-4 meters), such as San Francisco Bay, which is mesotidal during spring tides. I recently did a Google Earth tour of the coastline of Australia, during which I spotted many ebb-tidal deltas that matched the model quite well.

Flood-tidal deltas, the general model of which is illustrated in Figure 20A, tend to show more variation globally than ebb-tidal deltas. As shown in Figure 20B, the larger bedforms on the flood-tidal delta are typically flood oriented.

FIGURE 18. Vertical image of Essex Bay tidal inlet (Figure 12) at low tide, showing the relatively equally developed ebb- and flood-tidal deltas. Orthophoto acquired April 2001, courtesy of the office of Geographic Information (MassGIS), Commonwealth of Massachusetts Information Technology Division.

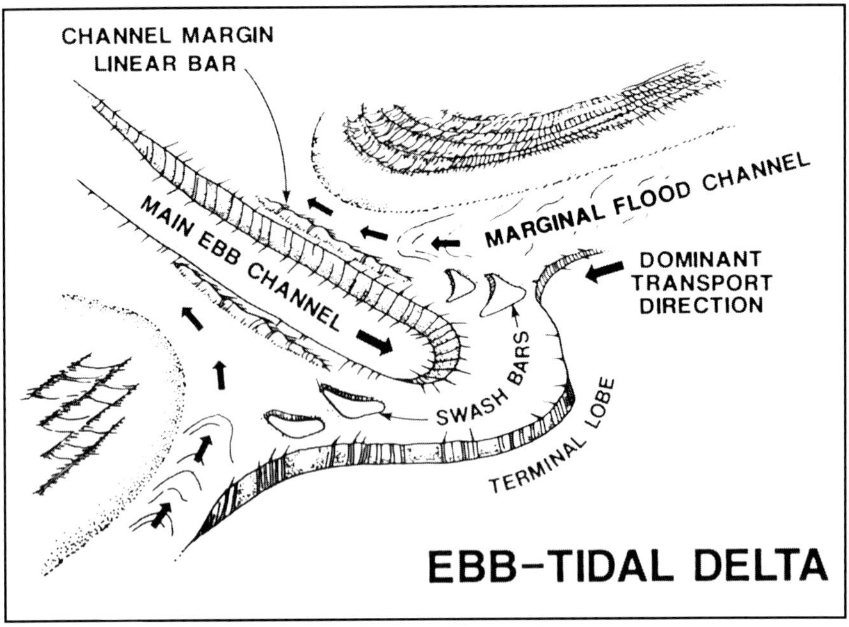

FIGURE 19. Typical morphology of ebb-tidal deltas in mesotidal settings (after Hayes, 1980a). Arrows indicate dominant direction of tidal currents. The model was derived for the ebb-tidal deltas of the New England area (see example in Figure 18), but it applies equally well to many other barrier-island systems around the world.

Based in part on these inlet studies with an assist from new research on the South Carolina coast, in 1975 I published a paper entitled *Morphology of Sand Accumulation in Estuaries* (Hayes, 1975), which has been cited some.

The spectacular tidal inlets of New England are still under intense scrutiny by one of the major COASTAL HEROES of Chapter 10 – PRICE INLET, Duncan FitzGerald, who is presently a Professor at Boston University.

Story One

(March 1967; early in the morning across the river from Newburyport, Massachusetts)

It was barely daylight as the driver pulled our limousine up in front of the diner across the Merrimack River from Newburyport, Massachusetts (Figure 12). We had just completed a two-hour drive from the UMass campus in Amherst to Newburyport with plans to survey a number of our permanent beach profiles along the coasts of Massachusetts and New Hampshire. Four or five graduate and undergraduate students accompanied me on this trip, during which we intended to repeat what we had been doing during low spring tides every two weeks for a couple of years. And, yes, we were riding in a limousine, none other than the one Richard Burton and Elizabeth Taylor had used when they filmed *Who's Afraid of Virginia Woolf?* in Northampton, Massachusetts a few months earlier. At that point, I did not yet have enough pull, nor money, at UMass to obtain some *bona fide* field

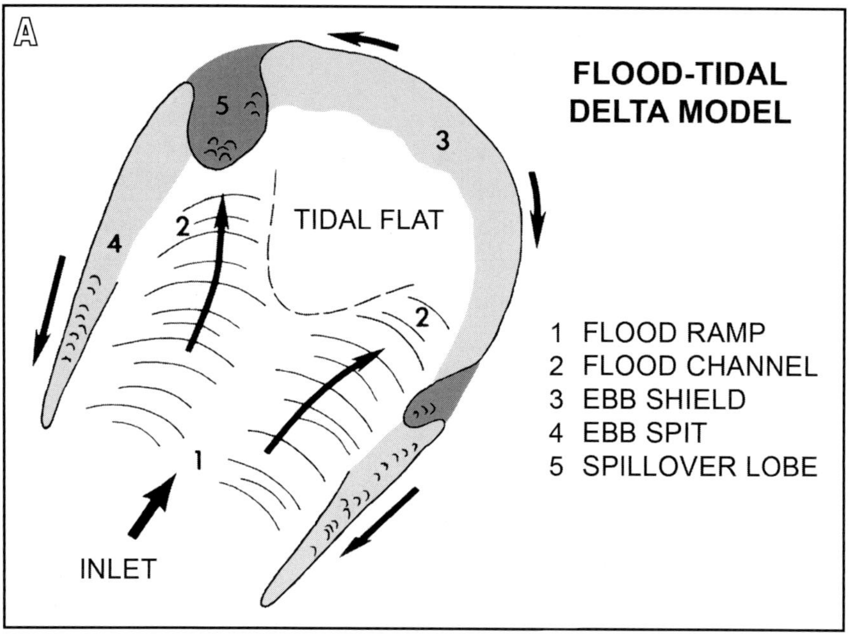

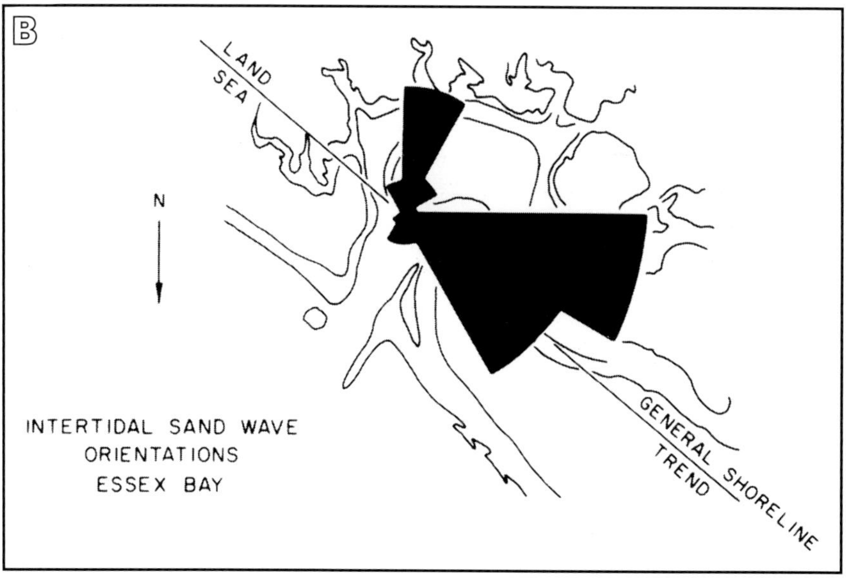

FIGURE 20. Flood-tidal deltas. (A) General model. Arrows indicate dominant directions of tidal currents. From Hayes and Kana (1976). (B) Summary of intertidal sand wave orientations, Essex Bay (see Figures 12 and 17). The dominant direction is landward with a strong component parallel with the shoreline. After Boothroyd (1969).

vehicles. That would come later. I cannot remember exactly how we managed to persuade the Department of Motor Vehicles to lease the limousine, but we surely were riding in style.

Inside the diner, the trim, 35-year old waitress took all five of our orders without writing anything down. She was amazing, never making a mistake.

But something about the diner was new this time. The owner/cook, a short, jovial mustached Italian fellow had plastered all of the walls with large banners, proclaiming –

RED SOX WILL WIN THE PENNANT AND WORLD SERIES IN '67 !*!

YAZ TO LEAD THE TEAM TO GLORY!

GO SOX, THIS IS THE YEAR!

And so on and so forth. I'm doing this from memory.

A couple of us, probably Jim Wessel and I, who were the most baseball savvy, gave him a hard time.

"You've got to be kidding," I said, "They don't have a chance! What was their record last year?"

"Uh, they won 76 and lost 90," Wessel informed us. Then everyone in our group laughed. The cook just grinned, shaking his finger at the banners, "Just you wait and see."

[NOTE: When we had this encounter, spring training was just starting for major league baseball. In one of the most stunning turnarounds in major league baseball history, the Sox record in '67 was 92 wins and 70 losses, virtually the opposite from the year before. They won the American League pennant, all right, but lost the World Series in seven games to the St. Louis Cardinals. Carl Yastrzemski was phenomenal, winning the Triple Crown. Want to know a good baseball trivia question? Here it is. Who was the last major league batter to win the Triple Crown (season leader in home runs, batting average, and runs batted in)? Answer: Carl Yastrzemski in 1967. I personally saw several of the home games in Fenway Park that year, no doubt one of the greatest in Red Sox history.]

[ANOTHER NOTE: It was about 7:30 pm on a Saturday night, 16 April 2011, and I was in my season-ticket-holder seat in the new baseball stadium of the USC-East Gamecocks, who were involved in a critical three-game series with the Vanderbilt Commodores. I believe these two teams were tied for first place in the Southeastern Conference standings at that time. Anyway, leading off the top of the second inning was the Commodore right fielder, who was batting in the 5th position in the Vanderbilt batting order. The batter's name was Mike Yastrzemski. I had heard a rumor that he is the Great Yaz's grandson (later confirmed). Not surprisingly then, he batted from the left side of the plate, looking a whole lot like grandpa. I believe on the third pitch to him, he hit the ball far over the right field wall and into the street–just like grandpa used to do it! Same swing, same trot around the bases. Young Yaz got another RBI on a walk later in the game, and he also made a great catch against the right field wall. Vandy won that game 6-4, but the Gamecocks won the other two games. It was the only SEC series loss so far this season for the Commodores (two more series to go), and at this time, they lead the SEC by one game and have been voted number 2 in the country by most of the writer's polls (I am writing this on 13 May 2011). The Gamecocks are ranked number 4. Actually, these two teams ended up being co-champions of the conference at the end of the regular season (along with the Florida Gators). I checked on Google and found out that grandpa had taught Mike to hit, because Mike's father had died at the age of 40. It certainly remains to be seen whether or not Mike will make the big leagues, where grandpa starred for 23 years, playing in 18 All-Star games, winning seven Gold Gloves (while playing in left field; Mike was playing in right when I saw him), and, as noted earlier, being the last player in the major leagues to win the Triple Crown (in 1967).]

Be that as it may be, after our delectable-delight breakfast that morning in March 1967, we were excited to go measure the beach profiles. This was a field trip, which means we were out in the field, probably the primary reason I became a geologist in the first place. How do I explain how it feels? Looking several years ahead, at 9:00 p.m. on 6 May 1974, I was sitting in my motel room near Cattaraugus Creek, New York after spending a day working on the shoreline of Lake Erie with Leita Jean Hulmes and Rob Clemens. I wrote in my journal–"I wish I could really explain how it feels to be in the field. It's the wind on your face, with your mind racing to keep up with the spectacle of nature you're trying to observe. It's companionship. It's plotting and scheming. It's dreaming the big dream of the great masterpiece to be written–that's already forming in your mind. It's working out a precise plan for tomorrow's work. It's reminiscing over the successful completion of today's plan."

Of all the different activities I ever did in the field, I enjoyed running topographic beach profiles the most. Starting from the back stake, we would stretch out the tape and I would sight on the horizon to determine whether the position where the front profile rod was placed was higher or lower than the last point. At the end of the run, we were usually out to at least knee-deep water in our waders. Seems like most of the profiles I ever ran were in cold water. Then I had the fun of making a sketch of the beach, a very important part of the process, because by sketching the beach, you finally really observed it in all of its component parts.

Seems like I have been leading field trips for most of my professional life, culminating in 2005 as the 30-year stint of leading Modern Clastics Seminars on the South Carolina coast for the American Association of Petroleum Geologists (AAPG) and other clients came to an end (discussed in Chapter 14 – THE ROCK RECORD.)

Speaking of field trips, let us skip ahead to May 1969, when our team led a three-day field trip along the shoreline of northeast Massachusetts and New Hampshire that was sponsored by the Eastern Section of the Society of Economic Paleontologists and Mineralogists.

It all started in October 1964, when Al Scott and I organized a trip on the depositional environments of the south-central Texas coast, the area I studied during my dissertation project (Scott et al., 1964). The leaders on that trip, besides Al and me, were Peter B. Andrews, Walt Siler, and Bill Behrens. It was a good learning experience, as we became acquainted with several things not to do the next time.

The two primary sponsors of much of the research discussed during the field trip on the Massachusetts/New Hampshire coasts in 1969 were the Office of Naval Research (Geography Branch) for the estuarine work and the Coastal Engineering Research Center, U.S. Army Corps of Engineers for the beach work.

I think about 35 or 40 people showed up for this field trip. There were a few logistical hurdles to jump. Special permission was needed to take such a large group on a drive into the Parker River National Wildlife Refuge, but the personnel of the Refuge furnished vehicles to do so. Getting to traverse down Crane Beach, a beautiful, somewhat isolated natural system, was accomplished with a number of four-wheel drive vehicles supplied and driven by members of the Massachusetts Beach Buggy Association, a fun trip. We had three boat rides engineered by some of my most skillful graduate students and last, but not least, was the overflight of the entire area visited on the ground. We visited 29 individual stops on the ground, with vigorous discussions of the points made by yours truly or one of the graduate students being common.

At the end of the last day, as we were getting ready to disperse, my long-time friend, the engineering guru and thinker, Cy Galvin, who at that time worked for the Coastal Engineering Research Center of the U.S. Army Corps of Engineers in D.C., walked up to me and said, "I think you would have done well in the military." I think he meant that he was impressed with the organization of the trip. We were lucky. The weather was good. None of the vehicles or boats broke down. We had no major problems with the planes. Yes, we were lucky.

As time passed, we received a lot of plaudits for this field trip. William T. Fox of Williams College, a modeler destined to write many papers on beach processes with my friend and ex-classmate at UT, Skip Davis, wrote me a letter in which he said our trip was "The best field trip I have ever been on."

George deVries Klein, another long-time friend, who was then at the University of Illinois Geology Department, also wrote me, saying how impressed he was with the participation and understanding of all the graduate students who had helped lead the trip.

A few weeks later, another letter came, this time from the famous future nemesis of the Corps of Engineers and modelers, Orrin Pilkey of Duke University, who said in his letter that he had heard good things about the trip, and that he would like to have a copy of the guidebook (i.e., the "White Elephant") now that he had "buttered me up." Okay, so we sent him one, for free, too.

That field trip was one of those unique experiences that could never be repeated in the same way, no matter how hard you might try.

Story Two

(Summer of 1970, coast of Massachusetts)

The following account, in reference to Skip Rhodes' M.S. thesis on the subsurface stratigraphy of the Essex Bay/Parker River Estuary area, was provided by Stew Farrell, who worked in the field with Skip on the project.

In the saga of the Rhodes seismic study, this was a technological first for the Coastal Research Group. We had big "Explosives" signs posted on the crew cab of our truck. One day we went to lunch in Exmore at a Burger King, parked the truck, and went in. Outside a crowd gathered at the truck, but the focus was Steve McQueen, not the signs. He was filming the "*Thomas Crown Affair*" and they shot footage of the dune-buggy scenes on Crane Beach (Figure 12). McQueen came into the Burger King and asked about the explosives truck outside, and we told him, yes, we are carrying dynamite and plan to use it that day. However, we were banned from the beach during the movie shoots, so we worked on the ebb-tidal delta of the Parker River Estuary instead. The A-frame on our research boat, the *RV Sorby*, can be seen in the distance in one scene in the movie.

The seismic effort was done using a relatively primitive Texas Instruments reflective seismic unit that employed geo-phones in a linear array each side of the shot hole. The shots were done using dynamite and fulminate of mercury electric blasting caps. The seismic unit set off the explosive down a hole or buried in the sand with the results projected onto a Polaroid camera print mounted on the unit. It was then Skip's job to reduce each set of data to derive the reflective layers and determine the depths to each.

I (Stew Farrell) was the primary dynamite handler since I had once been a chemist and allegedly knew the most about this material. In 1970, one could get a dynamite handler's permit in Massachusetts by filling out an address form. Dynamite was sold at most serious farm supply outlets to blast stumps and large glacial erratics in fields. Dynamite is a mixture of saw dust and tri-nitrotoluene in percentages between 40% TNT and 60% saw dust up to 75% TNT and 25% saw dust. It is wrapped in red paper with the ends tucked over. One never used a full 8-inch stick of dynamite in one shot hole, so there were times when I used an eighth of a stick, up to a quarter of a stick. The mercury caps were very powerful by themselves, so in sand we usually took a stick of dynamite, opened one end, and used a plastic butter knife to scoop out the gelatin like so much peanut butter and "grease the cap" with it. That learning experience followed an intertidal sand shot in Essex Bay where I used an eighth of a stick and buried it in a foot of sand. Skip triggered the shot and the blast 600 feet away rattled our teeth. While we looked at each other in amazement, suddenly 1 to 3-inch rocks began to rain down all around us. We all three dove under the crew-cab's tailgate, leaving the UMass seismic box running on the tailgate (this was Geology Chairman Bromery's pride and joy instrument), and watched 25 rocks thud down on the sand where we had parked. No damage done, but learning had occurred. The marshes, however, absorbed blast energy to the point where whole sticks of dynamite were required to get a return from the till or bedrock below. For these holes, we loaded the holes with an oar to push the dynamite about 6 feet below the surface. The blast usually produced a 30-foot geyser of marsh mud but very little noise.

So these were the tales related to Steve McQueen over burgers one day in the summer of 1970 while Miles and crew were discovering many things on the beaches, marshes, and glacial streams of Southeast Alaska (discussed in Chapter 4 – ALASKA: *SUZANNE,* EARTHQUAKES, AND TSUNAMIS).

The second phase of the project was to ground-truth each series of shot lines with a physical bore hole. For this task, we had borrowed the UMass Civil Engineering trailer-mounted wash-boring rig. This ancient contraption was readied for service (not easy since it had not been started in years), but it needed to be moved over water too. The trailer could be dragged along the beach, but the tidal deltas and marshes were a problem. The answer was to simply build a big raft. Rhodes, Richard Cahallan, and I worked behind Morrill Hall at UMass to construct a 16 by 16-foot raft that could be trailered to Plum Island, launched, and assembled to carry the drilling rig. This raft, with the drill rig, the 250-gallon water supply tank, the pipe rack, the gasoline supply, and gallons of green-head fly repellent, was pushed to each site by a 16-foot aluminum skiff and a 20 HP outboard (one of the same vessels used during the 1969 field trip that is discussed in this chapter). The rig would be positioned at high tide over the proposed drill hole, anchored, and drilling would commence as the tide dropped. The rig was capable of driving 150-feet of 4-inch casing and reaching inside with 2-inch rods with a variety of tools for sampling the material below.

The happy threesome, Rhodes, Callahan, and myself, proceeded to drill and blast our way around the Parker River, Essex Bay marshes, sand and mud flats and flood- and ebb-tidal deltas. The work continued into the fall of 1970.

With respect to the science, the information obtained during that summer of 1970 on the depth to bedrock and the nature of the variation of the deposition on the Pleistocene materials was awesome.

I will conclude by describing the last day of work on the seaward edge of the Parker River ebb-tidal delta in late October, which was during the bird-hunting season. Done at last, but with 3 sticks of dynamite left and not wanting to drive around to the Essex Town Dynamite Storage Locker yet one more time, Skip and I elected to try an air-blast for effect. The old oar was wired up, we got 1,200 feet away and hit the electric plunger. OMG, the shock wave knocked us both down and the echo rattled around the hills and estuary for 30 seconds. All the bird-shooting ceased for over an hour while we executed a classic "get-the-hell-outta-there" maneuver.

Skip took notice several days later of an article in the Essex *Gazette* about a mysterious sonic boom that had broken windows and rattled the local citizens of Essex on the very date of the tidal delta work. The blame was laid at the feet of the U.S. Air Force whose F-4 Phantoms were exercising from Pease Air Force Base that day. We took no steps to remedy this notion.

4 ALASKA: *SUZANNE*, EARTHQUAKES, AND TSUNAMIS

There are strange things done in the midnight sun
by the men who moil for gold;
the Arctic trails have their secret tales
that would make your blood run cold;
The Northern Lights have seen queer sights,
but the queerest they ever did see
was that night on the marge of Lake Lebarge
I cremated Sam McGee.

ROBERT SERVICE in THE CREMATION OF SAM McGEE

Music is the tree that never was,
falling in the imaginary forest,
heard by all that listen.

TOM FREEMAN

The Science

As far as I can recall, the reading by my teacher Mr. Wells of *The Cremation of Sam McGee* while I was in high school was my first reason to think much about Alaska. However, in the summers of 1969, 1970, and 1971, while still a professor at UMass, I led field excursions to the southern Alaska coast that changed my life.

In Alaska I learned something new about life. I learned to feel. I felt the wind, and I saw what the Indians called the "Spirit Father" or the "Great Spirit." The natural things, the earth god–pure, simple, cold, wet, real. I loved the god of nature that I had rediscovered there, because it brought back the child in me. And in so doing, I was able to take pictures superior to any I had taken before, and to decipher nature apart from pure science.

As a practicing geologist, I have by now worked on the Alaska coast for over four decades, kind of living out a part of the life described by Robert Service (no cremations, however). Some of these adventures will be discussed herein. In fact, as I write this, my wife and business partner, Jacqui Michel (who, as noted earlier, will be known as JM throughout this book), and I are in the beginning stages of writing book four in a series on the shorelines of the different states of the

U.S. This one is titled *A Coast Beyond Compare: Coastal Geology and Ecology of Southern Alaska*. Thus, I am still deeply immersed in the study of the landscape that "changed my life."

Once at UMass, except for the brief foray on The Ice, my graduate students and I first focused on the New England coast. However, in the summer of 1969, we initiated a major project on the southern Alaska coast.

[NOTE: Although the field work on this Alaska project began in the summer of 1969, I had an abdominal operation that summer and was able to spend only a fairly short time in the field. However, significant strides were made to get the project off the ground by three of my graduate students, Skip Rhodes, Stewart Farrell, and Jon Boothroyd, who did a somewhat historic trip to the area, initiating the studies that would follow during the next two years. Because I was not there while much of this work was being done, I asked Stew Farrell if he would do a write-up on their adventure that summer, which he did. The rest of the discussion in this NOTE was written primarily by Stew.

The early spring of 1969 was one of earnest anticipation by the Coastal Research Group for doing research work along parts of the Alaskan coast out of a fishing port called Cordova. None of us at the UMass Coastal Research Group had ever heard of Cordova, but we looked it up on the maps to see where we were headed. The initial mission was conducted by Eugene (Skip) Rhodes, yours truly Stewart Farrell, and Jon Boothroyd, along with two undergraduate field assistants, Bob Henry and Dennis Wilkins. We expected Miles to join us when the camp was established.

Getting ready was a series of logistical exercises in what to wear, what to have at hand, what to bring to eat, and what else to worry about. Since Skip had graduated as an Army Second Lieutenant from ROTC, we moved to acquire as much military help as possible. Skip secured MATS transport for food supplies and two amphibious COOTS (an all-wheel drive, all-terrain vehicle we figured we needed to have). We bought many supplies on base using Skip's military purchasing power.

Finally in early June, four of us departed UMass for Alaska in a rented station wagon. We had been told we needed "bear protection," so Jon borrowed Dr. George McGill's 44 Magnum revolver and I purchased a "sporterized" .303 WWI Enfield rifle. These and many other things were crammed into the station wagon and off we went. Being good geologists, we stopped at multiple places, reaching Seattle three weeks later via Grand Canyon, Yosemite (where Jon and I took a one evening course taught by Ansel Adams; neither of us bought any of his prints at $15.00 each, damn!), Devils Tower, Las Vegas, Mt. Hood, and Mt. Rainier.

In Seattle, the four of us boarded an Alaska Airlines flight to Juneau, hoping to be met there by Skip, who had driven an aging Geology Department International "suburban" and an over-loaded 2-axel U-Haul trailer to Vancouver, B.C. He planned to take the ferry from there to Juneau.

As we boarded the plane in Seattle, the stewardess requested that all weapons be surrendered at the door. The Enfield was packed as baggage, but Jon had purchased a nice leather holster for the .44. He carefully handed over the gun, expecting to see the girl flinch, but she took it by the grip, looked at it briefly, spun it around her finger and promptly stuffed it into the waistband of her uniform.

"Don't worry, sir, I will return it in Juneau," she said.

Now it was Jon who looked surprised. On the way back to our seats, I said, "Jon, you just met your first REAL woman."

After dealing with a few issues in Juneau, and successfully meeting up with Skip there, the five of us took the ferry to Haines and drove the ALCAN Highway to Anchorage. The ALCAN is now fully paved and more urbane, but in 1969 only the southern 500 miles was asphalt with the 1,000 remaining miles being composed of dirt and graded gravel. We managed about 250 miles per day, and about 2 flat tires per day, on either the International or the trailer.

Finally, we neared the Alaskan border at a town called Destruction Bay, Yukon. The "settlement" consisted of a restaurant, a serious garage/repair shop, and a single family home. Mother and two daughters ran the restaurant and father and son ran the shop. It was raining and the road was slick, so the 13th and final flat tire was slid along the muddy road for several miles to the "settlement." We arranged for a new tire, sat for dinner, and chatted with the family for several hours. Feeling sorry for us, they let us "camp" in the restaurant after it closed for the night. They gained the fee for breakfast for five in return the next morning.

That day we reached Anchorage and collected the MATS transported field equipment and food. We also rented a flat-bed truck for the COOTS and otherwise made preparations to head to Valdez in order to take the "ferry" to Cordova (see Figures 21A and 22). We all went to a movie that night in Anchorage, and afterward found it odd being chatted up by the local prostitutes in broad daylight at midnight. After all, it was June in the Arctic.

Jon drove the International to Valdez and Skip took the flat-bed. The descent into Valdez is quite steep and the International was almost pushed by the trailer off a cliff.

The ferry was a 12-vehicle vessel named the *Chilcoot*, and the captain was from New England originally, so I spent the night on the bridge drinking bloody Mary's with the Captain and 1st Mate as they tried to spot bear and moose with the ferry search light. Cruising about 50 feet from the near vertical fjord walls was disconcerting at night, but the depth sounder did say 250 feet most of the time. I also doubt that the U.S. Coast Guard would have approved of the non-charted course we were following. Meanwhile, the two field assistants had discovered that a truck driven onto the ferry at Valdez was loaded with the summer's liquor supply for Cordova. We had no lack of beverages for the rest of our stay.

The arrival in Cordova was uneventful except for when Skip went to open a bank account for the project. He presented the nice green U.S. government check issued to the Coastal Group and opened the account, asking for $500 in walk-around money. The $83,000 balance was gratefully accepted. Somehow, I was only briefly surprised when Skip and I walked into the marina to purchase a boat and motor for the field work to be immediately addressed as Mr. Rhodes and Mr. Farrell by the manager, who I guarantee you, neither of us had ever seen previously. Word of the "researchers" spread quickly in Cordova.

While we were in the field, Jon and Wilkins spent many happy hours mapping

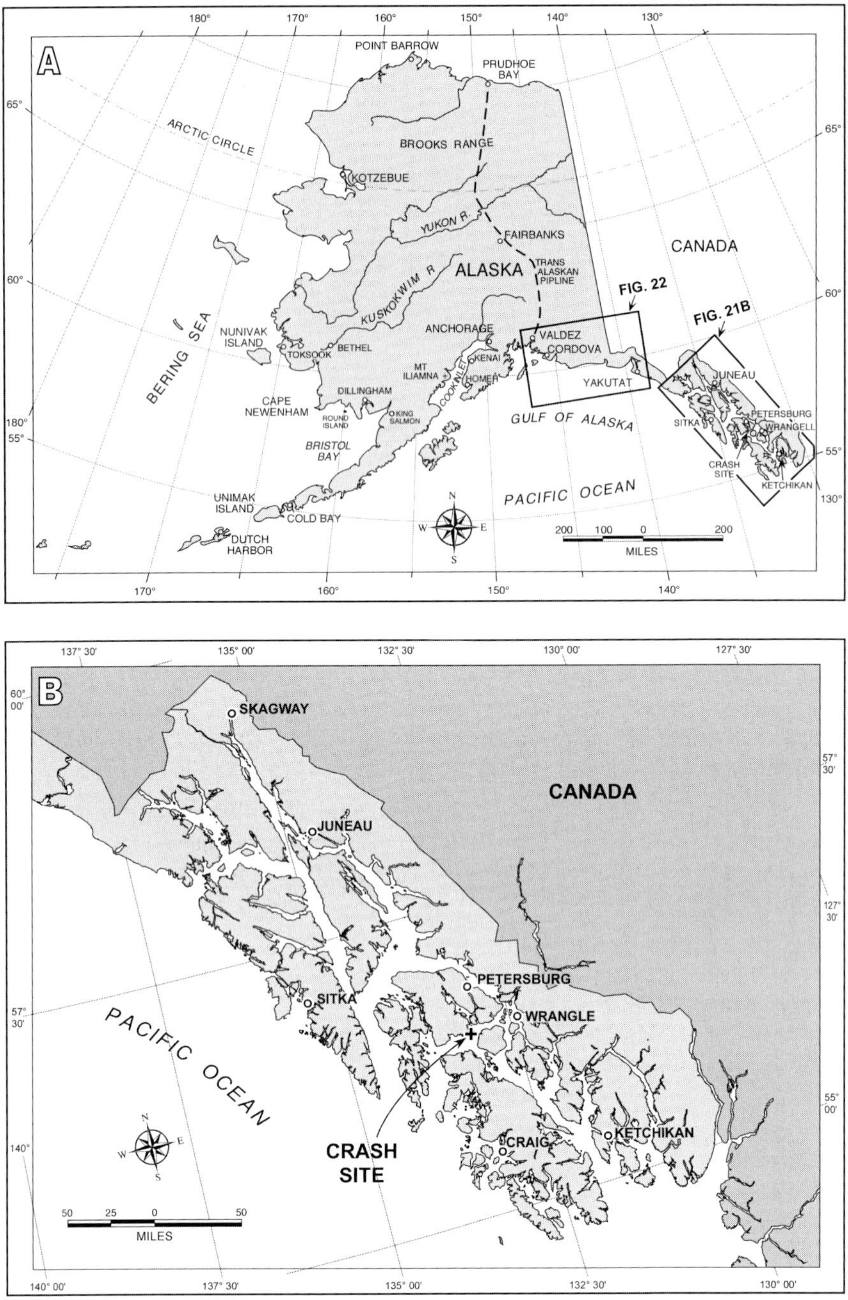

FIGURE 21. Location maps for the southern Alaska coast. (A) General location map. Boxes indicate the areas shown in Figures 21B and 22. (B) Far southeastern coast of Alaska, showing the location of my airplane crash in August 1990, which is discussed briefly in Chapter 15 – BY FOOT, BY CAR, BY BOAT, BY PLANE and in Chapter 16 – DOOMNESS.

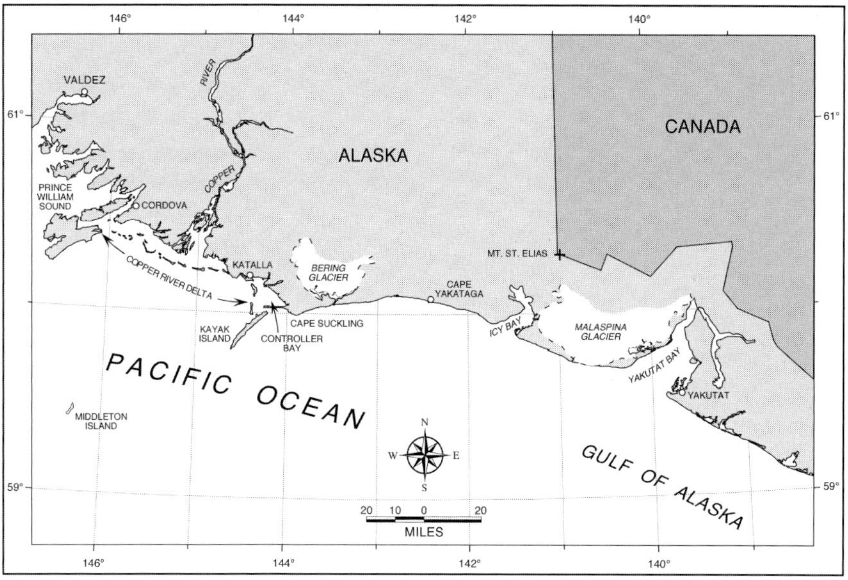

FIGURE 22. Location map, Cordova/Yakutat area.

the complexities of the Scott Glacier and its outwash plain, while Bob Henry and I walked miles and miles along the barrier island shoreline establishing beach profiles, measuring dunes, washovers, and photographing everything in sight. An aircraft was leased to gather the unprecedented views of coastal and glacial Alaska.

Miles arrived by plane in less than ready-for-field-work condition. His interest, however, was off the charts and, as best he could, he followed the flow of data collection until forced by the discomfort and outright pain to depart back to the lower 48 to recuperate. His time in the Alaska "sun" was to come in 1970.

One night in August, Bob Henry and I took a ride in our open, 18-foot skiff back to the camp from about 35 miles to the east, navigating through the tidal channels and marshes. Bob appeared to be certain that we had no chance of surviving. Later, I told Bob how ancient natives navigated using notches in the mountain ranges as aiming points and that I had set one up as we departed that morning. In August, it actually got nearly dark from midnight to about 2:30 a.m., so we worked from noon to midnight, because that was low tide at that time of year and we could survey the ridge and runnel systems. We did work long days.

In late August, the five-man crew packed up and drove back to Haines together and took the ferry to Juneau, where we all re-discovered civilization for several days. Then we boarded the RV *Matanuska* to travel the Alaska Marine Highway back to Seattle, and then drove across the U.S. to Amherst and the ordinary once again.

The expansion in the scope of understanding of coastal processes in a totally natural setting, under a 16-foot tidal range estuary exchange over the course of nearly 3 months of 22-hour daylight provided insight into every aspect of the coastal science and the sidebar of research logistics that could never, ever come from classroom experience or work along the developed Atlantic U.S. shoreline. I treasure every moment (we were there) greatly.]

These studies on the Alaskan shoreline were the first opportunity I had to carry out field studies of what are known as *leading-edge shorelines*. Up to that time, my personal fieldwork had been concentrated on *trailing-edge shorelines*, namely in Texas and New England. These terms are derived from the concept of plate tectonics, whereby a tectonic plate made up of a continental crust is riding over a sinking oceanic plate. For example, the western shoreline of the North American continent where the continental and oceanic plates collide, or the leading edge of the tectonic plate, is characterized by: 1) High mountains caused by relatively recent orogenic (mountain building) activity; 2) Bedrock of variable age but dominated by relatively young sedimentary and volcanic rocks; 3) Active tectonic uplift [e.g., with old beach lines (raised marine terraces) lifted up the sides of the mountains thousands of feet in some places]; and 4) Mostly short, high-gradient (steep) rivers emptying into the sea. The eastern side of the North American Plate, or the trailing edge of the continental plate, which is moving westward at a rate of around 4 inches/year (Monroe and Wicander, 1998), is dominated by depositional coasts (Amero-trailing edge coasts as defined by Inman and Nordstrom, 1971) characterized by: 1) Coastal zone made up chiefly of broad coastal and deltaic plains; 2) Bedrock composed generally of Tertiary and Quaternary sediments (young geologically speaking); 3) Tectonically subsiding areas (e.g., thousands of feet of sediments have accumulated in some areas); and 4) Many large and long rivers emptying into the sea.

With reference to the leading-edge Alaskan coast, this is also where we conducted our first field work on a shoreline where glaciers carry heavy loads of sediment to the shore.

One of the most noteworthy products of that early Alaskan research in 1969 and 1970 was a tone poem or slide show, called *Suzanne's Lament*, which dealt with the environmental preservation of the Alaska shore, among other things. I had thought about doing something like this before we started the fieldwork in the summer of 1970, using slides Barry Timson and I took during a brief survey of the Alaska coast in February 1970. For example, I put together a set of slides set to the Beatles song *Let It Be* that compared the pristine Alaskan shoreline with highly developed shorelines, such as the one in New Jersey. I showed this vignette at the end of a lecture that I gave on a speaking tour around the country later in the spring that year. This went very well. Accordingly, during the summer of 1970, we took a large number of photographs with the idea of expanding upon this earlier start. Paul Hague, my friend and co-worker on the 1970 field project in Alaska, and I created the show in the fall of 1970. It was presented all over the country and was eventually converted into an educational movie. According to our records, during my remaining days at UMass, from December 1970 to June 1972, we showed this program to audiences of over 28,000 people. We showed it to the students at 12 elementary and high schools in the western Massachusetts area, with audiences of up to 500 students being fairly common. In the spring of 1972, Frank J. Raffaldi, Jr., also a member of our field team in the summer of 1970, and I took the show on a tour of 10 major universities that included UCLA, University of Montana, and University of Illinois, showing it to approximately 4,000 college students.

Thirteen other universities and colleges were also visited during my remaining time at UMass. After I moved to the University of South Carolina (USC-East), we presented the show many more times, but I did not keep a record of them. Also, the movie was purchased by a pretty large group of universities and colleges; no doubt, a significant number of students saw the movie.

In October 2010, I was one of two "Heroes of the Profession" at a national meeting of the American Shore & Beach Preservation Association (ASBPA) in Charleston, South Carolina, which meant that the other "hero" (a coastal engineer named Dave Basco) and I had to sit on the stage in front of about 200 people and field questions solicited at random from the audience. One question was: What was the apex (i.e., greatest accomplishment) of your professional career? Without hesitation, I said *Suzanne's Lament*. Quite a few people in the audience had seen *Suzanne*. Even at this late date, forty years later, I still run into people at scientific meetings and elsewhere that bring it up. What more is there to say?

Paul Hague, my co-creator of *Suzanne*, and I often wondered what element of the show was most important–the basic plea for the environment, the imagery, or the music. The photography (see example in Plate 8) was spectacular, if I do say so myself! I think Paul probably favored the music, being the primary music consultant, with contributions like *Suzanne* by Leonard Cohen, *Both Sides Now* by Dave Van Rock, *Let It Be* by the Beatles, and so on providing the background for the imagery. No doubt the music provided the emotion that fed the imagination and made the message more real.

In addition to the difference (for us) between this leading-edge coastline and those on the east coast and along the Gulf of Mexico, as well as the preponderance of glaciation, another added attraction of the southern Alaska coast (illustrated in Figures 21 and 22) was the frequent occurrence of earthquakes and their associated tsunamis. We arrived on this coast for the first time in 1969 only five years after the devastating Good Friday Earthquake of 27 March 1964. Consequently, we were able to observe the effects of this phenomenon up close in many localities along the shore. The shorelines in the Prince William Sound area were uplifted as much as 30 feet during the earthquake. These uplifted shorelines will be discussed in more detail later (Chapter 15 - GRAVEL BEACHES AND FAN DELTAS). Another interesting aspect of the 1964 earthquake was the tsunamis it created.

A tsunami, a Japanese word translated as *harbor wave*, is a large ocean wave usually generated by a submarine earthquake, although they may also be generated by volcanic eruptions, landslides (both those initiated on shore and submarine), and even by a comet or asteroid impact (not by meteorites; they are too small). Historically, these waves have been called tidal waves, but astronomical tides are not involved in their formation. *Seismic seawaves* is another term that scientists commonly use. These are typically exceptionally long waves that move very fast in the open water at speeds measured in hundreds of miles per hour (mph); however, as they approach shore, they usually slow down to 40-50 mph. Like their smaller, wind-generated cousins, when they enter shallow water, they abruptly peak and break, reaching heights of well over 100 feet in some cases.

As noted in the Wikipedia encyclopedia, tsunamis are most commonly generated by earthquakes where converging plate boundaries generate faults in which at least one side of the fault line undergoes abrupt vertical movements that displace the overlying water. It is very unlikely that major ones can form at divergent plate boundaries or along transform faults because those types of features do not typically have significant vertical movement. Therefore, faults generated by these types of movements do not usually displace the water column vertically. Subduction-zone-related earthquakes generate the majority of all tsunamis (i.e., where major plates converge as along major oceanic trenches around the Pacific Ocean "ring of fire," obviously including the southern Alaska shoreline). As noted by Pararas-Carayannis (2010) in his detailed account of the "Good Friday" Alaska earthquake, that earthquake generated catastrophic tsunami waves that devastated many towns in Prince William Sound, along the Gulf of Alaska, along the West Coast of Canada and the U.S., and in the Hawaiian Islands. In Alaska, the tsunami run-up measurements varied from 20 feet at Kodiak Island, 30 feet at Valdez, 80 feet at Blackstone Bay, and 90 feet at the native village of Chenega in Prince William Sound. This 1964 earthquake caused 119 deaths in Alaska alone, with 106 of these due to tsunamis that were generated by tectonic uplift of the sea floor, and by localized upland-generated and submarine landslides. According to Pararas-Carayannis, there were two different types of "tsunami generation mechanisms" associated with this earthquake: one along the continental shelf bordering the Gulf of Alaska; and the other in the Prince William Sound region. Tectonic movements that uplifted some of the shoreline over 30 feet and the resulting landslides "were responsible for most of the local tsunami waves which caused the destruction and deaths within the Prince William Sound region." These tsunamis destroyed local towns and fishing villages, killing 82 people. The maximum wave height recorded in the sound was over 200 feet.

Back to the fieldwork in Alaska during the summer of 1970, we were faced with surveying several hundred miles of shoreline within a relatively limited amount of time. In the summer of 1969, we established a number of permanent beach profiles to survey at repeated intervals. However, during the more detailed coastal survey in the summer of 1970, we had to expand the study to include the whole area. To do so, we ended up applying something we eventually called the "zonal method," which allowed us to do detailed surveys of blocks of the shoreline within distinct geomorphological units, using the detailed surveys to characterize the overall unit in general (described in detail in Hayes et al., 1973). The field team that carried out this survey consisted of Paul Hague, Bob Henry, Frank J. Raffaldi, Jr., Woody Hobbs, and me. An outstanding bush pilot, Pat James, was also with us the whole time. This application of the zonal method, plus several detailed thesis projects, allowed us to learn a lot about a number of topics, including:

1) The effects of glaciation on the shoreline. In 1967, I published a paper in which I recognized that the common occurrence of gravel in continental shelf and shore zone sediment budgets in the higher latitudes (>40 degrees north and south) is primarily the result of glaciation at those latitudes (Hayes, 1967c). This was based on the findings of my worldwide study of inner continental shelves sponsored by the Defense Research Laboratory at the University of Texas. This fact regarding gravel distribution was also noted later by Davies (1973) and by Orford, Forbes, and Jennings (2002). One of our earliest observations on the southern Alaska coast was the correlation of gravel sediments on the beach with the presence of glaciers near the coast, as is illustrated in Figure 23. Figure 24 shows the coarse gravel on the beach at the margin of the Malaspina Glacier. This

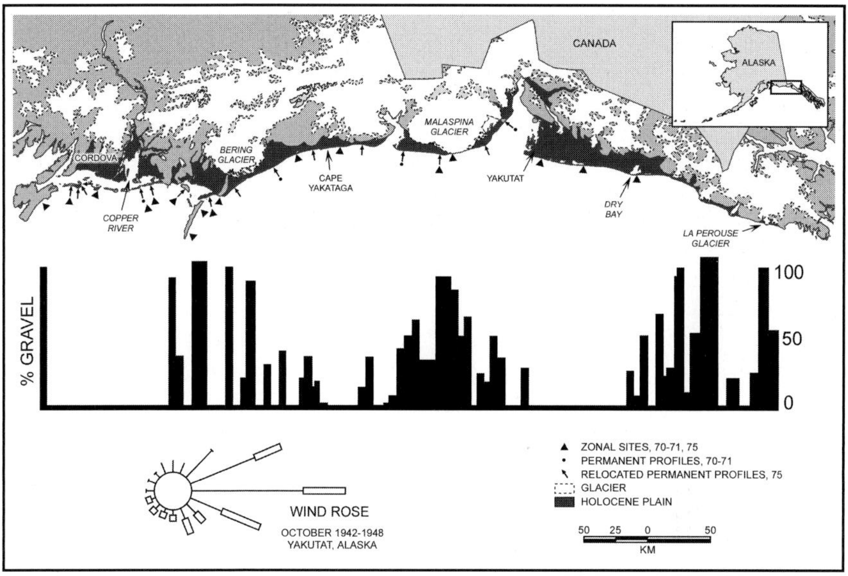

FIGURE 23. The south-central coast of Alaska, showing the distribution of coastal plain sediments (in black). The sediments that compose this 700-kilometer long coastal plain were derived mostly from the largest ice field in North America. Maximum width of the coastal plain, which is made up primarily of glacial outwash plains and fans, beach-ridge plains, and fan deltas, is about 35 kilometers. The histogram plots show average percent gravel occurring in 3 samples at each of 90 sampling stations along the outer beach. The high gravel percentages occur in areas where glaciers or rocky cliffs border the shore. The wind rose at Yakutat demonstrates the dominance of easterly winds generated by the numerous extratropical cyclones that occur in the area. Locations of some of our study sites are also shown. From Hayes, et al. (1973).

photograph was featured in *Suzanne's Lament*, as well as in several of our publications (e.g., Hayes and Ruby, 1994).

2) We began a long-term study of the somewhat unique barrier islands along the outer margin of the Copper River Delta on this leading-edge shoreline. According to Stutz and Pilkey (pers. comm.), only 16% of the barrier islands in the world occur along leading edge shorelines. The Copper River Delta, and its accompanying suite of barrier islands (see Figure 25), is there because of an overwhelming amount of sediment brought to the shoreline by the Copper River, considering its relatively small size (annual sediment load of 100 million metric tons; Reimnitz, 1966). Much of this sediment is fed into the river by drainages from major glaciers.

3) This study also gave us a chance to study on the ground a variety of gravel beaches (discussed in some detail in Chapter 15 – GRAVEL BEACHES AND FAN DELTAS).

4) Because of the dominant wind and wave approach out of the east/southeast (see the wind rose in Figure 23), we observed a number of features that indicated strong longshore transport of both sand and gravel sediments

FIGURE 24. The beach at Sitkagi Bluffs, located along the southern margin of the Malaspina Glacier, the largest piedmont glacier in the world (Figure 22). The COASTAL HERO kneeling is Paul Hague and the one standing is Frank J. Raffaldi, Jr. Photograph taken in the summer of 1970.

from east to west, such as the sand and gravel spits illustrated in Figures 26 and 27.

As part of this program, a separate study of the outwash fans associated with the Scott and Malaspina glaciers was carried out. My friend and colleague at UMass, Joe Hartshorn, the glacial/Pleistocene geologist, took on an advisory role for the project that included supervision of a dissertation by Tom Gustavson (discussed in Gustavson and Boothroyd, 1987). Gail Ashley, another one of Joe's students, also participated in the fieldwork. Jon Boothroyd, my student at both UMass and USC-East, took a leading role in the project. This fieldwork was only the beginning of Jon's long-term study of braided streams and alluvial fans that eventually resulted in a number of highly regarded publications (e.g., Boothroyd and Ashley, 1975; Boothroyd, 1976; among others).

[NOTE: As mentioned earlier, the coastal part of this work resulted in a publication that outlined the study technique we used in the field called the zonal method (Hayes et al., 1973). We later adopted this method of study as a primary field technique for the assessment of the initial extent and persistence of oiling at major oil spills (publishing several papers and manuals on this subject; e.g., Hayes and Michel, 1992). This modified zonal method consists of the following seven steps:

1) Collect and study available literature, aerial photographs, maps, and charts.
2) Conduct an aerial overflight of the entire area during low tide to observe maximum exposure of the oiled intertidal area.

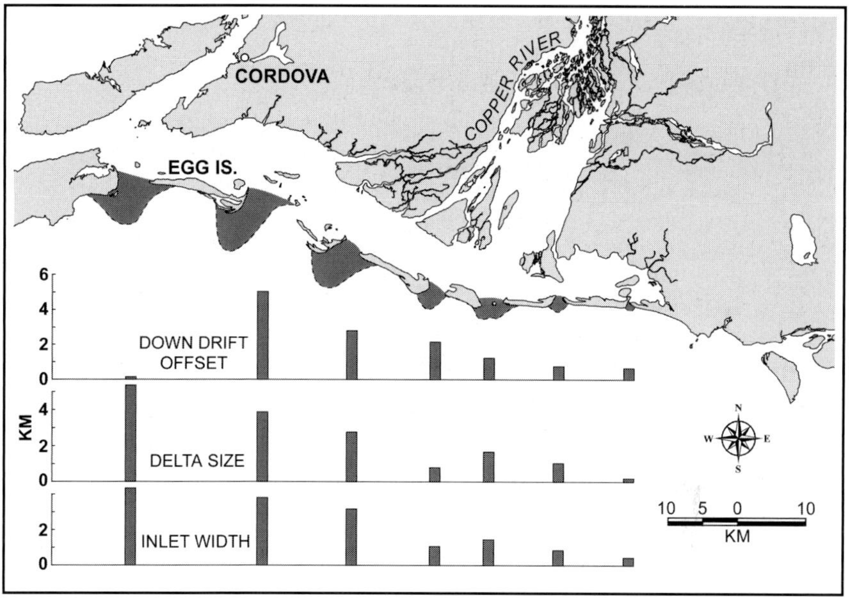

FIGURE 25. Barrier islands on the outer shoreline of the Copper River Delta. Lower graphs show increase in downdrift offset at tidal inlets, ebb-tidal delta size, and inlet width in a westerly direction, the direction in which nearshore sandy sediments move along this delta front (see wind rose in Figure 23). After Hayes and Ruby (1994).

3) Select specific areas for detailed study, based on the aerial survey, local maps, and the oil distribution. Those areas selected are representative of the local geomorphology, habitat type, and oil distribution.

4) Setup two types of stations: (a) Visual inspections stations, sites to be surveyed rapidly to determine coastal type, aerial extent of oil coverage, and the thickness and depth of buried oil layers as observed in small trenches dug across the beachface. These stations are spaced a consistent distance apart (e.g., 250 meters); and (b) Zonal stations, more detailed study sites chosen to represent all of the oiled intertidal habitat types. It is important that a hand-drawn sketch is made of a wide zone surrounding a central profile line (again, a standard width, such as 50 meters) to force the geomorphologist to carefully observe all aspects of the site, including oil distribution, depth of oiling, and so on.

5) Determine the grain size of the oiled sediments, using standard methods.

6) Categorize the coastal zone into its geomorphological components and oil distribution, superimposing this information onto base maps. Oil distribution may be categorized as light, moderate, heavy, etc. (using clearly defined criteria).

7) Conduct follow-up surveys and overflights to determine longer-term oil retention, cleanup effectiveness, and geomorphic changes.]

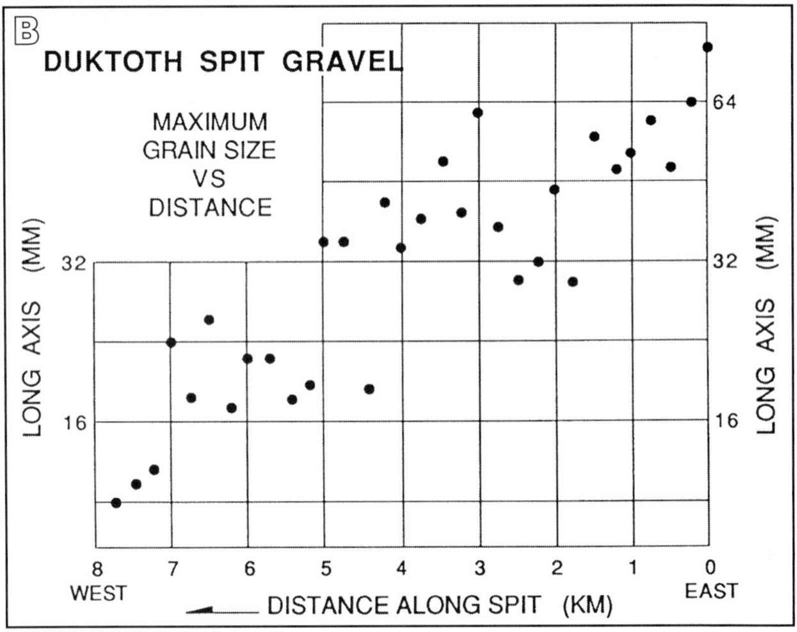

FIGURE 26. Morphology and grain size trends of Duktoth Spit. (A) Aerial view looking east at the prograding beach-ridge plain west of Cape Yakataga (Figure 22) in the summer of 1970. The sand and gravel beach sediments that compose the Duktoth Spit are derived from the Duktoth River, which had cut through the neck of the spit a few days before this photograph was taken. Arrow A points to the new outlet of the Duktoth River, and Arrow B points to the eroded beach-ridge plain scarp shown in Figure 27. (B) Plot of maximum grain size vs. distance along the 8 kilometers length of the Duktoth Spit as a result of the predominant easterly wave approach direction. From Hayes et al. (1973). These data are from a Senior Honors Thesis by Bob Henry (Henry, 1971).

FIGURE 27. Vertical section of eroded beach ridge (located by Arrow B on Figure 26A) showing dominance of flat-bedded, spit-top overwash deposition. Face of cut is parallel with beach. View looks landward. Sand at top of section is an aeolian deposit (wind-shadow dune). Photo taken in the summer of 1970 by Frank J. Raffaldi, Jr.

The Story

(Summer 1970, Cordova, Alaska)

Six years after the tsunami wrecked the harbor at Cordova, Alaska (on the southeastern shore of Prince William Sound; Figure 22), Paul Hague and I sat in the living room of a professional fisherman whose house overlooked the Cordova Harbor, as the fisherman told his story about the day the tsunamis of 1964 hit the southern Alaskan coast.

First the harbor itself, which was in plain view far below the elevated terrace upon which sat the town and the fisherman's house. According to Pflaker (1969, 1971), the harbor was uplifted about 6 feet during the earthquake. There are various reports on the height of the tsunami wave that hit the harbor, but possibly it was up to as much as 20 feet high. Some reports noted that there was $1.7M in damages to the harbor. It is hard to imagine there was less than $2M in damage, but again, this was 1964.

This is the fisherman's story. They had taken his boat on a fishing or crabbing expedition about 35 miles south of Cordova on the outside in the Gulf of Alaska. When the earthquake struck, the boat was in a large estuarine complex in the lee of Kayak Island called Controller Bay (Figure 22). The first unnerving thing that happened after the quake, which, of course, they could not feel on the boat, was a huge drop in water level that lowered the boat to near the bottom of the channel where they were located. When the first large tsunami wave came, the boat was lifted up and transported well back on top of a wide marsh/tidal flat system at the back end of the estuary. When the wave receded, the surface of the marsh and tidal flats, which were many hundreds of feet across, erupted in a series of Old Faithful-like geysers, all over the place! And, of course, the boat was left sitting high and dry on top of the marsh. Worried about what would happen when the next wave struck, a crew member jumped off the boat onto the marsh and–WHOOSH!!–disappeared without a trace down deep into the sediment that had been turned into a gel by the quake. Then they started tearing boards from the side of the boat and laid out a wide boardwalk across the surface of the marsh along which they individually crawled on their bellies to the landward edge of the marsh where it abutted against a steeply sloping hill. Once all of the survivors made it to land, they walked a good distance up the slope where they sat all night listening as wave after wave flooded the estuary.

5 KIAWAH ISLAND

One does not sell the land upon which the people walk!

CRAZY HORSE

The Science

As mentioned earlier, while I was completing my Ph.D. study of the south Texas coast, which has a small tidal range, my ambition was to eventually carry out similar studies of coastlines around the world with intermediate and large tidal ranges. Therefore, the very first research grant I applied for, with the U.S. Navy research program while I was still working at the Defense Research Laboratory at UT, was to study the depositional coastline of South Carolina with its intermediate-range tides. That research proposal was turned down flat. Also, there were no job openings requiring my skills in any of the geology departments of the universities in the southeastern U.S.

Fortunately, shortly thereafter I was able to secure the job at the University of Massachusetts. The relatively unstudied coastline of New England, which has a tidal range similar to South Carolina, was at the disposal of me and my versatile crew of graduate students. Some of that work was discussed in Chapter 3 – THE FIELD TRIP.

It was a great experience working on the New England coast, and we learned a lot, but the South Carolina coast still beckoned. Therefore, in the fall of 1972, a group of my students, technical assistants, and I migrated to Columbia, South Carolina to join the Geology Department of the University of South Carolina (USC-East). We eventually called our program the Coastal Research Division of the Geology Department. I also spent a four-year stint (fall of 1973 to summer of 1977) as the Chairman of the Geology Department.

Several major personal events occurred during my years at UMass. My two daughters, Joy Elizabeth and Mya Suzanne, were born before the 1970 Alaska trip discussed in the previous chapter, in February 1966 (Joy) and in July 1968 (Mya). Also, during the late fall of 1970, my first wife, Barbara Anne, and I split up and we were eventually divorced. I also entered into a three and a half year personal relationship with Leita Jean Hulmes, which was ending a few weeks before our group started the field work on Kiawah Island, South Carolina in the summer of 1974, which will now be discussed.[10]

Research on the origin, morphology, and stratigraphy of barrier islands was a part of our efforts in New England, where much of our work was conducted on a barrier island called Plum Island (Figure 12), as well as on the southern Alaska coast (see Figure 25). However, our first major and totally focused study of these characteristics of barrier islands was initiated at Kiawah Island, South Carolina (located on Figure 28) in the summer of 1974. After that, my graduate students at USC-East carried out many projects on the stratigraphy and evolution of barrier islands. Those studies, along with studies of the barrier islands of the Copper River Delta in Southeast Alaska carried out in 1975 and later (Hayes and Ruby, 1994), allowed us to develop some of the most complete publications on the geomorphology and stratigraphy of barrier islands carried out up until that time (Duc, 1981; FitzGerald, 1977; Hayes, 1977, 1979, 1994, 2005d, 2009; Imperato, Sexton, and Hayes, 1988; Moslow, 1980; Ruby, 1981; Sexton and Hayes, 1996; Tye, 1984).[11]

One of the results of these studies was the production of the barrier island "drumstick model," which is illustrated in Figure 29 (Hayes, 1976, 1979, 1985b; Hayes et al., 1976b; Hayes and Michel, 2008). This general model, which develops more readily along mixed-energy shorelines [i.e., those subject to moderate wave and tidal conditions (usually mesotidal - tidal range = 6-12 feet), the conditions that prevail on the coasts of South Carolina, Georgia, New England, and the

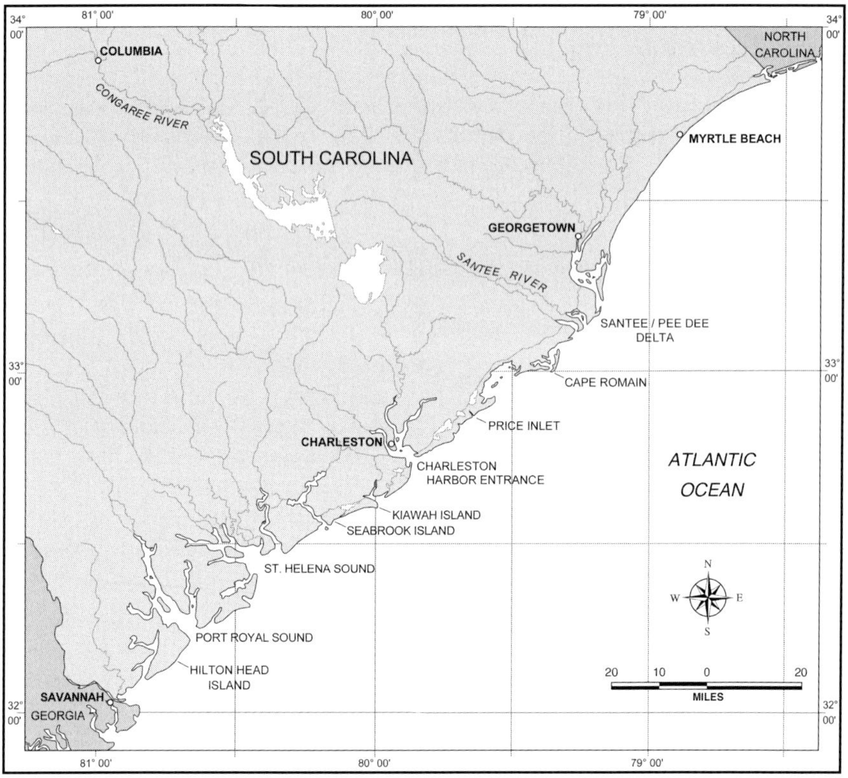

FIGURE 28. The South Carolina coast.

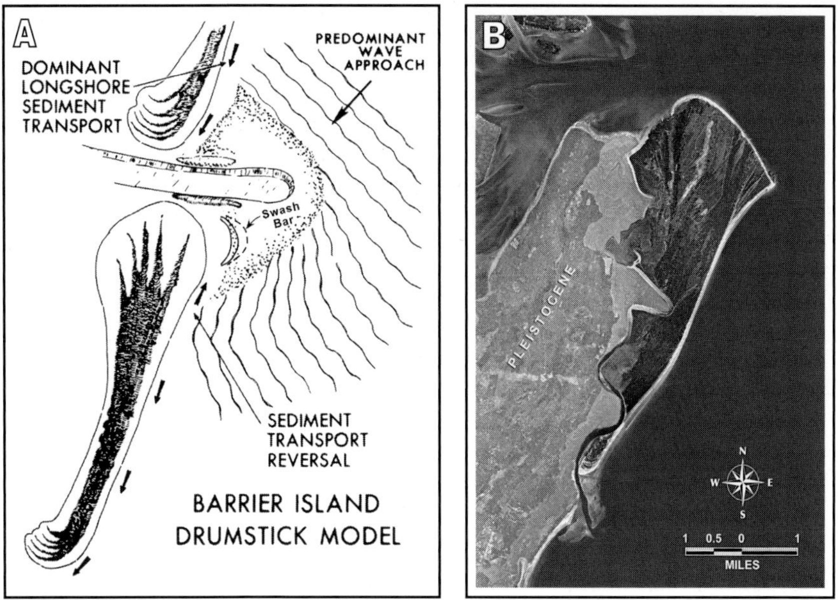

FIGURE 29. Typical morphology of prograding barrier islands on the South Carolina and Georgia coasts. (A) Barrier island drumstick model, primarily the result of welding of masses of sand derived from the updrift ebb-tidal delta in the form of large swash bars (see example in Plate 3). From Hayes (1976). (B) Blackbeard Island, Georgia, a seven-mile-long prograding barrier island that reflects the drumstick configuration. The updrift end (direction the sediments are coming from) of the barrier island is at top right of the photograph. Orthoimage acquired June 2008, courtesy of U.S. Geological Survey National Map Seamless Server.

Copper River Delta region of Alaska], is still cited frequently in the literature.

The geological history of Kiawah Island, which is about 10 miles long and 1.5 miles across at its widest point, is fascinating in all its many aspects. The island began to take shape about 4,500 years ago, when sea level stopped rising after the last Ice Age and stabilized at near its present level. For the most part, the shoreline of the island has been building out in a seaward direction ever since. Unraveling this history was one of the projects we initiated in 1974, culminating in a dissertation project by a graduate student of mine, Tom Moslow. Tom's work produced one of the most complete 3-D pictures of barrier island stratigraphy available. His stratigraphic cross-section of Kiawah Island is illustrated in Figure 30 (Moslow, 1980).

After a group of Kuwaiti investors purchased the island in 1973 (more details on the island's human history will be presented later), Charleston County applied rather strict development regulations. These regulations required an environmental impact statement, which incorporated the development of construction set-back lines for all the beachfront property. A fellow faculty member at USC-East, John Dean, and his partners formed a new consulting company and landed the environmental project. John's company contracted our group at USC-East to develop the required set of construction set-back lines for the island, among other

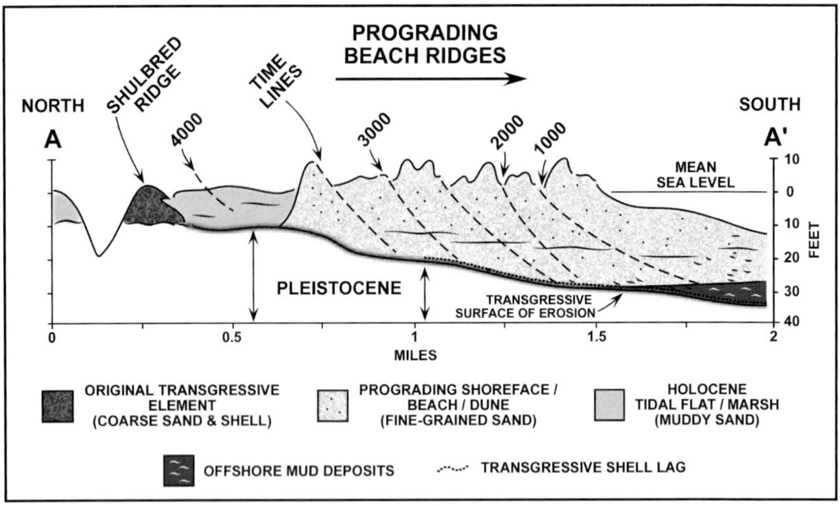

FIGURE 30. Generalized stratigraphy of Kiawah Island. Modified after Moslow (1980). Time lines represent thousands of years before present (BP). This cross-section is located (approximately) on Plate 2.

tasks, which we did for the whopping sum of $14,000. Despite this rather meager paycheck, we also did a basic geomorphological study of the island, with the map in Plate 2 of the island's geomorphic provinces being one of the products of the study (Hayes et al., 1975).

The set-back lines we drew were based primarily on two criteria: 1) A detailed analysis of the shoreline changes on the island, starting with historic charts dating back to 1661 (six maps and sketches and five coastal charts) and five sets of vertical photographs taken between 1938 and 1973 (data summarized by Stephen et al., 1975); and 2) Historical beach erosion rates during hurricanes elsewhere, relying heavily on my work on Hurricane *Carla* (1961) (Hayes, 1965, 1967a), which was discussed in Chapter 1 – THE HURRICANE. The goal of the set-back concept was to draw a line beyond which the beach was not expected to erode within the next 50 years or so. For the record, it is now 37 years since those lines were drawn, and significant erosion has not occurred on the island where the original setback was adhered to, and to this date, no structures have been lost due to erosion. There is presently concern about the erosion of the beach along the Ocean Course at the northern end of the island (T. Kana, pers. comm.), but this golf course was built seaward of the original set-back line. If the eye of Hurricane *Hugo* (1989) had crossed the shore to the south of the island, no doubt severe wind damage would have occurred, but it is doubtful that any beachfront lots would have been lost, as was the case on parts of the Isle of Palms, located 16 miles to the north. Erosion of the Kiawah beach was minor during *Hugo*.

One of the reasons for this success of the set-back lines at Kiawah Island, besides the fact that the managers of the island's development projects enforced adherence to the concept, is that the island is building seaward at a relatively rapid rate, especially on its northern end. This process, which accelerated in the early 1900s, is illustrated by the historical data given in Figure 31. A large swash bar

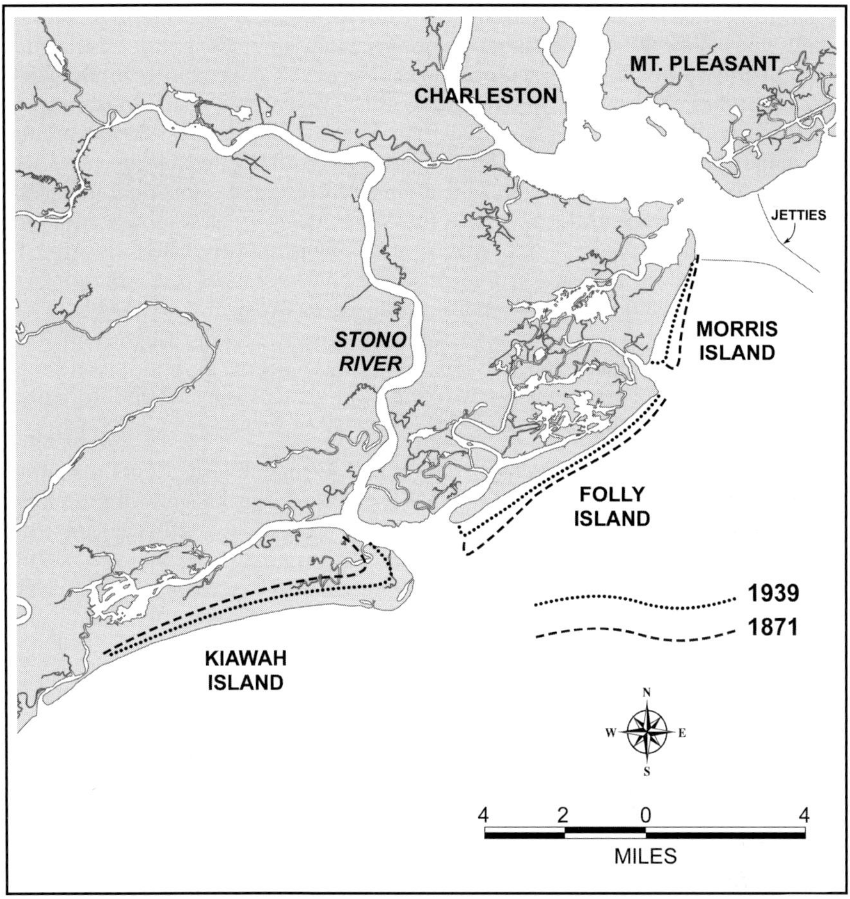

FIGURE 31. Historical shoreline changes, Morris Island to Kiawah Island, South Carolina. Morris and Folly Islands eroded hundreds of feet and Kiawah Island prograded more than 3,000 feet during this time period (1871-1974). This diagram was created in 1974, but this trend of shoreline changes shown continued until the 1980s. Since 1993, over 5 million cubic yards of sand have been added to the Folly Island beach (artificially). From Hayes and Michel (2008).

welding to the north end of the island is pictured in Plate 3.

When we compiled the shoreline data showing such a trend, we were curious why this had happened, considering the fact that Folly Island, which is located only a few miles to the north, suddenly began to erode dramatically around the 1930s (see Figure 31; later noted by Lennon, 2000). Accordingly, we examined the historical impact to the shoreline of the jetty construction at the entrance to Charleston Harbor, which was begun in the late 1870s and completed in 1898. A 1780 map of Charleston Harbor made by the British military showed that the huge ebb-tidal delta associated with the harbor entrance was deflected far to the south of the harbor entrance at the time the jetties were constructed. With the establishment of the new east-west oriented channel straight off the harbor entrance created when the jetties were constructed, this large sediment mass (south deflected ebb-tidal delta) was no longer built out by the ebb-tidal currents. Consequently, a significant part of this mass of sand was driven onto the shorelines of Morris and Folly Islands by wave action. This initial burst of sand accumulation was eventually eroded away and much of it has made its way to Kiawah Island. Consequently, the dramatic erosion of these islands illustrated in Figure 31 was the result of the eventual erosion of the sand of the historic ebb-tidal delta that was originally welded to the beach. In the 1970s, we made several presentations in which we related the erosion of Morris and Folly Islands to the construction of the jetties. However, the U.S. Army Corps of Engineers office in Charleston was reluctant to accept that premise, noting that the beach to the southwest of the jetties had accreted significantly soon after the jetties were built. That observation was true, of course, because of the landward migration of the huge, abandoned ebb-tidal delta. Ultimately, however, with no more sand passing around the jetties, the shoreline began to retreat.

In any event, I made many trips to Folly Beach (on Folly Island) at the request of the mayor, Richard Beck, to give lectures, *pro bono*, on this and other matters related to the erosion at Folly Beach. Long after we had departed that scene, the powers that be agreed with at least some of our original conclusions about the jetties, and programs to nourish the beach with sand derived from elsewhere were adopted. Over 5 million cubic yards of sand have been dredged and added to the beach at Folly Island since 1993 (Tim Kana, pers. comm.).

With the science of our project on Kiawah Island completed, the contractors took over and the development of the island flourished. The construction set-back lines we developed for beachfront development worked out well, and have become something of a model for that kind of thing, I'm proud to say. I still lead field trips to the island and continue to be amazed at the rapidity with which the beaches and inlets change. Each trip reveals something new and exciting.

Several years after the completion of this Kiawah Island project, some interesting issues concerning the development of the adjacent barrier island to the south, Seabrook Island, allowed our new company, Research Planning, Inc. (RPI), which was established in 1977, to become involved. During this work, which will be discussed in Chapter 9 – A RELOCATED TIDAL INLET, we applied some of the information gathered during these early studies of Kiawah Island.

The Story

(June 1974 - Kiawah Island, South Carolina)

Kiawah Island is named after one of the many small Indian tribes that occupied the land now known as South Carolina before the white man started arriving in force around 1670. The Kiawah tribe is now history, as are most of the other lesser tribes, such as the Congarees, Sewees, and so forth. During the First Civil War, some people call it the Revolutionary War, the British captured the island fairly early in the War and held it until the end. Sometime thereafter, a Dutch man named Vanderhorst purchased the whole Island, where his slaves grew cotton and indigo. During the Second Civil War, the island was captured by the Yankees, and a black Union regiment was encamped on the island for some time. When we did our work in 1974, the old Vanderhorst mansion was in rather poor shape, but the walls on the second floor were covered by graffiti written in gorgeous large script by the Union soldiers, such as FUCK YOU GENERAL BEAUREGARD! (the Confederate general in charge of Charleston at that time), and things like that.

In 1948, a rich doctor named Royal bought the whole island for $250,000, and then he cut down the trees, which were worth almost that much. Dr. Royal allocated a few beachfront lots upon which were built roomy and elegant houses for his wealthy doctor friends, I assume. About ten bungalows were located down the road from the larger dwellings, presumably built for the help.

Sometime around 1973, the old man passed away and his heirs, as we might have expected, sold the island–this time to the aforementioned Kuwaiti investors for $17 million! At that point, the second largest barrier island in South Carolina was owned by investors from a foreign country. This action prompted the requirement of such things as the set-back lines discussed above.

I personally carried out the fieldwork for the set-back line project during the summer of 1974, assisted by a brilliant undergraduate student in geology named Steve Wilson, who decided to go to college after his stint as a green beret in Vietnam. Tall and thin, with a prominent crooked nose, which was severely sunburned most of the time, his demeanor could best be described as smoldering. I thought he might explode at any time, but he never did, not around me anyway.

Steve and I spent several weeks on the island living like kings in one of the huge houses on the beachfront. We claimed the entire second level of the elegantly furnished mansion, to which we invited numerous weekend guests, who will not be discussed any further here.

One night along about the middle of our stay, while lounging on the boardwalk that led from the house out to the beach, I was writing a letter to my friend Debbie, who had helped with the surveys a few weeks earlier. In the letter, I said, "The moon is nearly full and I can see it reflecting off the water (the shoreline faces east). The waves are high and their roar is steady and melancholy. .... Heavy clouds hang near the horizon, but the moon is above them and keeps out in the open most of the time. It's late, about midnight, and I am alone writing by the glow of the moon. I can see a light from a lone shrimp boat far offshore. The breeze is strong and cool, sometimes shaking the walkway, and it feels good after a hot day in the sun.

"This morning I made a speech for the Sea Pines people, the developers of the island, and I think it went well. And this afternoon we took them to the beach where I gave them a few coastal geology lessons. They seemed interested. I hope and believe they will take most of our recommendations seriously." And then I fell asleep there on the walkway under the moon and dreamed about Hoojer Smith and Darlene and my Uncle Luster.

For that presentation I mentioned in the letter, we had prepared page-sized maps of the island, colored yellow, on which I drew zones of erosion and deposition and a proposed set-back line concept, though it was as a new conception for me at that time. The next day after the presentation, I was standing out on the beach by myself as Steve went back to the house to retrieve some equipment we had left behind. I was far out on the beach near the low-tide line. Taking out a data sheet, I wrote on the back of it: "I'm freaked out by the notion that the island is more than eleven inches long, not a yellow blob on a piece of paper, and not so simple to analyze as I implied at that meeting. I am standing here on the beach and looking three miles down to the end of the island. I can't see across the foredunes back onto the central portion of the island; however, looking across the wide intertidal zone, I now realize that I don't really understand any of this! How the island was made, how the set-back should work. But, it is the clearest, and sharpest, and most exquisite day I've ever seen, which is why I came in the first place. Life seems beautiful today."

The lower level of the mansion we stayed in was occupied by a bevy of crazies from Georgia we called the "herps," part of the team assessing the environmental quality of the island. Remember, except for about ten acres or so on the southern end, this ten-mile-long sanctuary was in a completely natural state.

Anyway, the "herps" were experts in herpetology, a branch of zoology dealing with reptiles and amphibians. Steve and I were up at all hours, because our surveys had to be carried out at low tide. The "herps," however, rose at the gentleman's hour of 8 or 9 o'clock, had a leisurely breakfast, and then started drinking can after can of beer. When the pile of beer cans reached a certain critical level, they gathered together their burlap bags and headed for the woods and ponds to capture snakes. Late in the afternoon, they would return with several heavy bags labeled timber rattlesnake, cotton-mouth water moccasin, red-bellied water moccasin, and so on. I'm not sure how they studied the alligators, but at least there weren't bags full of 'gators waiting on the front porch to greet us when we returned from the field.

About mid-July, I went to Columbia to pick up a new purchase that had finally arrived, a brand new four-wheel-drive Jeep, something I had dreamed of owning ever since the Padre Island days. It was painted white with COASTAL RESEARCH DIVISION neatly lettered on each door. When I arrived back on the island, Steve jumped in and we went on a driving tour in our new toy. A few old logging roads composed of sand encircled the entire island. For the most part, Steve and I and the herps and another research crew studying the nesting habits of loggerhead turtles were the only people driving along those roads. At some time in the past, an old Dodge Power Wagon truck, one of the larger models, had broken down on the road about half way between the big house where Steve and I stayed, and the bungalows where the managers lived and had their offices. The Power Wagon had been pushed aside, catty cornered with two wheels in the ditch beside the road and two up on the shoulder of the road. It was the only other vehicle on that part of the island, except those belonging to the research personnel.

One Friday morning in late July, with our field surveys for the summer having been completed, I made a little speech in which I summarized our progress to a managers' meeting down at the bungalows. I should point out that the straight engineers and consultants from Hilton Head, who had been hired by the Kuwaitis to do the dirty deed and, yes, even the USC-East academic crew that had hired us, looked askance at the boys living in the big house because of our unorthodox ways of conducting business, among other things. They didn't necessarily approve of some of the visitors Steve and I entertained, nor of the "herps" drinking habits, and so forth.

In any event, with the task of delivering our summer's progress report completed, Steve and I went back to the mansion for one last time and packed our field gear, data, and other belongings into the new Jeep for the trip back to

Columbia, about a two-hour journey. We pulled out of the drive, with Steve behind the wheel, and headed down the one-lane, sandy road toward the bungalows.

As he was driving along, Steve starting rummaging through his duffel, which was resting on the seat behind him, looking back as he continued down the road. He pulled a can of antiperspirant out of the duffel and started spraying under his armpits, still not looking where he was going. I was laughing at him.

"Forgot to spray this morning," he chuckled.

WHAM!

I was thrown out of my seat, no seat belts at that time, and crashed into the right side of the windshield with the top of my head, shattering the glass, which didn't hurt in the least because of the heavy mop of hair I had grown over the summer. As I bounced back into my seat, I looked ahead and observed what had happened. Steve had run straight into the Power Wagon in the ditch, the only other vehicle on the island that morning, except for those belonging to the people at the managers' meeting at the bungalows. The left front tire of the Jeep straddled the giant left forward tire of the Power Wagon, which was facing us, and the front end of my precious new Jeep was folded up like an accordion. The Jeep was cocked up at an angle of about 50 degrees, dipping toward the middle of the road.

Seeing the irony of the situation, namely that we had managed to crash into the only other vehicle on the island road, never mind that it was well over to the side and even half-way into the ditch, I let go my grip on the side of the door and rolled out of the Jeep into the middle of the road, where I lay on my back and laughed until I couldn't laugh any more.

Steve finally crawled out of the Jeep and helped me up so we could both stand and stare at what we had done so he could go odor-free for the day.

After unsuccessfully trying to pry the Jeep loose from the Power Wagon, Steve said, "Well, what do we do now?"

"Guess we will have to walk down to the bungalows and see if we can get some help getting these vehicles unstuck, and maybe call for a wrecker, or something."

"Boy, what are they going to think of us now?"

"Probably, that we are high on something. Kind of wish we were, don't you?"

"Might make this a little easier."

The folks at the bungalows didn't seem to be the least bit surprised that we had crashed into the Power Wagon, but many of them wanted to see it, so we led the parade a few hundred yards down the road to the wreck site.

After a certain amount of consultation on how to get David and Goliath unstuck, one of the Hilton Head engineers suggested that if we let the air out of the left front tire of the Jeep that might do the trick. Which Steve did and the Jeep rolled down off the huge tire of the Power Wagon into the sand with a heavy thud. The engine of the Jeep was totaled, that much was clear, so everybody started to walk away.

"Hey, we could use some help pushing this thing," I called to the retreating backs of the crowd.

A few volunteers joined us, and we pushed the wounded Jeep slowly down the road to the bungalows, where the rest of the attendees of the meeting came out to admire our handiwork.

I went inside to use the phone, reaching a towing company located on the north side of Charleston, about 20 or so miles away, finally being assured that a wrecker was on its way. Maybe two hours later, the wrecker arrived, manned by two burly black guys named Isaac and Chester. They backed the wrecker right up to the mangled Jeep and attached a long cable to what was left of the front bumper. Isaac was barking out the orders and Chester was obediently following them. Of course, a cheering section had gathered, carefully following every move they made.

They turned on the winch and lifted the nose of the Jeep up at an angle of about 45 degrees. Then Chester, following Isaac's orders, turned off the winch, at which the Jeep came crashing down on the ground while the cable continued to spin off the winch. The cable instantly became a tangled mess that looked just like what happens when you throw a fishing plug too hard with a bait-casting reel and forget to feather the spool with your thumb. In fishing terminology, it's called a backlash. The cable looked like a plate of rigid, rust-colored spaghetti. Much foul language followed, from both the participants and the cheering section. About twenty minutes later, with the backlash finally untangled, they tried again, with the exact same results.

After one more failed try and backlash, Chester decided he was the boss of the operation and started ordering Isaac around, with Isaac complacently following his orders. Three more backlashes. Finally, about two hours after they had arrived, and with the help of some of the Hilton Head engineers, the Jeep was finally attached to the cable and winch with the front end lifted about a foot off the ground. We were then ready to head for North Charleston. I went back inside the bungalow to call my secretary, Jill, asking her to start the two-hour drive from Columbia to Charleston in order to give us a ride back home.

With the front of the Jeep riding only about a foot off the ground, we had to drive very slowly, not more than 10 mph, on our 20-mile trip to North Charleston. Steve and I jammed ourselves into the front cab of the wrecker with Isaac and Chester. It was a hot day, and after all that pulling and pushing and cursing and sweating, our position in the front seat was pretty hard on the old olfactory nerves.

The four of us were facing a long interval in which to talk with each other, but we had a hard time breaking the ice, you might say. After about 20 minutes, I said to Chester, who was driving, "Hey Chester, why don't you pull over at that store up there and I'll buy us all a beer."

That was all it took, the cold beer hit the spot, and we were the best of friends after that. They told us the short versions of their life histories, a good deal of which had been spent in the slammer, from which they had both been recently released. We laughed and joked as we crept along. They thought our running into the Power Wagon with the now-disabled Jeep was a story good enough to earn us all another round of beers, which I was glad to supply.

The people living along the road out to the Kiawah Island were mostly low-country black folks. Any time we passed a yard with a pretty young lady in it, either Chester or Isaac would lean way out of the window and call out, "Oh Mama, don't you worry, don't you worry now, we'll be right back now, you hear?"

When we finally reached the repair shop, Jill was waiting for us. She had been there for almost an hour. We told Chester and Isaac good-bye, wished them the best of luck, and headed for Columbia, with no more trips to the island in the works until fall.

Kiawah Island is now a famous tourist and golfing resort; they have held the Ryder Cup there twice, and in 2012 the island will host the PGA Championship. Also several years back, the Kuwaitis sold the island for $125 million, not to mention the return from the development itself.

6 BRIDGES OVER TROUBLED WATERS

Night after night I kept for her
half of my quilt
in expectation of her spirit
coming back to me.....

LI PO, Tang Dynasty

Just because you're paranoid,
it doesn't necessarily mean that
they are not out
to get you.

JEFFREY DAHLIN

The period between around the middle of 1971 until toward the end of 1975 composed perhaps the most unsettled episode of my entire life. Having to visit my two daughters in faraway Massachusetts, the added stress of trying to be an effective Department Chairman, splitting with Leita Jean in early 1974, and other issues added to the chaos. Also, trying to keep up with a major project in Iceland which, unfortunately, I will not cover here because of the relatively minor role I played in the fieldwork, also added a certain amount of stress. From a scientific perspective, this stretch of about four years was without doubt my least productive time as far as publications were concerned. However, I did have a great and productive group of graduate students and research scientists in my program, and the Coastal Research Division at USC was flourishing, so all was not bad news. In total, nineteen graduate students finished their Ph.D. dissertations under my supervision at USC (Appendix A). In this chapter, I will discuss the dissertation projects of Patrick McLaren, Ed Owens, and Chris Zabawa, all of which were carried out far from the confines of the South Carolina coast.

McLAREN – The Science

Patrick McLaren's dissertation, completed in May 1977, was titled *The Coasts of Eastern Melville and Western Byam Martin Islands: Coastal Processes and Related Geology of a High Arctic Environment* (McLaren, 1977). The Magnetic North Pole was located about 80 kilometers east of his study area (see Figure 32) when he

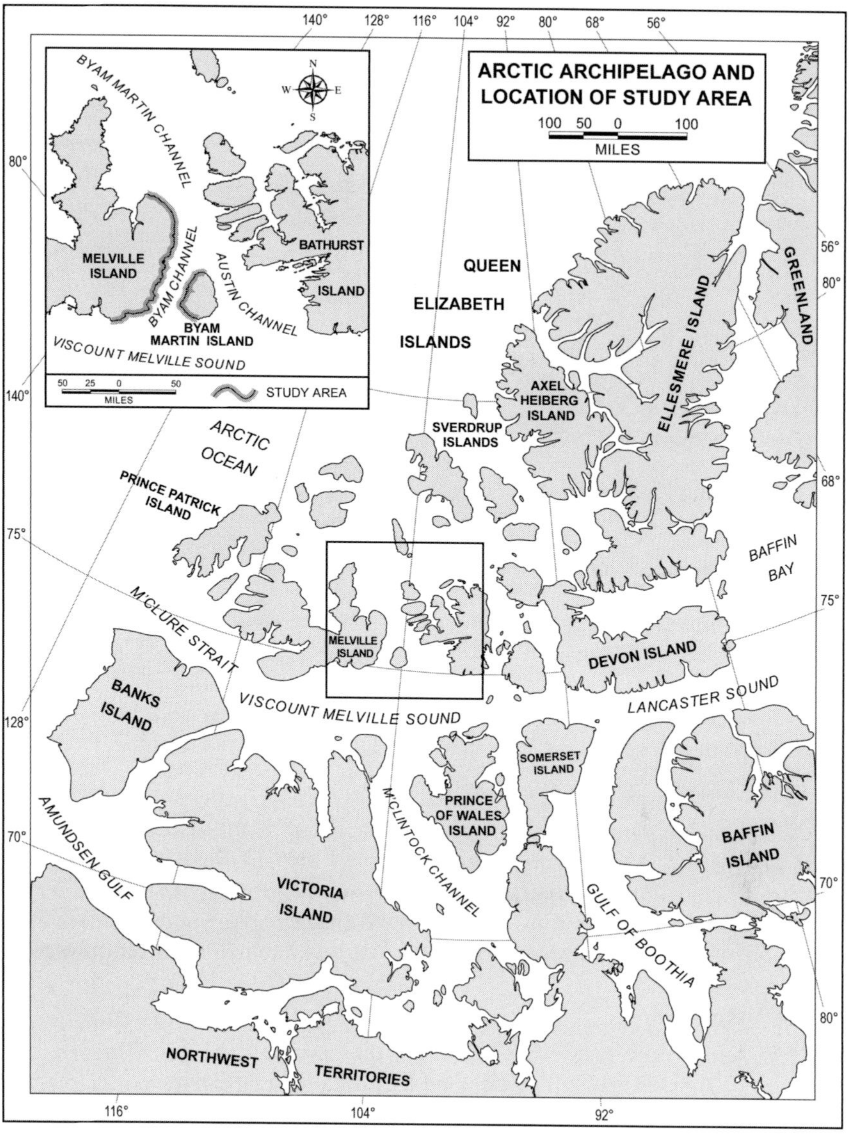

FIGURE 32. Pat McLaren's study area in the vicinity of the Magnetic North Pole.

completed his fieldwork in the summers of 1973 and 1974. By now, however, the pole has moved to over 200 kilometers north of there.

I had only been at USC-East a few months when Patrick joined our team. He secured his own funding, which was significant, from the Canadian Geological Survey. He did a phenomenal amount of fieldwork, which included sessions of SCUBA diving under the ice in the Byam Channel. A total of 93 dives during 50 underwater hours, carried out at depths ranging from 3 to 25 meters within 4 kilometers from shore, were completed. The goal of the diving was to investigate

FIGURE 33. Grounded ice on the beach at Melville Island in the High Arctic. Note the abundant gravel in the sediment. Photograph taken in July 1973.

the "important role of grounded nearshore ice on the coastal environments." A photograph of this grounded ice is given in Figure 33.

Some of Patrick's principle findings follow:

1) "The coasts of eastern Melville and western Byam Martin Islands have emerged (been elevated) approximately 100 meters during the Holocene, and are still undergoing isostatic recovery at the rate of approximately 0.35 cm/year." This has resulted in the creation of the phenomenal raised beach ridges shown in Figure 34, the best example of this I have seen anywhere.

2) Along the shoreline, he also found sediments deposited in Gilbert-type deltas (presumably fan deltas; see discussion of these in Chapter 17 – GRAVEL BEACHES AND FAN DELTAS) and sand flats.

3) "SCUBA observations showed that moving ice blocks impinging on the substrate leave long linear scour tracks approximately 23 meters deep, which are flanked on either side by embankments 1 meter high. The ice scour tracks are terminated by a crater, which is surrounded by an embankment composed of excavated substrate." Maybe I should add that this work was carried out in an area of active oil-and-gas exploration. These findings had obvious implications regarding the modes of transportation of these petroleum products, such as pipelines laid along the ocean floor.

4) Where ice scouring is sufficiently deep, the resultant ice-push deposits on the beach contain gravel deposits with clasts as large as boulders (see Figure 33).

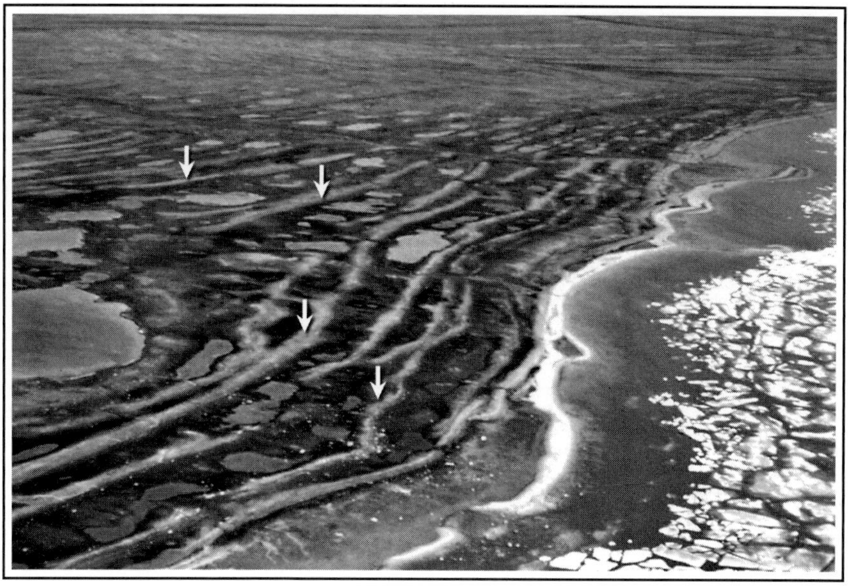

FIGURE 34. Raised beach ridges (arrows) on Melville Island in the High Arctic. Photograph taken in July 1973.

5) "Suspended sediment analysis and SCUBA observations indicate that little or no present day sedimentation is occurring beyond the 7 meters isobath."

In the field, the team had an impressive array of field instrumentation. There was one problem, however. Occasionally, the polar bears would come onshore at night and play games with the instruments. Very costly. They also did that with one of Patrick's compatriots the next year but, somehow, he survived. Tricky place to work. Diving with polar bears? Hmmm.

After returning to Canada, Patrick founded his own scientific consulting company that is now called SedTrend Analysis, Ltd. According to his web site, his method of sediment analysis, called Sediment Trend Analysis (STA), "has been applied to a large number of international projects encompassing port design, ocean dumping, coastal erosion, dredging, contaminated sediment remediation, and environmental impact assessment."

McLAREN – The Story

(29 July 1973 – Rea Point, Canadian Arctic Islands. Sitting in a Panarctic Electra waiting to take off for a trip to Edmonton, Alberta)

Panarctic Oils, Ltd. had a base of field operations, including an airfield, at Rea Point on Melville Island, which was where we flew in and out of to visit Patrick's field area. I paid a visit to the site for a few days in late July 1973. As best I can tell, the oil-and-gas exploration activities in the High Arctic were beginning to start in a serious way about then. However, this particular Panarctic base was closed down when the oil business went to hell and gone in 1986.

In my journal, I had a brief description of the four days I had spent in the field with Patrick and his field crew. As best I can tell from these notes, we examined the beaches and sediments along the shore near the main camp during the first two days, but during the last two, we spent most of our time exploring a long ways along the shoreline and in the valleys that drained onto the coast. Patrick had a fleet of 3-wheeler ATVs that we all rode along in single file past herds of musk oxen and other spectacular sights. I had never ridden such an ATV before, and it took some getting used to.

During day three, on a clear, warm, exhilarating day, I had felt like a member of Rommel's Afrika Korps patrolling the desert wastes. But, on that particular day, we didn't take any daredevil rides down the steep slopes into the valleys. However, I reminisced to myself while sitting in the plane waiting for the takeoff, "Yesterday we did, and I just mostly held on for dear life and prayed for the end of the ride." We went down one very steep hill and had to jump over the little gully at the bottom of the slope. Oops! Not a very good jump. I ended up on my back in the bottom of the gully with the three-wheeler directly on top of me with the wheels spinning wildly in the air.

So there I was sitting in the plane with all my muscles aching from that bone-jarring Honda ride with its three back flips and 10,000 jolting bumps. At that time, I was one year short of my 40th birthday and, after that ride on day four, perhaps for the first time, I finally realized that I was getting old.

Back on the plane, it was now 11:40 a.m., and the pilot informed us it would take four hours and fifteen minutes to fly to Edmonton. We had been scheduled to leave at 3:00 a.m. As we were getting on the plane earlier, I heard one of the oil-field workers comment, "Another day of my life lost in the Arctic, I'm getting mighty tired of that."

It was nice to feel and see the white smooth concrete runway on the flat plains of Edmonton, Alberta and know we were back to "civilization" again. As the plane slowed to a stop, I thought, "Oil is king in Edmonton, and soon it will be in the Arctic as well."[12]

OWENS – The Science

Ed Owen's project was aimed at determining the factors controlling short-term changes on the beaches of the Magdalen Islands, located in the middle of the Gulf of St. Lawrence, eastern Canada (Figure 35). He had detailed instrumentation set up in the surf and on the dunes at different localities to monitor the changes in wave and wind conditions. A number of permanent sites were established along both the east and west shorelines, where topographic surveys were carried out routinely in order to determine how the beach morphology responded to the changes in wave regime measured by the instrumentation. Current measurements in the nearshore zone were also made at times. This was one of the most detailed studies of this type that had been carried out on the east coast of North America up to that time. The Geological Survey of Canada provided the funds for this dissertation through a grant to the Bedford Institute of Oceanography, Dartmouth, Nova Scotia. The Institute also provided manpower and equipment for the project. Ed secured this support himself.

The title of this dissertation was *The Geodynamics of Two Beach Units in the*

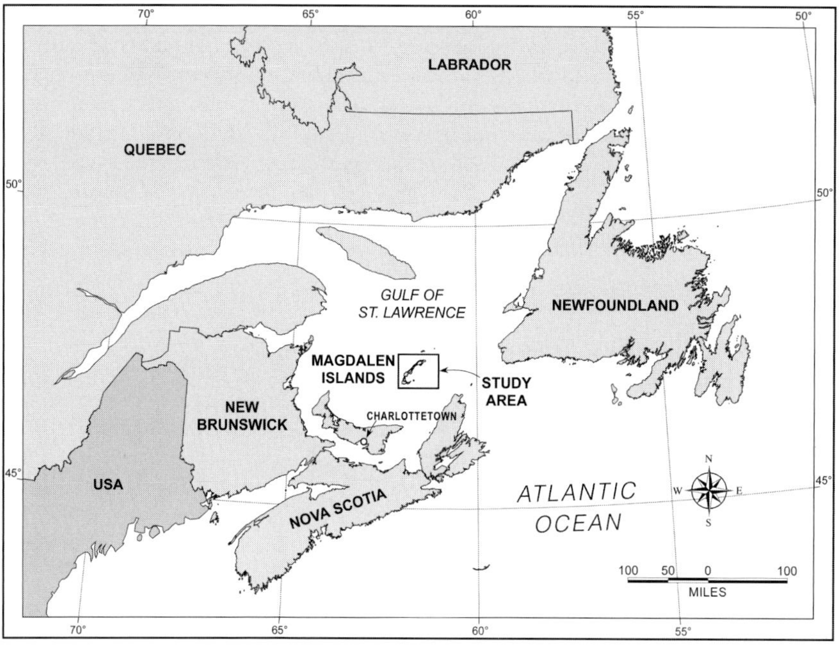

FIGURE 35. Location of the Magdalen Islands.

Magdalen Islands, Quebec, within the Framework of Coastal Environments of the Southern Gulf of St. Lawrence (Owens, 1975), and at 650 pages in length, it was the longest of all the theses and dissertations that I supervised.

A few of Ed's major conclusions follow:

1) "The Magdalen Islands are a series of small bedrock outcrops which have been connected by barrier beaches to form an extensive tombolo system oriented approximately northeast-southwest."

2) This microtidal environment is "dominated by locally generated storm waves out of the westerly quadrant," which are associated with the passage of low-pressure systems across the region.

3) One of his main conclusions was that he was dealing with two distinct "energy environments;" a) The west-facing barrier beach, which is characterized by a net longshore sediment transport to the north at the rate of over 200,000 cubic meters per year, and by "a complex pattern of rhythmic nearshore bars;" and b) A similar sediment dispersal mechanism on the more sheltered east-facing barrier, with a net longshore sediment transport to the south at a rate of 100,000 cubic meters per year. In this area, "primary wave-induced currents result in frequent modification of the inner-bar morphology." Figure 36 shows these sediment transport directions. In a paper published in 1977, he said–"The exposed west coast has a distinct seasonal wave climate and the sheltered east coast is a storm-wave environment."

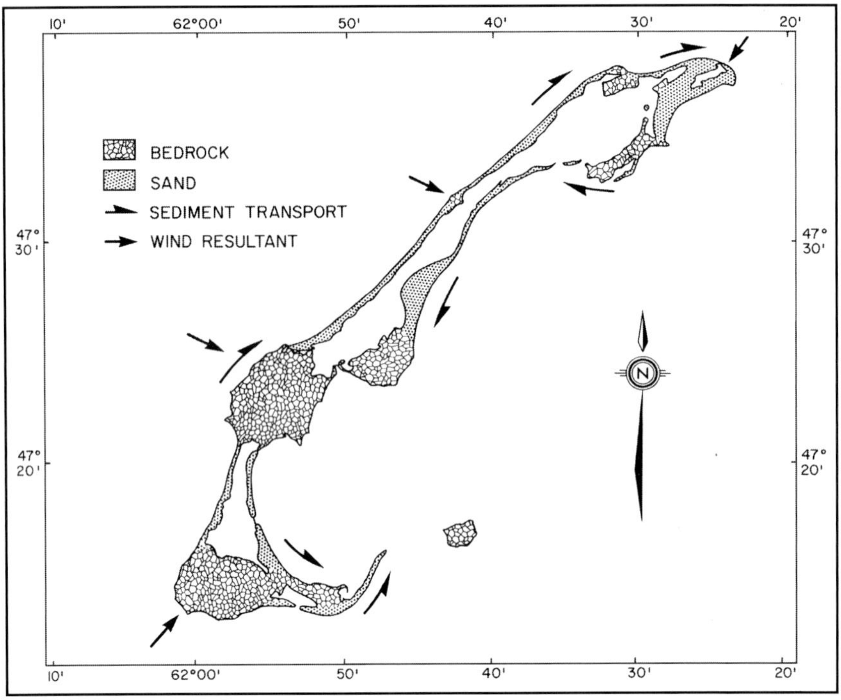

FIGURE 36. Onshore wind resultants and nearshore sediment dispersal around the Magdalen Islands, Quebec. From Owens (1975).

4) It is difficult to synthesize such an overwhelming amount of data into a few short notes. For example, his information on the changes of the nearshore bars was especially noteworthy, but the details are beyond the scope of this discussion.

After leaving USC-East, Ed went on to a research position at Louisiana State University. Later, he was involved in scientific consulting firms, with his present position being as a primary partner in Polaris Applied Sciences, Inc. on Bainbridge Island, Washington. Over the years, Ed has become undoubtedly the prime responder to oil spills for industrial clients. He publishes prodigiously and offers a very popular training course in oil spills for both industry and some government clients. As I write this, he is now in Houma, Louisiana sitting across the table from JM as they work out ways to deal with the *Deepwater Horizon* oil spill in the Gulf of Mexico, his client being BP and JM's being NOAA/U.S. Coast Guard.

Ed Owens was one of the most productive graduate students I ever had. His outstanding professional career speaks for itself. He was perhaps the very first coastal geologist to tackle the oil spill issue. He certainly preceded me. In fact, when he came to UMass and wanted to discuss his work on spills, I showed no interest whatsoever. He was always great at organizing complex projects and managing groups of people. His work organizing the *Exxon Valdez* scientific response, the SCAT[13] surveys in particular, is a legend in itself, as far as I am concerned.

OWENS – The Story

(16 August 1974, Friday - On Eastern Provincial Airline flight from Iles de la Madeleine to Charlottetown, Prince Edward Island, Canada)

It was rainy and cold as the propeller on the right engine began to turn, signaling our departure from the Iles de la Madeleine (Magdalen Islands), which are located in the middle of the Gulf of St. Lawrence off the coast of eastern Canada (Figure 35). Having arrived three days earlier as worn and depressed as I had ever been; now I felt relieved, wind-blown, refreshed, and happy. These islands are so beautiful and unique, with the most striking aspect about them being the combinations of colors that constantly engulf you–blue-green water, green grass on the hills, red rocks and red soil, white sand on the beaches and in the shallow water. A number of individual islands composed mostly of red sandstones are joined by multiple long, linear sandy spits that create a circular, atoll-like structure in the middle of the Gulf (Figure 36).

As already mentioned, I was there to visit the study area of Ed Owens. Unfortunately, because of some last-minute issues I had to solve in the Geology Department at the University of South Carolina, I missed my plane. The new connections resulted in a tortuous trip–Columbia to Atlanta to Boston to Montreal.

"Have a pleasant stay in the city of Montreal," said the college-boy bellhop as he delivered me to my $10 room in the Queens Hotel in the not-so-upscale portion of the city. Later, I wandered the dark streets, ending up in a bar populated by gigantic black ladies looking for a date. I decided I was too tired for that, so I retreated to my cheap hotel room.

Next morning, when I boarded the plane headed for Charlottetown, Prince Edward Island, I sat in the middle seat, something I seemed to do a lot in those days. On my left was a tractor salesman, who was in the process of moving to Birmingham, Alabama, interrogating me about what it is like to live in the South. On my right was a nice-looking lady in her early thirties named Margaret, who agreed to let me place my famous orange camera bag underneath her feet in our crowded space. She said she had noticed the unusual bag as we had stood in line at the ticket counter. I assured her that all I had in the bag at the moment was two cameras and 20 rolls of film.

Margaret worked in the Library of Congress in D.C. and was a devoted classic literature buff. She said she would send me a book of poems by William Butler Yeats, her favorite poet, which she did a couple of weeks later. After reading it, I could understand why she was so attracted to his work.

"Yeats lasts forever and has written many small gems that always spill out of my mind in obscure situations," she said. That particular situation may have been obscure, all right, but we managed to make the time fly with a totally engaging conversation during that flight. She said I was an October Scorpio, I was born on 20 October,[14] and proceeded with the following analysis of October Scorpios, which are, according to her:

1) Deeper than most

2) Love a challenge

3) Can get deeper into shit than most and still somehow pull themselves out of it and be better off for the experience. I assured her that I was in pretty deep at the moment and thanked her for the encouraging words.

4) Their poetry would hit you right between the eyes. After such hype, I didn't have the nerve to show her any of my poems.

When we were getting ready to say good-bye, I was thinking that maybe we could spend some time together when I came back through Charlottetown but, anticipating what I was thinking, I presume, she informed me that she was meeting her lover there.

"Oh well, she was too smart for me anyway," I thought as I walked away.

Facing a 2.5-hour layover in Charlottetown, I went for a long hike all the way around the airport to the other side and laid down for over an hour in the middle of a huge field filled with flowers. Finally, with only a few minutes left before take-off, I jumped up and hitchhiked my way back to the terminal.

On the plane to the islands, I sat beside one Roger the Photographer, who was also going to see Owens to make a photographic record of his rather complex project for his sponsor, the New Bedford Marine Institute, a Canadian government facility and Roger's employer. Roger was about 45 years old, a handsome French Canadian with a handle-bar mustache. He turned out to be a serious, hard worker of the sensitive type, compared to the rest of the field party anyway.

Ed's field team consisted of eight to nine souls, who lived in a small beach cottage on top of the dunes. Like most such field camps, it was a minor disaster area. I stayed in the master bedroom, which had one bed surrounded by shoes, clothes, wrenches, and other tools, all in general disarray. My room was the most comfortable one there, however. Most of the crew slept in a trailer parked out back, and there were two tents on the lawn in front of the cottage in case things became too crowded elsewhere.

Some of the key players on that particular field expedition included:

1) <u>Owens</u>, a 32-year-old, ex-rugby player from England, who was full of energy and gained and lost weight at prodigious rates. Seemed to enjoy swimming in the nude in that cold water, for some reason. At that time, he was also working on getting his head straight, recovering from a broken marriage, but was very much into his work, which he valued highly. He also drank a lot of beer and participated in a project to swallow dollar bills in order to see if they could easily be recovered in a day or two, a strange after-dinner activity.

2) <u>Heiner</u>, or <u>Yoselite</u>, a tall, blonde-haired and handsome, 25-year-old technician working on a degree in environmental science at Dalhousie University. He was an extremely fit skier, jogger, rock climber, and glider pilot. Very hard worker, who liked to hang on the refrigerator door. Yoselite was the name he gave people when he was so drunk that he couldn't remember his real name.

3) <u>Sparky</u>, a 45-year-old gentleman from Newfoundland. He was tall and thin, bespectacled with a hooked nose. Mostly quiet and slow sometimes,

they said there were 90 seconds in his minutes–Newfy time. In case you don't know it, the clocks in Newfoundland are set one half hour earlier than those in the other parts of eastern Canada. Newfys are the equivalents to the Pollock's and Aggies in the jokes of our culture. Ex-drunk.

4) Billy Whiskers, a 39-year-old science bum from the mountains of western North Carolina. About a year or so earlier, Leita Jean and I had become good friends with another professor at USC-East who was born in Sweden, B.J. Kjerfve, and his wife, Lucy. Lucy's parents visited one weekend and invited us to dinner with the four of them. Upon meeting us, Lucy's father, a native Georgian, took one look at my beard and said, "Oh no, a Dumb Swede and a Billy Whiskers both in the same night!" Thus, Billy Whiskers became my adopted name for the next few years.

Then there were the other five field assistants, mechanics, and technicians, plus one guy's wife, two cats, and a dog. That field camp was a bustling place, never a dull moment.

Because of the complexity of Owen's project, with numerous activities going on almost continuously, a modest sort of chaos reigned. As time progressed, a new language evolved in order to relieve the monotony of the repetitious surveys, I guess, or for whatever reason. Radio messages filled the airways, with the three main stations being given the following names:

a) Walla Walla - The trailer at the main field site, where much of the recording equipment and data records were kept. It was a rest station and main communication center. This radio was manned and the equipment was watched over twenty-four hours a day. The name "Walla Walla" was modeled after the words that Owens and some of his wild associates yelled as they streaked the main thoroughfare in Quebec during a recent scientific meeting.

b) Daisy - The main cottage. Named after the principal flower on the island.

c) Beaver patrol - The mobile radio transmitter located in the truck used principally for transporting the crew back and forth to the field sites, and for viewing all the fair young damsels camping and living on the island.

More of the new language:

a) BWC - Big white chief, Owens' handle.

b) That's fuck all you see - It really isn't important or doesn't amount to much.

c) Rambling - A number of things: 1) looking around for girls; 2) going to the bar or generally screwing off; 3) leaving; or 4) simply going from one place to another. (e.g., Are you ready to ramble?)

d) Growlies - Groceries.

e) Bad case of the growlies - Very hungry.

f) Humper - Assistant too dumb to think. A low-grade technician who only reads the numbers. (As in, don't send me any more humpers!)

g) Do the job - Have sexual intercourse.

h) Nookie - Guess.

Anyway, when Roger and I arrived at the Magdalens on that Tuesday night in August 1974, BWC met us at the plane and we went to Daisy where he cooked dinner. It was late, and we all had a bad case of the growlies, thus the food tasted good, even if BWC did cook it. However, Roger's assessment of the meal was that he would man the cooking duties for the remainder of his stay.

As usual, during the after dinner conversation, someone started making disparaging remarks to Sparky about his origins in Newfoundland. Roger said the highest car insurance rates in all of Canada were assigned to a small island in Newfoundland on which only two families lived, each owning a single car. Driving back and forth along the one road on the island, the drivers of the two cars were continually running into each other, hence the high insurance rates.

I said, "What's so unusual about that?" and told them about Steve smashing my new Jeep into the Power Wagon on Kiawah Island, the only other vehicle on that part of the island. I also filled them in on the details of the ensuing ride in the wrecker with Isaac and Chester. They all thought I made that story up.

[NOTE: In case you missed it, the Isaac and Chester saga is told in Chapter 5 – KIAWAH ISLAND.]

Next morning, BWC, Roger, and I drove over the hills to Walla Walla to inspect the instrumentation set-up. We also visited the west beach, where Roger took pictures of the crew running a beach survey oriented perpendicular to the shoreline designed to measure a topographic profile of the nearshore zone. This was accomplished by two persons wading into the surf with a tape and wooden survey rods so that the most landward surveyor could sight along the top of one of the stakes to the horizon in order to measure the vertical change between the survey points.

[NOTE: This beach profiling technique, which was originally invented by Emery (1965) and later modified and upgraded by our group, was also discussed briefly in Chapter 3 – THE FIELD TRIP.]

When the measurements were completed, the data were entered into a computer, which printed out the shape of the surface of the intertidal and shallow subtidal portions of the beach. Short-term changes in the configuration of the beach, as well as of the volume of sand present during each survey, could thus be determined.

As that particular activity was taking place, BWC and I walked down the beach a ways, busily engaged in a conversation about the techniques he was using and the different aspects of the beach morphology at the study site. A thick fog bank that extended far out to sea hung low against the ground ahead of us. As we approached to within a few yards of the edge of the fog, two bikini-clad, stunning French-speaking girls suddenly emerged from that opaque wall of mist and stopped about 10 feet or so right in front of us. As they walked up, they smiled warmly and said "*bon jour!*" Not knowing how to speak a word of French, I was tongue-tied, and I don't know what was the matter with BWC, who spoke a little French, but he was tongue-tied as well. The four of us stood gazing at each other for a good 30-40 seconds, while I waited impatiently for Ed to say something. Nothing being said, the ladies gave us a puzzled look and walked on by. We never saw them again.

That afternoon, BWC, Yoselite, and I ran the east-beach profiles. The east side of the island is exposed to the dominant winds in the Gulf, thus the waves were consistently huge on that beach. I wrote the following in my journal about this survey:

"They profile into depths up to their ears! The waves frequently splashed up over my head. Sometimes, Owens would go completely under the waves. The water was cold, but it was fun. I could feel the sand waves move under my feet. I wore my cut-offs and a wet suit jacket toward the end (against the cold)."

We returned to Daisy about five o'clock, and Roger said he would cook his famous macaroni dinner for us, but that it would take a while. With time on our hands, Yoselite, BWC, and Billy Whiskers decided to go *cherchez la femme*, which is accomplished in the late afternoon on the island by meeting the ferry that arrived once a day during the summertime.

When we drove up, a huge crowd was gathered at the ferry landing. It looked as if half the island's population was there. There were so many people milling about, I thought a festival of some kind was going on but, no, they were only there to meet the ferry. A number of fishing boats were in port, and we talked with one fisherman who had been fishing off Labrador, "Catching salmon, my son!"

Yoselite told me that a large welcoming party like this one met the ferry every day, and that the ferry landing at that time was the best place on the island to pick up girls. As we joined the throng of people walking randomly about, the pickings seemed a little slim to me. I did notice two ladies in their twenties or early thirties, one quite lovely and one a little less so, sitting aside on a bench watching for the ferry. As we walked by them, the less beautiful one said "*bon jour*" to Yoselite, but we kept walking.

When the ferry arrived, the waiting horde formed two long lines to greet the arriving cars, something like the cheerleaders at a college football game lined up for the team to run between them onto the field. Each disembarking car was greeted with a loud cheer as it rolled off the ramp between the two lines. All but two of the cars had Quebec license plates.

After this flurry of excitement, we walked back by the two ladies on the bench, and I caught the eye of the more comely one. Wow! She had blondish-brunette hair, was kind of short with big eyes behind large, round-rimmed glasses. She had been reading a book the whole time we were there and didn't seem very interested in either the people wandering around or the emptying of the ferry. We fooled around with the car for a while and on our way out, we drove close by the two women on the bench. As we passed by them, I leaned out of the window a little ways, a technique that I had learned a few days earlier from Chester and Isaac, and said "*bon jour!*" They laughed and greeted us back.

I was concerned that Roger's dinner would be getting cold and he would be miffed if we didn't show up soon, so we finally made a date to meet the two ladies at the Hotel Central at nine o'clock and left. To make a long story short, this encounter eventually worked out as an evening together, which I will not discuss further at this point.

Thursday, August 15, was a very windy day. In the afternoon, BWC and I took a helicopter ride to take some photographs of the site. It had cleared up, the sky was blue, and the sun shining on the rocks in the cliffs was spectacular, as was the green grass on the dunes waving in the wind. Later on, we profiled the east beach again during a howling gale. Yoselite and one of the humpers crashed through the big, 8-10 feet waves breaking on the offshore bar. I took notes, staying out of the water this time. I wrote later, "The east beach, with its color and high dunes, is an exquisitely beautiful place."

When we arrived at Daisy, Roger was preparing another one of his much-publicized dinners with the flair of a professional chef. As we all sat around the kitchen table drinking a beer and watching his performance, that is, everyone except Yoselite, who was hanging on the refrigerator door eyeing the growlies inside, Roger began telling us about his morning out in the helicopter photographing the small islands north of the Magdalens. Seems that part of the photographic mission was to deliver the month's growlies, mail, and other supplies, as well as a new crew, to a lighthouse on an island near the outermost margin of the island chain. The lighthouse sits in the center of the flat-topped rock island, called Bird Rock, which is about 75 yards wide and is encircled by vertical cliffs several hundred feet high in the red Devonian sandstones.

As the helicopter approached the landing circle, which was located only 20 feet from the edge of the cliff, the wind was howling at 35-40 knots. Just as the pilot started to land, the lighthouse keeper's dog ran out and lay down in the landing circle. The pilot lifted up and down a little in the ferocious wind, but the dog, probably anticipating an improvement in his cuisine in the shipment from the big island, just turned over on its back and stuck its feet playfully up into the air. The cursing pilot tried the up-and-down trick a couple more times, but the dog refused to move. Then he had to lift the chopper away from the island in the turbulent wind so he could radio the people in the lighthouse to come out and remove the blessed dog. Roger banged the pots and pans loudly to simulate the turbulence in the plane.

Once the dog was removed and the chopper landed with much difficulty, the pilot told Roger to stay in the plane while he and the co-pilot unloaded the cargo. Then they left him alone with the rotor blades still turning at a high rate of speed as they delivered their goods and rounded up the returning passengers. There he sat all alone, 20 feet from the edge of the cliff, petrified that somehow the aircraft would lift up by itself into the wind and topple over into the ocean.

"I lost five years off my life sitting there in that helicopter," he said, as he delivered the tossed salad to the table.

But the worst part was the take-off. With much rattling, the chopper lifted up into the turbulent wind and was swept out a few feet off the cliff and, once away from the lift created by the wind blowing up the cliff edge, dropped about a hundred feet straight down toward the ocean. This caused a thousand or so of the seabirds nesting on the cliff, mostly puffins and kittiwakes, to take off all at once and fly over the helicopter, shitting as they went along, as birds always do when frightened. The helicopter hung about 150 feet above the water, enduring the seabird shit storm until, finally, the pilot was able to steer away from the cliff.

As they headed for home, just skimming over the surface of the chaotically breaking wave crests, the pilot said, "It's always the same, every time we come here."

ZABAWA – The Science

Chris Zabawa's dissertation, which was completed in 1978, was titled, *Flocculation in the Turbidity Maximum of Northern Chesapeake Bay* (Zabawa, 1978). Chris also procured his own funding, which was provided by the Chesapeake Bay Institute, carrying out his fieldwork in the bay aboard the Institute's research vessel, the *Maury*. In fact, the study was suggested and supported by Jerry Schubel, one of the leading scientists at the Institute. Chris also did a study of the sediments in the Winyah Bay Estuary in South Carolina, which was published in our course notes for the AAPG field seminars, *Terrigenous Clastic Depositional Environments* (Zabawa, 1976).

Chris was a very efficient worker, who had an immaculate laboratory on the Institute's boat equipped to filter the suspended particles out of the numerous water samples they were collected. His focus was to eventually examine the samples under a scanning electron microscope (SEM) in order to determine the detailed physical characteristics of the agglomerated suspended material, which is called floccules (see example in Figure 37).

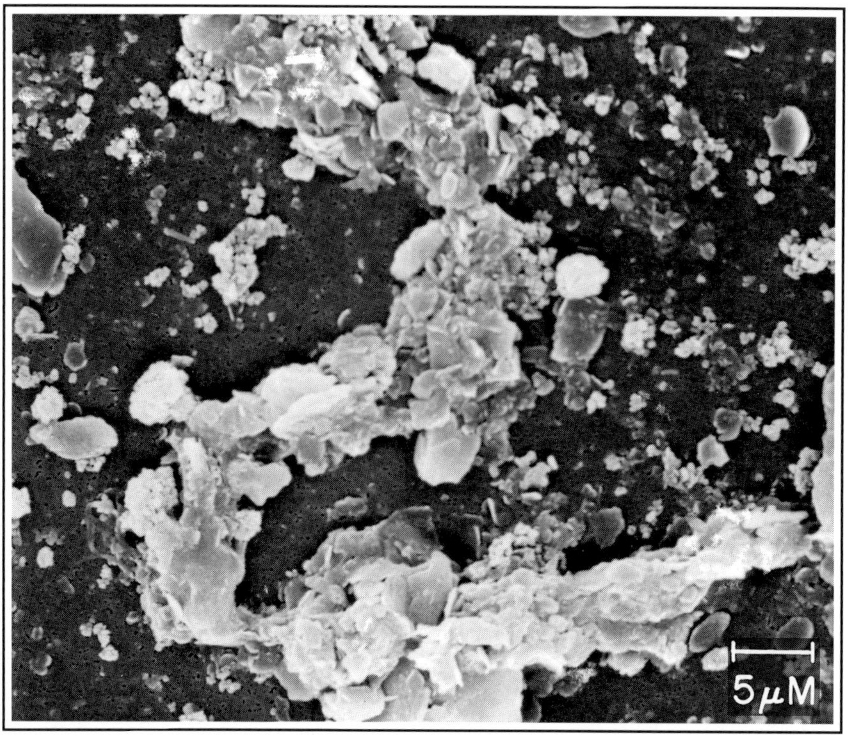

FIGURE 37. Flocculated mass of suspended sediments produced by adding filtered seawater to a suspension of dispersed fine-grained Chesapeake Bay sediments. From Zabawa (1978).

Some of Chris's major findings follow:

1) In Chesapeake Bay, "evidence of flocculation is contained both in the *in situ* appearance of agglomerated suspended sediments in the turbidity maximum, and in the *in situ* distribution of agglomerated sediments in the background and resuspended bottom sediment populations in the turbidity maximum."

2) SEM observations show that the suspended sediments in the northern bay are not single grains, but are rather composite particles "formed by several mechanisms of agglomeration: electrochemical flocculation, coagulation, microbial growth, and fecal pelletization."

3) "Pelletization is essential for the deposition of colloidal particles in the estuarine turbidity maximum, but flocculation and coagulation attach to the pellets many colloidal particles which would not otherwise be deposited from the background sediments."

[NOTE: The concept of the turbidity maximum should probably be elaborated upon a little more. As noted by Hayes and Michel (2010), in the middle portions of estuaries, tidal flow is typically increased and the flood tide reverses the river currents when the tide rises. This results in a complex two-way, two-layer circulation pattern, creating a zone of mixing of the two water masses. In that zone of mixing, an important process known as **flocculation** of the clay particles in suspension occurs. Flocculation takes place because the abundant clay particles in suspension brought in by the river flow slow down their seaward movement at the point of mixing as a result of the reversing of tidal flow directions twice a day. This allows an increase in the number of clay particles in the water column to the extent that they begin to collide with one another. These colliding particles tend to adhere together under the influence of the increased concentration of cations, such as $Na+$, $Ca++$ and $Mg++$, derived from the seawater. Therefore, the clay particles, which may have diameters as little as one micron, cluster into groups (**flocs**) that may have diameters measured in hundreds of microns. As a result, many of the flocs sink to the bottom at slack tide. Once on the bottom, some of the flocs are resuspended by ebb- and flood-tidal currents. The net result is the formation of a zone in the middle reaches of the estuary that contains abundant fine-grained sediment in suspension known as the **zone of the turbidity maximum.** This zone of turbidity can migrate up and down the estuary over time, moving downstream during periods of flood on the river and upstream during low-water periods. A side effect of this process is the creation of a "mud zone" on the tidal flats and in the tidal channels in the middle of the estuary.]

When Chris Zabawa left USC-East, he joined the Maryland Department of Natural Resources, Coastal Zone Management Program in Annapolis, during which time he continued to produce interesting scientific results. Presently, he continues his environmental work at EPA's Office of Wetlands, Oceans, and Watersheds.

ZABAWA – The Story

(*Saturday, 12 October 1974, airport in Charleston, South Carolina*)

Steve Wilson and Leita Jean dropped me off at the airport in Charleston for the flight to Baltimore. Chris Zabawa met me in Baltimore for a quick ride to the Chesapeake Bay Institute's research vessel, the *Maury*, which was at anchor in Annapolis.

We sailed out of the harbor at Annapolis early on Sunday morning, winding our way among the sailboats. It was regatta weekend at the Naval Academy. Out in the middle of the bay, as we headed north, there were sailboats all around us in the fog, which slowed our progress quite a bit.

Chris was running a very efficient operation on this first-class research vessel supplied by the Chesapeake Bay Institute, obtaining cores and suspended sediment samples to do some original research on sediments thus acquired, as mentioned earlier. His primary assistant, Jeff Boggs, was "a typical hippie," I wrote in my journal, with long hair and metal glasses. A graduate of some technical college in Florida, he worked exceptionally hard and was quite friendly.

As we moved along in the light fog up the center of the bay toward the first sampling station, the three of us were standing by the rail at the stern when Jeff pulled a joint out of his pocket, lit it, and we passed it around a few times. Then I sat down in a deck chair, while the other two went about their business. In a few minutes, Jeff came back to near where I was and climbed up on top of the rail at the stern, hanging onto two cables from the winch. He stayed in that position for a long time, bending the middle of his body out over the water with his back silhouetted against the fog and the scattered sails across the bay.

He climbed down, walked over to me, and said, "Hey, over there's the buoy for station number 2!" which was one of the permanent sites the Institute monitored for various environmental characteristics of the bay.

Then he walked back over to the rail, looked out at the buoy, and yelled in a loud voice, "Hello, there, station number 2!"

"I think he has become a little attached to the bay," I thought.

I was becoming attached to the deck chair I was sitting in, staying in that position huddled in a warm jacket for hours in a semi-depressed state and listening to the cacophonous mixture of sounds from the vessel and its surroundings the breeze carried back to me. Hearing their honking overhead, I would occasionally look up at one of the numerous Vs of Canada geese headed south. From the lab came the clinking of the glassware as Chris and Jeff worked with the samples, those sounds harmonizing with the music from Jeff's hard-rock tapes.

In the galley, the cook and the engineer were engaged in a rowdy card game. "I think I'll have a fucking orange!" said the cook, as he left the table after throwing his losing hand down in disgust.

My most vivid memory of this trip was of the long hours alone in the deck chair. That, and looking up at the curving, giant arches of the Chesapeake Bay Bridge as we sailed under it, one of the many bridges over troubled waters I crossed under, and over, in the trying years of 1974-75.

So it went through the rest of the fall. I went out and bought a small house in a quiet neighborhood in town, why I am not too sure. The last and final breakup with Leita Jean came in the late fall and I did not go with her to Massachusetts for the Christmas season. Instead, I spent the holidays alone in my new house, with an occasional visit from my friend Sally, who was also breaking off from a relationship with her ex-husband. Sally told some great stories. She had formerly circulated in the society of notables in the city, about whom she had all the inside poop. I laughed when she told me that one of the more prominent men in town spent the whole 24 hours of Christmas day secluded all by himself in a darkened clothes closet. There had been something missing in his childhood, I guess.

"Sometimes I feel like doing that myself," I responded.

Sally's assessment of the holy season: "It wasn't really that bad. I was surprised, thanks to you."

During that winter, I spent a lot of time creating another tone poem, called *A Day Just Like Today*, based on our two-year project on the coastline of Iceland. There was a good turnout for the first showing of this program at USC-East on the first Saturday night in May. At the end of the show, Jim Durig, the Dean of the College of Science and Mathematics, with whom I worked as Chairman of Geology, walked up to me and said, "I hope the science was as good as the pictures!"

In fact, this show was not near the success of *Suzanne's Lament*, but it gave me something to do.

The big news for me that spring was that, because I had participated in and directed research on the Alaskan coast since 1969, I was able to win a major contract from the National Oceanic and Atmospheric Administration (NOAA) to continue working there for the next few years along with a number of my graduate students at USC-East.

7 THE MEN THAT DON'T FIT IN

Thirty years ago I was born into the world.
A thousand, ten thousand miles I've roamed,
by rivers where the green grass lies thick,
beyond the border where the red sands fly.
I've brewed potions in vain search for
life everlasting.
I read books, I sang songs of history,
and today I've come home to Cold Mountain
to pillow my head on the stream and
wash my ears.[15]

HAN-SHAN in COLD MOUNTAIN

Then when as wolf-dogs fight we've fought, the
lean wolf-dog and I;
Fought and bled till the snows are red under
the reeling sky;
Even as lean wolf-dog goes down will I go
down and die.

ROBERT SERVICE in THE HEART OF THE SOURDOUGH

The Science

Needless to say, we were very excited over the opportunity to go back in the field along the coast of southern Alaska. Once we reached the field, we split into two field crews. Jon Boothroyd and his team of two, Ray Levey and Mark Cable, set up a camp by the Malaspina Glacier (Figure 22), where they focused their activities on the dynamic processes of the streams that drain the glacier. My field team, which covered the entire shoreline between Cordova and beyond Yakutat to as far south as the La Perouse Glacier (Figure 23), consisted of myself, a graduate student working on a M.S. thesis on the project, Chris Ruby, and Janie Zenger, an enthusiastic undergraduate English major.

The southern coast of Alaska is an exceptionally dynamic area with respect to physical processes. Intense tectonic activity, strong winds, large waves, strong tidal currents, and active glaciation interact to produce one of the most rugged and variable coastlines on earth. The Chugach, Kenai, and St. Elias mountains, extensions of the Cordilleran Mountain system, control the general orientation of the coastline, a southwest-facing, broad arc that extends from the Canadian border in the southeast to Dutch Harbor far out on the Alaska Peninsula. Our study area extended for over 400 miles along this shoreline.

Rapid advance and retreat of the numerous glaciers that extend down from the mountains have caused sudden and dramatic shifts in the loci of erosion and deposition along the coastline (Hayes and Ruby, 1994; Hayes, Michel, and Betenbaugh, 2010). Therefore, it is both a treat and a challenge to try to synthesize the coastal geomorphology of such an area.

My group arrived in Cordova in late May 1975 to start the surveys along the outer margin of the Copper River Delta. After a few days studying the delta, I wrote a letter to my friend Sally in which I said that our drumstick model for barrier islands had been confirmed during "A spectacular clear day as we were flying at five thousand feet taking pictures." Some of these pictures are still featured in my lectures on barrier islands. I also observed that "There have been some remarkable changes of the shoreline since 1971. So far, we have been lucky enough to find several of our old surveying stations, so we are getting good data on the changes that have occurred since we were last here."

Next, I commented about the growth of Egg Island and the other barrier islands along the delta front since the Good Friday earthquake of 1964, as is illustrated in Figures 38 and 39. Then I continued by saying: "Actually, the dominant theme here is the spectacular beauty of the place, and the joy of being away from the phone and other hassles that go with being back there. The science is good also and I have been getting a lot of ideas for papers to publish and new research projects."

In fact, I have always received a certain amount of grief about not publishing enough of this stuff. The work is summarized to some extent in two papers, Hayes et al. (1976a) and Hayes and Ruby (1994). Some of the points emphasized in those papers include:

1) The morphology of the barrier islands, with their predominant drumstick shape and other aspects of their beach processes and tidal inlet dynamics (see Figures 25, 38, and 39). The recovery of the barrier islands, with continued accentuation of the drumstick configuration in response to the fact that the shoreline was uplifted 2.5 to 3.0 meters in the 1964 earthquake, was a fascinating thing to watch. Details of the intertidal geomorphology at Egg Island are clearly shown by the three oblique aerial photographs in Figure 40.

2) The other major shoreline type was the outwash-plain shorelines, with their wide prograding outwash-stream deposits and downdrift beach-ridge plains, such as at the Alsek River mouth (see Figure 41). Large dune fields also are commonly associated with the beach-ridge plains, evidence of the large amount of sediment supplied to the shore by the outwash streams. Also, the strong east-to-west longshore sediment transport in the area, discussed in Chapter 4 (see Figure 26), has a strong influence on the development of this type of geomorphology.

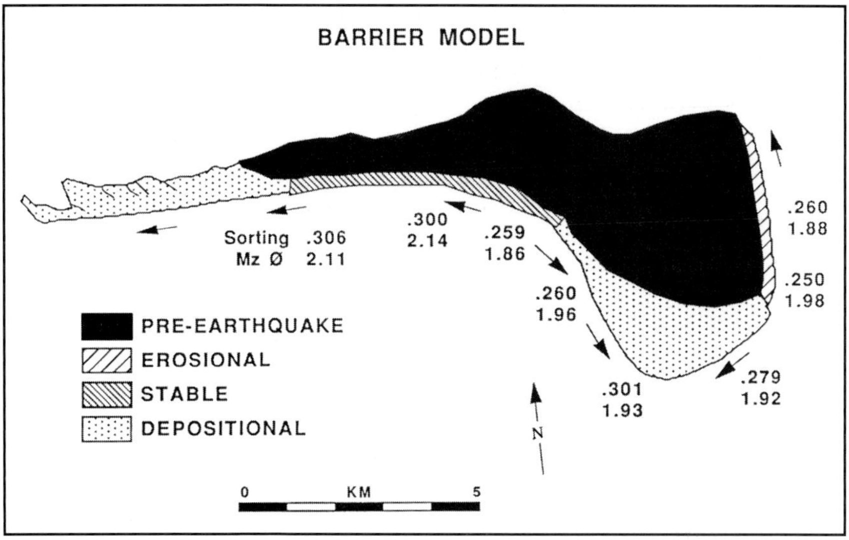

FIGURE 38. General post-earthquake erosional-depositional model of barrier islands of the Copper River Delta, based on changes at Egg Island between 1964 and 1975. The core of the island was subaerial before the Good Friday earthquake of 1964, which caused uplift of the land surface in this area of between 2.5 and 3.0 meters. After the uplift, the drumstick shape of the island was accentuated by rapid accretion downdrift (west) of the inlet throat (located along the easternmost side of the island) and westerly migration of the island itself. Numbers give values of mean size and sorting to the beach sediment samples. In general, sediments of the erosional and stable areas are slightly coarser than those of the depositional area. Arrows indicate sediment transport directions inferred from beach morphology. From Hayes and Ruby (1994).

3) Numerous aeolian sand dunes are present in this area, the general models of which are illustrated in Figure 42A. The Bering Foreland, the area along the seaward side of the Bering Glacier (Figures 22 and 23), has the largest development of coastal dunes in the area we studied. The orientation of this segment of the coast in an east-northeast/west-southwest direction allows the east-southeast winds to cross the shoreline obliquely. Because of the abundant glacial silt in the sediment, huge plumes of aeolian silt are sometimes blown off the beach in spectacular helical flow patterns (Figure 43A). Large fields of longitudinal dunes, mostly vegetated, have developed to the east of Cape Suckling (located straight off the end of the Bering Glacier; Figure 22) in response to these winds. These dune fields extend inland for several kilometers and attain heights of tens of meters (see Figure 43B). They are composed of silt-rich, medium- to fine-grained sand. Wind-shadow dunes, showing similar orientations, develop in great profusion on the tops of the overwashed spits and sandy high berms in this area. Logs and vegetation clumps form the baffles that initiate these dunes. Low foredune ridges occur on some of the barrier islands along the margin of the Copper River Delta. Wind-shadow dunes are quite common in this area as well. During periods of high winds, the upper intertidal flats and barren parts of the islands are covered with low-amplitude, dune-like mounds of sand which have barchanoid shapes and

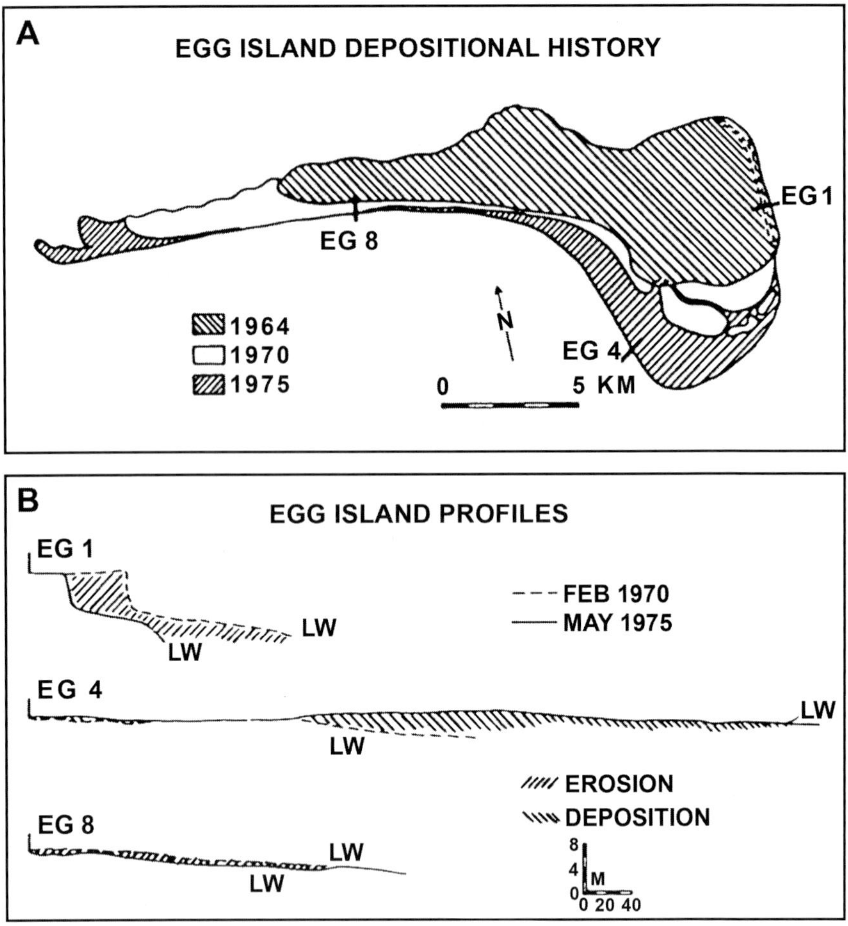

FIGURE 39. Evolution of Egg Island, Alaska, cont. (A) Changes of Egg Island after the March 1964 earthquake. Note continual accentuation of the drumstick shape of the island through time. (B) Beach profiles showing the post-earthquake changes on Egg Island at three stations, which are located on diagram A. From Hayes (1976).

lack discrete slip faces. In the field, we called these features *flat dunes* (see illustration in Figure 42B). These dune-like structures migrated slowly downwind and were usually oriented in response to either prevailing southeasterly winds or to the offshore-directed adiabatic winds blowing down the Copper River Valley.

4) We constructed general depositional models for the two main shoreline types, mixed-energy delta and outwash plains, which are illustrated in Figure 44.

5) During our studies, our team surveyed a little over 100 beach profiles, some of which were established in 1970 and resurveyed in 1971 and 1975. There were three basic types of profiles: a) Gravel beaches, which will be discussed in some detail in Chapter 15 – GRAVEL BEACHES AND FAN

FIGURE 40. Aerial views of Egg Island, Alaska. (A) Low tide view taken in June 1971. (B) Low-tide view of the eastern end of the island in May 1975. Arrow points to same sand bar as the one indicated by the arrow in A. This beach accreted 300 meters between February 1970 and May 1975. (C) Multiple intertidal bars welding to the beach near the position of the arrow in B. Low-tide photograph taken in summer of 1969 (by Stewart C. Farrell). From Hayes (1976).

DELTAS; b) Relatively wide and flat fine-grained sand beaches, which followed cycles of erosion and deposition somewhat similar to those surveyed in Massachusetts (Figure 13). These beaches commonly contain intertidal bars, such as the ones illustrated in Figure 40; and c) Coarser-grained sand and fine gravel beaches with relatively steep profiles. These types of beaches, such as the one illustrated in Figure 45, are commonly associated with migrating rhythmic topography.

[NOTE: The concept of rhythmic topography was discussed as follows by Hayes and Michel (2010). A mode of longshore sediment transport occurs when evenly spaced bulges of sand along the beachface move down the shore like the sinusoidal loops one might throw along a rope. These bulges of sand, called **rhythmic topography**, are typically spaced 10s to 100s of feet apart. They are created by waves that approach the beach at an oblique angle. The side of the sand bulge facing directly into the wave approach direction orients perpendicular to that direction, with the beachface being aligned parallel with the approaching wave crests, the most stable configuration possible. However, the distance the sediment on the exposed beachface of the bulge can extend offshore is limited by water depth in most situations. Sand is therefore either lost from the outer edge of the bulge into deeper water or is transported to its lee side. The result of this process is that the exposed face of the bulge erodes

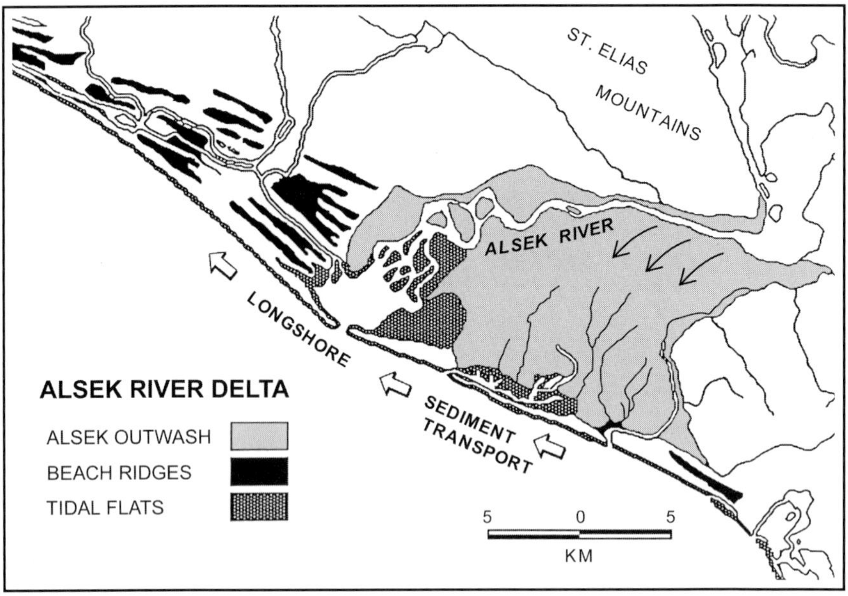

FIGURE 41. Depositional features associated with the mouth of the Alsek River. Note the apparent westward drift of the main channel of the river, as well as the development of a broad beach-ridge plain to the west, in response to the strong, east-to-west longshore sediment transport in this area. From Hayes and Ruby (1994).

while sand accumulates on its sheltered side. Therefore, the bulge moves along the shore in what is known as known as the **downdrift** direction (direction in which the individual sand grains are moving along the shore within the longshore sediment transport system).]

6) In his M.S. thesis, Chris Ruby described many of these details (Ruby, 1977). Chris was ably assisted in the field by Mike Stephen and Steve Wilson. Some of the results of this work are given in Hayes et al. (1973). Chris Ruby also eventually put together shoreline sensitivity maps for oil spill response in the area (Ruby and Hayes, 1978), using a new concept we had just recently developed based on studies in Lower Cook Inlet in the summer of 1976 (Hayes, Brown, and Michel, 1978). This mapping technique will be discussed in considerable detail in Chapters 8, 17, and 20.

7) During the summer of 1975, Jon Boothroyd and his team continued their long-term studies of glacial outwash fans, which were discussed briefly in Chapter 4. Inasmuch as this was not strictly coastal work, and I was not personally involved except in an advisory capacity, it will not be discussed further here. This is not meant to dismiss the importance of Jon's findings, as these are still considered to be some of the most important contributions to the early understanding of braided stream and outwash fan deposition (e.g., Boothroyd and Ashley, 1975; Gustavson and Boothroyd, 1987). Probably his most famous stratigraphic model is

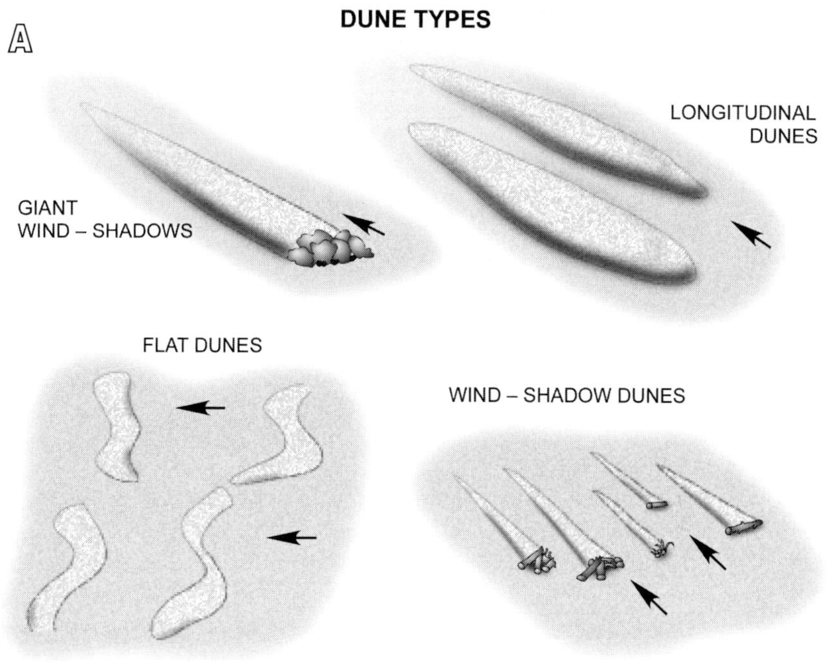

FIGURE 42. (A) Types of aeolian sand dunes that occur on the southern Alaska coast. (B) Flat dunes, Copper River Delta, Alaska. COASTAL HERO Barry Timson holds the tape. February 1970.

FIGURE 43. Aeolian features of the Bering Foreland. (A) Helical transport of wind-blown silt off the beach of the Bering Foreland (along the front of the Bering Glacier; Figure 22). Wind was blowing out of the southeast at approximately 50 km/hr. Photo was taken in July 1970. (B) Longitudinal dune field located a few kilometers east of Cape Suckling, which is located off the end of the Bering Glacier (Figure 22). View looks east-southeast. Photo was taken on 27 February 1970.

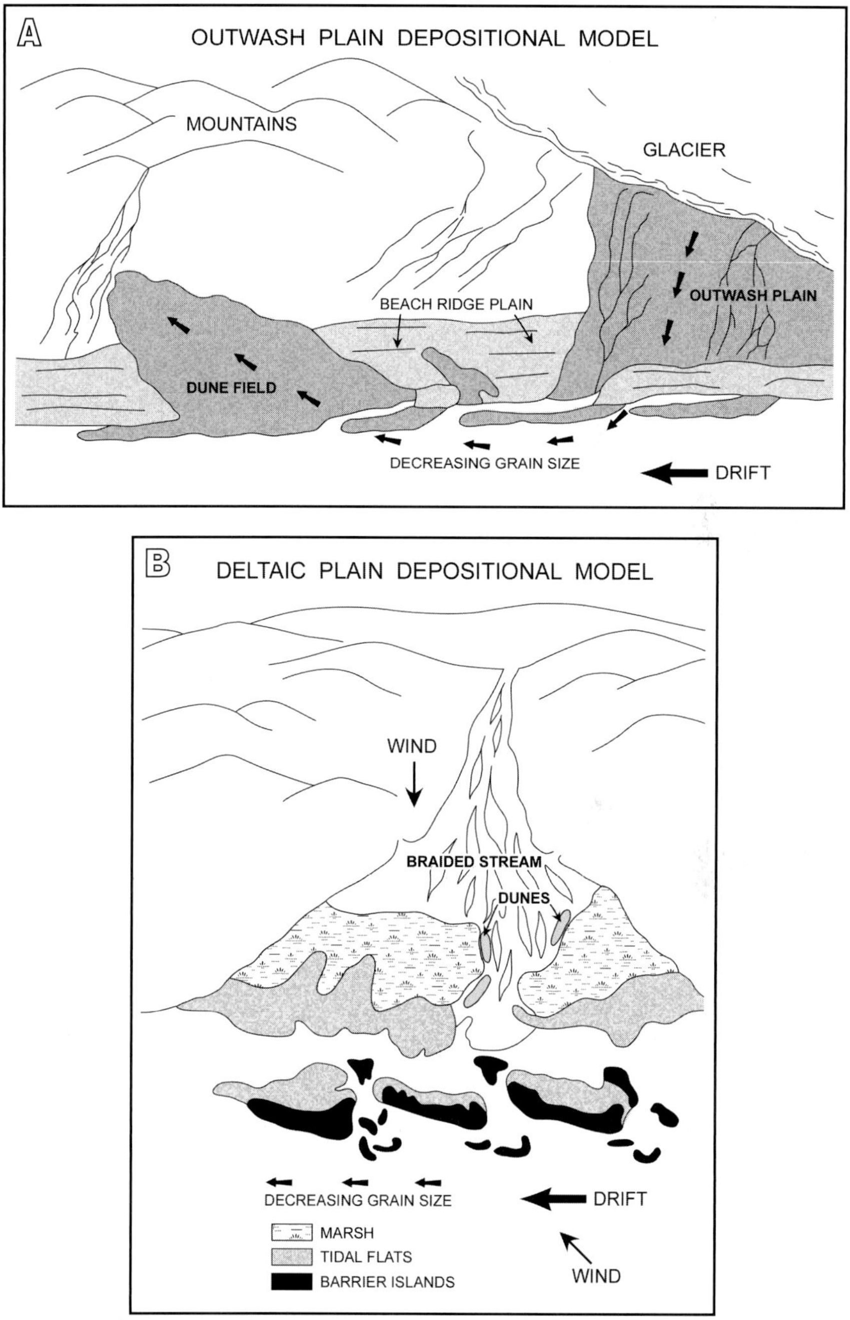

FIGURE 44. Depositional models for the outer coast of southern Alaska. (A) Glacial outwash plain depositional model. Principal components are outwash streams and fans, coastal spits and beach-ridge plains, and coastal dune fields. (B) Deltaic plain depositional model. Principal components are braided stream, salt marsh, tidal flats, subtidal estuarine, and mixed-energy barrier islands. From Hayes (1976).

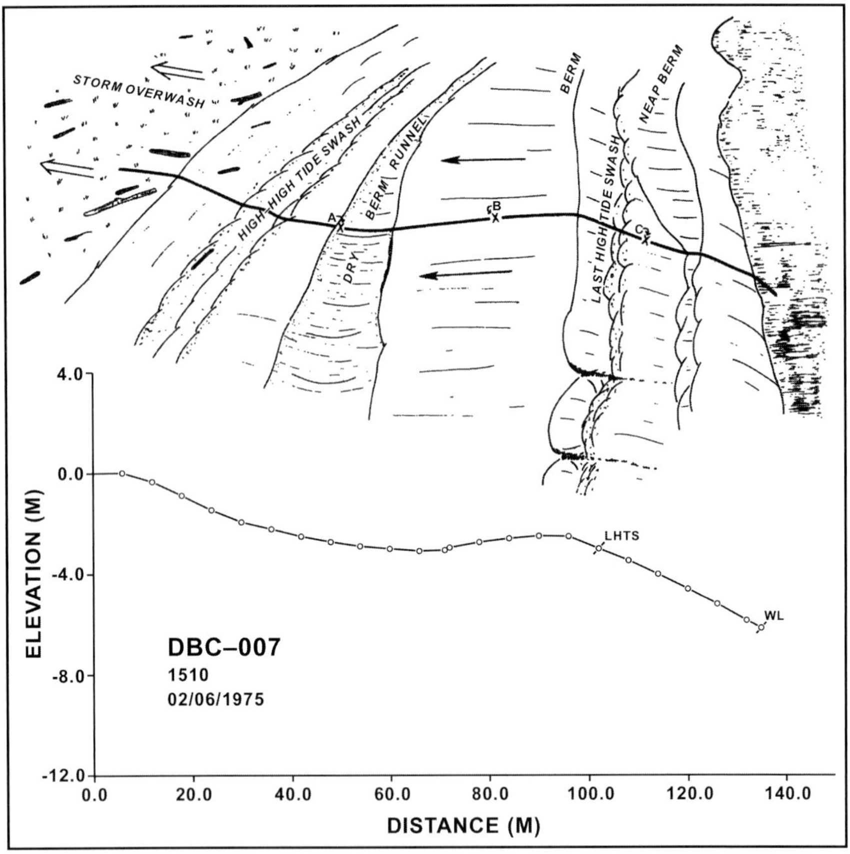

FIGURE 45. Beach profile site DBC-7 located about 40 kilometers northwest of Dry Bay (Figure 23). At this time, the beach consisted of a high, broad berm and berm runnel backed by a high overwash berm. The profile was convex upward, indicative of progradation. The topographic profile of the beach, plotted at a vertical exaggeration of 5:1, is given below the sketch. From Ruby (1977).

given in Figure 46.

Story One

Toward the end of May, we moved from Cordova to Yakutat and stayed there for a little under two weeks. On June 9, we moved on to Cape Yakataga (Figure 22), where we spent the rest of the field season, with me heading for South Carolina on June 29, leaving Chris and Janie to finish out the fieldwork.

One of the main goals of this book is to introduce the reader to the especially unique and anachronous people met during field projects such as this one. For example, while in Yakutat, we spent some time with a halibut fisherman named Halibut John, who was an amazing character. George the Mountain Climber was another of the unique characters we met and became friends with in Yakutat.

During one rainy night early in June, as we were lying around in the hut at the

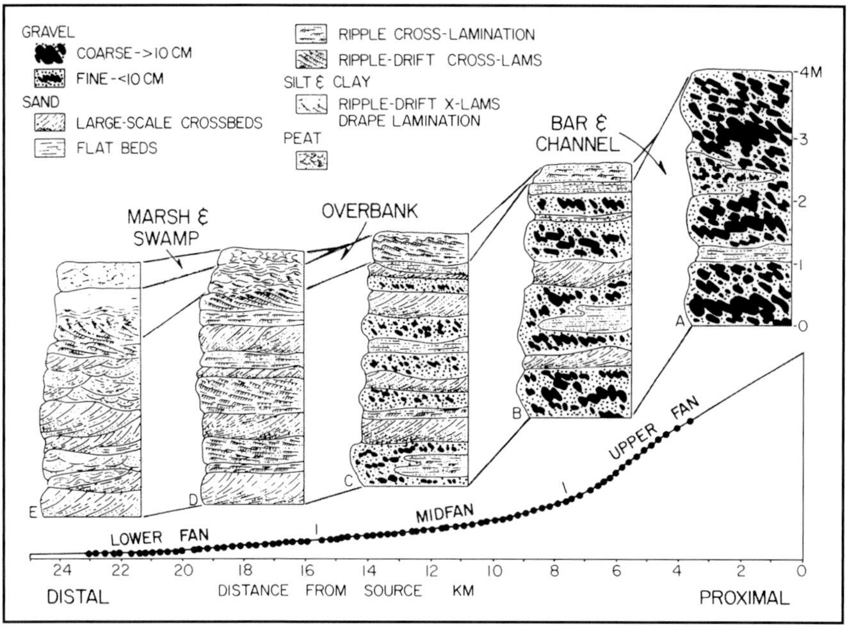

FIGURE 46. Downstream variation in facies and sedimentary structures for Alaska outwash fans. Bar and channel sequences do not fine upward, but are capped by a finer overbank facies that becomes more important in a down-fan direction. From Boothroyd and Ashley (1975; after Boothroyd, 1976).

Yakutat Lodge, our living quarters located along the side of the Yakutat Airport runway, a fuzzy blonde head projected through the door and said, “Has anybody seen Gordon the Game Warden?”

“He’s out in the x?#!x woods! What do you think?” I answered.

Then he stepped inside to join us, a good break from a rather dull evening. He was part of a group of mountain climbers who were living in tents on the other side of the hangar. They were gathering from all over to make an attempt at conquering Mt. St. Elias, an eighteen thousand feet high peak located a few tens of miles northeast of the entrance to Icy Bay (Figure 22). They were going up as soon as the weather cleared and the plane could get in to their first base camp. They would have to land on the ice, and expected the climb to take about sixty days. The mountain had been climbed eight times before, but the face they were going up had never been attempted. That was a group of true crazies. By the way, when Vitus Bering finally was able to explore the Alaskan coast on 16 July 1741, the first land sighting his party made was of the peak of Mt. St. Elias. Next time you fly from Seattle to Anchorage, if it is a clear day, you may spot the isolated, pyramidal shape of this spectacular mountain.

We eventually came to know our new friend, George Ochenski, as George the Mountain Climber. He grew up in Detroit and went to Michigan State, where he first majored in Biology, and then English, and then dodging the draft. He was twenty-five years old. He ran away from Michigan to live in the woods in Montana, where he had gained a reputation as a mountain climber. He was about five feet, eleven inches tall, thin and blonde, with penetrating eyes, and he was an incessant talker. His biggest kick, after girls and climbing mountains, was getting stoned and driving his dog sled through the Montana woods at night. He brought over a book called *Cold Mountain: 100 Poems by the T'ang Poet Han-Shan*, a Chinese Buddhist poet-recluse who lived 900 years ago. It was an amazingly appropriate book for us to read there in the rain in Alaska (see introductory quote).

Anyway, George was worried about this last outpost of Tlingit Indian culture, Yakutat, being destroyed by the town being made into an oil port. We talked about that frequently as the days passed. My contention throughout our discussions was that Captain Cook and the Russians that followed had already destroyed what Indian culture had originally existed, and that it really didn't matter anymore. I later changed my mind about that. Maybe George had something to do with my change of heart, I don't know.

The next evening after our heated discussion of Tlingit culture, George visited again and told us that an old Indian fisherman had given him a piece of salmon in honor of their sacred mountain, St. Elias, and told him to be good to the mountain. I think that pleased George immensely.

We talked a lot again that night, especially George. He was concerned about the police shows on television and about the country turning into a police state. He used his gun to hunt with and didn't want to give it up. I expressed my feeling that the grand bureaucracy would grind to a halt in about ten years. What were we going to do?

A few days later, we moved to Cape Yakataga, where I sometimes thought about George the Mountain Climber, whom we had left sulking in the rain back in Yakutat. He eventually sent us a note in one of the planes that came up from Yakutat.

Among other things, he said in the note, "In any event, your kindness and hospitality offered at such a time and in such a place to such a bum as myself means a great deal to me. …… I shall think of you all in both my spare moments of reflection and at the summit, should I be fortunate enough to reach it." Whether or not they reached the summit, I haven't a clue. But, for the record, George went on to eventually become a major player in the Sierra Club in Montana.

Story Two

(23 June 1975 - Monday; Cape Yakataga, Alaska)

By the 23rd of June, our team was staying in a somewhat dilapidated building at the seldom used airport at Cape Yakataga, which is located a little over a hundred miles southeast of Prince William Sound and the town of Cordova, on the open shoreline of the Gulf of Alaska (Figure 22). After dinner that day, I was on the beach by the landing strip talking with Gayle Ranney, who was our pilot throughout the stay at the Cape. I had noted that whenever we landed the tail-dragging Cessna fixed-wing on the beach at one of our study sites, she would walk away from the plane, find a comfortable log to sit on, and calmly observe the waves and the shorebirds while we took the usual thirty to forty minutes to carry out the survey.

I asked, “What are you thinking about while we run the profiles?”

“Good thoughts. Just getting my head straight. I love it out here. It’s been a good–a beautiful time for me.”

Then I asked her to take a walk and we went down along a dirt track a couple of miles from the end of the airstrip and our abode. At the end of the track, where the bridge had fallen into the Duktoth River (this area is pictured in Figure 26), we saw a cabin just off to the right in the edge of the forest. We looked in the window, which was dark, but we thought we smelled smoke. Shortly, an old man came out of the cabin and asked us if we wanted to come in for some coffee.

We had a long talk. His name was Paul Shoner. He had been in Alaska forty years; he came up from the state of Washington. He had been a prospector and a trapper. He enumerated the animals that he had trapped–lynx, beaver, fox, and wolves. He was 70 years old, but still walked his trap lines. He had cabins all through the woods that he built himself. This one, which was made from logs, had two rooms, with bunks lining the back room. He had built the stove from an old oil drum. In the middle of the door was a single picture of a black-haired woman, just the face.

He said the glaciers had retreated since he came to Alaska, and that the plains in front of the ice had grown up in trees. He was concerned that the game would be depleted by over-trapping. One of his best friends had been an old Indian native who used to live in Katalla, a small, abandoned village located about 70 miles west of Cape Yakataga (Figure 22).

“He killed at least three of them,” he said, “that’s why he couldn’t go back to Yakutat.”

“Why did he do it?” I asked.

“Because they had it comin’ to them. He was the best friend I had around here.”

Then he sort of spaced out for a while, as these old timers often do. We waited silently for him to talk again.

He talked about his son in California.

Then I asked him if he knew Charley Bilderback, an old prospector I had met on my earlier trip to Cape Yakataga. Charley had a cabin right up against the dune at the tip of the Cape, a point where vertical rocks project straight out into the Pacific Ocean. I had visited Charley several times in 1970, and had many discussions with him about the geology of the Cape Yakataga area.

Charley once told me, “If you want to know where the sediment goes, son, just ask me. The sand goes where the wind blows!” A basic truth that I have repeated often to my students. I guess I was young enough to be his son in 1970.

Charley had a phenomenal collection of classical geology texts; *Field Geology* by Lahee and *Structural Geology* by Billings numbered among his collection. He knew all the terms, and he could really talk geology. He showed me a collection of postcards from all over the world that had been sent to him by the different geologists who had visited the Cape. You see, the rocks around Cape Yakataga had long been known to be potential source rocks for oil; hence, they had been studied often by many different groups. It was rumored that an experimental well was to be drilled on the continental shelf just off Charley's cabin in 1976 but, as far as I know, this never happened.

Paul Shoner, something of an infidel on the subject of geology, laughed about some of the discussions he had had with Charley. "Charley didn't know his geology, heh, heh," he said, showing the stubbornness that characterizes these old men that live alone in the woods.

Charley was there in his cabin during the Good Friday earthquake of 1964, and he made some of the most accurate measurements that anyone had recorded of the run-up of the tidal waves (seismic sea waves) generated by the earthquake. He retreated up the hill as the waves came ashore and then ran back down it while the water level was low so he could mark the highest elevations the waves had reached. He kept this up all night as wave after wave came in. Meanwhile, about 60 miles up the shore, the fishermen, whose story was presented in Chapter 4 – ALASKA: *SUZANNE*, EARTHQUAKES AND TSUNAMIS, were sitting up on the side of that mountain listening to the "tidal waves" come in.

I had nearly forgotten all about this until in early July 2008, I started writing the first draft of the section on tsunamis in the book JM and I were writing on the Central California coast (Hayes and Michel, 2010) and found the following quote in the Pararas-Carayannis write up cited earlier in Chapter 4:

"Cape Yakataga: Outside the immediate generating area, the record of Cape Yakataga, as constructed from the personal account of C.R. Bilderback, a resident of the area, is the next most reliable record. This record is the only one obtained outside the generating area that shows an initial drop in the water level. It is quite possible therefore that the first waves to arrive at Cape Yakataga had a different origin from that of the first waves to arrive at Yakutat. It could very well be that the Cape Yakataga waves traveled over the shallow portion of the shelf, whereas the Yakutat waves came from the open ocean."

"It is concluded that the waves arriving at Cape Yakataga had their origin in the shallow coastal area near Bering Glacier, whereas the waves arriving at Yakutat traveled from a different path through the deeper waters of the Gulf of Alaska."

The actual data that Charley Bilderback collected were also presented in this write-up by Pararas-Carayannis. Guess old Charley was a pretty good geologist after all!

As Gayle and I departed from this pleasant discussion with Paul Shoner, he invited us to come back again for some of his patented dandelion wine.

Next day, as we were getting ready to fly out to work the early morning tide, we saw Paul walking along the road with a back-pack, heading for one of his cabins up the shore.

He asked, "Where you going?"

"Oh, we're going down to Okalee Spit and then come back by the Bering Glacier," I answered.

"Well, if you find any of that yellow stuff, just keep it to yourself," he laughed, as he waved and walked away.

Then, he turned and said, "You can always come back, you know."

We went back to visit Paul's cabin a week later but, by then, he was many miles down the shore in another one of his cabins.

At the end of that day's run, we decided to go over to visit the ghost town of Katalla (Figure 22), which had been featured in many photographs in *Suzanne's Lament*. We found that the picturesque old hotel had been raped. Boards had been used to make campfires that had been built inside the rooms. The antique piano had been smashed and thrown out into the growing weeds.

As we went back to the plane, Les Knew, an old man who had lived alone in Katalla for many years, came out and took a seat on a log. Janie walked up to him and sat between his legs, while I took some pictures. He was tall, 6' 5", and gangly. He took us to his cabin, which was somewhat like Paul's, with a drum stove and a single pin-up, a nude black girl, in the middle of the wall. Paul had built Les' cabin in 1937, a replica of several others he had scattered throughout the woods.

Les was a native Texan and had lived in Alaska 20 years, sometimes spending weeks alone in his cabin buried to the roof in snow. I wondered how he coped with the loneliness. He told us that a year or so earlier, a visiting group of fishermen had left a dog with him, a beagle. The past spring, the dog was eaten by a big wolf.

He poured Chris and me some whisky and the ladies some coffee. Then he talked about his dog.

"Couldn't house-break him. The little son-of-a-bitch eat so much that he'd have to go a few times a night, and he didn't wake me up all the time, you know. And finally, we had a little disagreement and I kicked the hell out of him and told him there won't be no more of that. If I can't break ye, I'll kill ye. And I'll just put you out and you'll have the whole world to crap on."

We all laughed.

"I built him that dog house and I put a full fur bear hide in it, a wool blanket, and there were three feather pillows in it. Oh, he loved it out there."

He looked at his hands, slowly turning them over as if inspecting them for blemishes, and then he ran his left forefinger around the open top of the whisky bottle.

"He woke me up that morning, the morning the wolf got him. If there was anything around that didn't belong he was ready for tearing it up. He was just more dog than he could produce, that's all."

He stopped and thought for a second, seeming to want to avoid continuing with the story.

"Yeah. He went down toward the old house down there. I could hear him. He bayed. It was in March. I thought it was a bear, about time for them to start coming out. I didn't go check on him."

He spoke in measured tones.

"And after a while he quit barking. And ordinarily when he did that he came back to the house. And I set his breakfast out there and he didn't come back."

When he didn't come back, Les went out and followed the tracks to what was left of his dog.

"They cleaned him right up. Yow, I promised that big wolf I'd get him if I lived long enough. And I will."

Back in the shack that night, the four of us had a long talk about religion and our personal backgrounds. I talked about my childhood in North Carolina, about our gospel singing group, and about my friendship with Darlene and Hoojer Smith, and my best friend, Adrian, who eventually joined the Marine Corps and then became a preacher. Later, I was thinking, asking myself, what is this thing we all have in common–Les and Paul, Halibut John, George the Mountain Climber, and yes, maybe even Billy Whiskers? Are we all losers? And concluded, yes, the losers go to Alaska. They don't really want to get rich, they want to lose. Robert Service said it this way in a poem called:

THE MEN THAT DON'T FIT IN:

There's a race of men that don't fit in,
a race that can't stay still;
so they break the hearts of kith and kin,
and they roam the world at will.
They range the field and they rove the flood,
and they climb the mountains' crest;
theirs is the curse of the gypsy blood,
and they don't know how to rest. ....
He has failed, he has failed; he has missed his chance;
life's been a jolly good joke on him,
And now is the time to laugh.
Ha, ha! He is one of the legion lost;
he was never meant to win;
he's a rolling stone, and it's bred to the bones;
he's a man who won't fit in.

8 OIL SPILLS

I saw the oil
cemented cormorant's wing.
On the beaches,
in the gravel,
. . . .
So with anger
total full and
completely overrun,
I suddenly concluded
we had just
run out of fun,
in the Great
Patagonian Oil Spill.

BILLY WHISKERS in BLACK TIDES (August 1975)

Shit happens!

ANONYMOUS BUMPER STICKER

The Science

I remember well the first time I saw that bumper sticker. It was as Bruce French and I were driving along on the freeway that circles Atlanta after just spending two full days marketing investment firms in the big A. At that time, the late 1980s, our company (RPI) was in its all time worst slump. Dan Domeracki and I were going around the country closing down our service companies for the oil-and-gas industry. The environmental arm was holding its own, thanks to the efforts of JM, but that was not enough to get the company growing again. Therefore, we made a major effort to diversify by getting into the aquaculture industry, emphasizing the technology for growing marine shrimp. We had some success with this activity in El Salvador, forming a company to do this called RPI/El Salvador, and Bruce and I mapped out some sites for shrimp farms along the coast of the Red Sea in Saudi Arabia, which later became a success, or so we were told.

Bruce French was one of the biologists we hired to get into the aquaculture business. Before joining us, he had been in the Peace Corps, building fish ponds on an island in the Philippines. He had a favorite saying when things went sour– "Nothing is as bad as the Philippines!" Bruce was also diabolically entrepreneurial, coming up with new ideas on how to make a lot of money about once a week.

Despite some success as consultants to groups wanting to get into shrimp mariculture, we realized that in order to make a real profit, we would have to be involved with the actual shrimp growers, which meant becoming at least part owners of some of the shrimp farms we consulted for. We had interests in doing this in Mexico and other places, even in South Carolina. This trip with Bruce was one of my first meetings with investment bankers. We tried to sell them on the fact that we were on the leading edge of this business, technically, which I do believe was true thanks to the experts on our team, such as Paul Sandifer and Linos Cotsapas, and Bruce, of course. However, the investors always asked us the primal question–What else are you bringing to the table? Meaning how about a couple of million from RPI, an utter impossibility at that time. We were also asked more than once if we knew the Golden Rule. No, not that one, this one–HE WHO HAS THE GOLD MAKES THE RULES.

Accordingly, Bruce and I had put in a couple of humiliating, non-productive days knocking on doors throughout Atlanta. We were pretty down and out as I drove along, probably in my little blue Chevy S-10 pickup. All of the company station wagons, our new Mercedes, and other company vehicles were history by that time.

As we moved along silently, Bruce said, "Hey, look at that!" He was pointing to the bumper sticker on the rear of a car that had just passed us by.

"See that. Shit happens! Oh, ha, ha, ha!" I joined Bruce as we both surrendered the day in total amusement.

We still had a 3.5-hour drive ahead of us before we would get back to Columbia. About every thirty minutes or so, one of us would say–"Shit happens. Ha, oh, ha, ha, ha." Thus, the ride home wasn't too bad.

No, the aquaculture thing did not work out for us, not in any major way, that is. But, the same cannot be said about the oil spill response business.

Over the years, our research groups became heavily involved in responding to major oil spills, to the extent that in the past three decades or so, our scientific efforts have been directed to solving problems in that field more than to any other. Also, this activity has been a significant contributor to the success and longevity of our science technology firm, RPI (also discussed in some detail in Chapter 12 - THE TRADE SHOW).

This field of scientific endeavor started in March 1967, when a supertanker named the *Torrey Canyon* was shipwrecked off the west coast of Cornwall County, England, giving scientists their first taste of trying to respond to a major tanker oil spill in the western world. A couple of years later, in January 1969, an offshore rig in the Santa Barbara Channel suffered a blow-out, eventually spilling between 3,360,000 to 4,200,000 gallons of crude oil that oiled the coastline from Goleta to Rincon, and all four of the northern Channel Islands (Wikipedia Encyclopedia). Thus, the scientific study of and response to oil spills was begun in a big way in the U.S.

Meanwhile, I was a Professor in the Geology Departments at the Universities of Massachusetts and South Carolina at that time (late 1960s to early 1970s), happily directing the research of an army of graduate students on the coasts of Alaska, South Carolina, and elsewhere, without any thoughts of consequence about oil spills. However, presumably because of my previous experience of working along the Alaska coast, in August of 1975 I received a grant from the National Science Foundation (NSF) to do a brief study of a large oil spill by the tanker *Metula* in the Strait of Magellan on 9 August 1974. I was informed of the availability of this project during a phone call from an NSF project manager named Ralph Perhac that I received shortly after I returned to South Carolina from participating in the southern Alaska study discussed in the last chapter. The purpose of this new project was to see if any lessons could be learned from the spill, which had received no significant clean-up during the year following it, that could be applied to any potential future spills along the Alaska coast. A newly arrived graduate student of mine, Erich Gundlach, who had recently returned from a two-year stint with the Peace Corps in Chile, and I conducted a reconnaissance study of the *Metula* spill site shortly after receipt of this grant (during the last few weeks of August 1975). This was my first oil spill, and I was so affected by it that I wrote a poem called *The Great Patagonian Oil Spill* (see introductory quote).

Based on this quick initial survey, we concluded that there were, indeed, many lessons to learn that could be applied to the Alaska shore, because of similarities between the two areas with regard to their climate, tidal range, and recent geological history. Therefore, a much more extensive study was carried out in January-March 1976, with a crew of seven Americans and a number of Chilean scientists.

Also, shortly after Erich and I returned from the first trip to the strait, we formed a small study unit within the Geology Department at the University of South Carolina called the Oil Spill Assessment Team (OSAT), in order to become more knowledgeable in this field. This study group included a graduate student working on her Ph.D. project under Dr. Billy Moore, Jacqui Michel (JM). As previously noted, JM and I were eventually married, and today, she is one of the leading experts on oil spills in the world. In fact, as I write this in January 2011, she is in Houma, Louisiana (under the sponsorship of NOAA) heavily involved in the management of the response to BP's *Deepwater Horizon* oil spill.

By the end of the fall semester of 1975, we were making rules and becoming smarter about oil spills, but at the time of our second field study, the *Metula* spill was 1.5 years old thus, we had never been to a "live" oil spill. This deficiency ended when we invited ourselves to conduct a study of the active spill of the tanker *Urquiola* at La Caruña, Spain in mid-May 1976, mostly using the same field team that did the *Metula* study.

Although we were eventually asked to leave Spain, we learned a lot at this "live" spill, because of the variety of shoreline habitats that were heavily oiled. Two of these heavily oiled beaches are illustrated in Figure 47 and Plate 4. As a result of studies at these first two spills, we began to compile a list of what we called the RULES OF OIL SPILL BEHAVIOR. The first ones were:

1) Salt marshes are subject to long-term impacts from oil spills. This was made abundantly clear by the long-term impacts to the heavily oiled marsh at Punta Espora in the Strait of Magellan (Figure 48).

2) On exposed, high-energy, rocky coasts, cleanup of spilled oil by natural processes is rapid (hours to days).

FIGURE 47. COASTAL HEROES Larry Ward (right) and Bob Stein running a beach profile across a heavily oiled beach at the *Urquiola* oil spill site near La Caruña, Spain. Photograph taken in May 1976.

3) Depth of oil penetration and burial increases with sediment grain size: a) Fine-grained sand – very little penetration or burial (see Plate 4); b) Coarse-grained sand – moderate penetration (up to 10 centimeters); potential for considerable burial; and c) Gravel – maximum penetration (up to 100s of centimeters); potential for considerable burial.

Both rules 2 and 3 were based on observations made repeatedly at the *Urquiola* spill site in Spain. These and other lessons learned in Chile and Spain allowed us to devise a coastline mapping system, now called the Environmental Sensitivity Index (ESI), which provides planners a systematic database outlining the most critical areas and resources along their shoreline that should be protected in the event of an oil spill. The ESI is discussed in more detail in Chapter 20 – ENVIRONMENTAL SENSITIVITY INDEX (ESI).

In March of 1978, we responded to the major spill of the tanker *Amoco Cadiz* on the coast of France, this time at the request of the French government. By then, we were fairly well entrenched into the NOAA oil spill response group, which was headed by John Robinson (see DEDICATION). We wrote several scientific reports and papers on these three major early spills (e.g., Hayes and Gundlach, 1975; Gundlach et al., 1978; Hayes, Gundlach, and D'Ozouville, 1979; Gundlach, Domeracki, and Thebeau, 1982).

On 15 July 1979, our team showed up in Corpus Christi, Texas to begin several months of response to the spill caused by a blowout of the offshore platform *Ixtoc I* off the coast of Yucatan, Mexico. This spill released 13 times as much oil as the *Exxon Valdez* spill in 1989, oiling 265 miles of shoreline in Texas. The volume of this spill was recently surpassed by another famous offshore well blowout, the *Deepwater Horizon* spill off the shore of Louisiana on 20 April 2010.

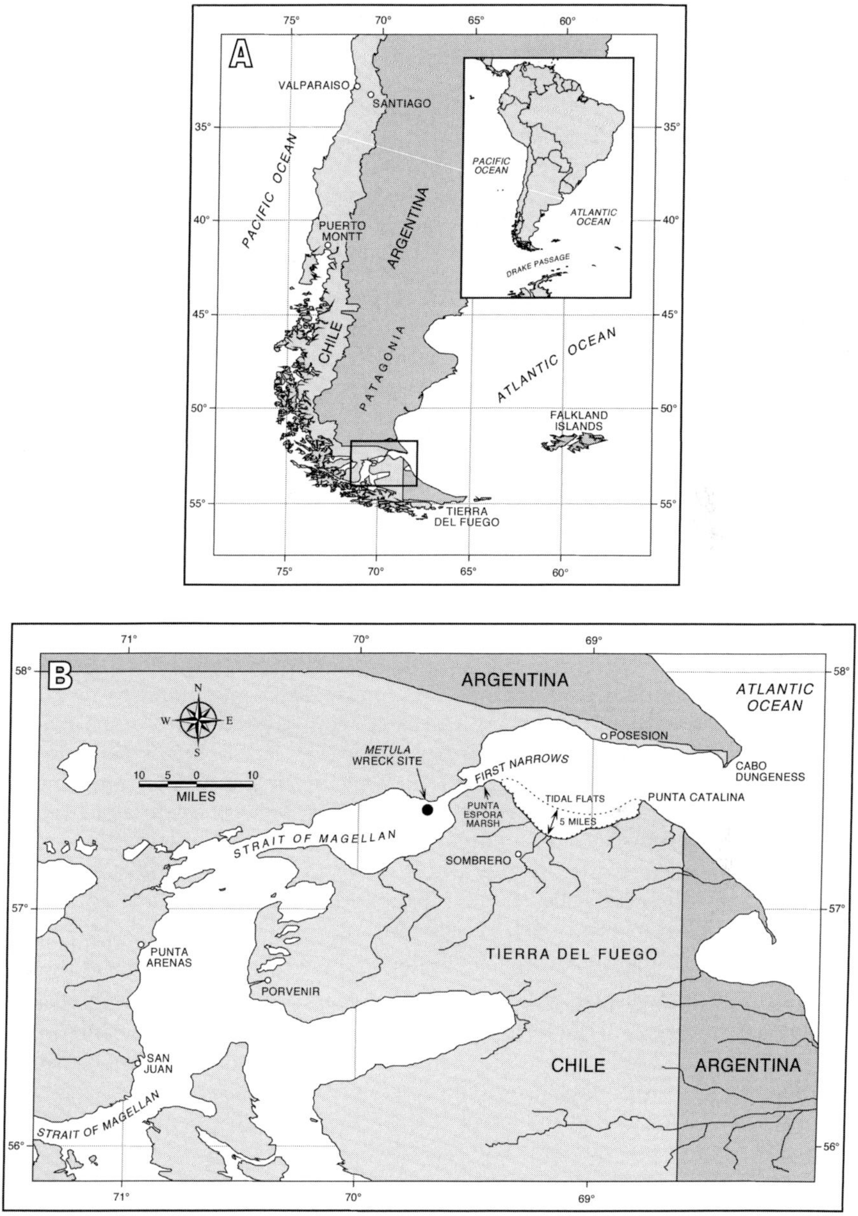

FIGURE 48. Location maps of South America. (A) Lower box indicates location of B. (B) Location map, Strait of Magellan.

Under NOAA's sponsorship, we were able to utilize our ESI mapping scheme with considerable success in the preplanning phases before the *Ixtoc I* oil reached the Texas coast. That, plus some success with inlet protection strategies (discussed in more detail in Chapter 19 – INLET PROTECTION STRATEGIES), helped establish our group as a major player in the science of oil spill response.

The Story

(19 February 1976 - Posesion, Chile)

We were in the middle of our second expedition to study the impacts of an oil spill by the tanker *Metula* in the Strait of Magellan. There were three of us working on the north side of the strait (Figure 48) out of our somewhat unreliable Argentine jeep that day, myself, Cristobal, the same young man (Chris Ruby) that had worked with me the previous summer in Alaska (discussed in Chapter 7 – THE MEN THAT DON'T FIT IN), and Natalia, a Chilean biologist.

[NOTE: The particular team being discussed here was the zonal study team, whose job it was to "Survey the entire oiled area to obtain a complete picture of the oil distribution." We did this by sampling and describing the oiling at 70 locations. In addition to that, at 15 locations, we surveyed two parallel topographic profiles a couple of hundred meters apart across the entire intertidal zone along which we made detailed notes and collected samples of the oil. We also did a complete geomorphological description of the intertidal zone and mapped the distribution of the oil between the two transects. These were called zonal study sites (Hayes et al., 1973). The fifteen sites where these types of surveys were completed were chosen to represent all of the different major habitat types present.]

On this particular afternoon, we went to the north entrance of the strait to work on the cuspate foreland located there, which protrudes into the strait for about ten kilometers. It is obvious from the geomorphology that this foreland, called Cabo Dungeness, is migrating in a westerly direction at a relatively rapid rate. Waves from the open Atlantic Ocean erode away the older beach ridges located on the east side, which are composed of gravel, each of the ridges marking a former strait-side boundary on the western side of the spit. Currents generated by these waves transport the gravel around the south point of the foreland to the west side, where it is deposited again in a series of new beach ridges. I made the field sketch given in Figure 49 and sent it to JM in one of my letters to her. A similar cuspate foreland, called Punta Catalina, is located at the south side of the entrance to the strait (Figure 48). The spectacular Punta Catalina is shown by the oblique aerial photograph in Plate 5, my all time favorite coastal photograph.

It is probable that some of the pebbles and cobbles that make up the regenerated ridges on the west side of Cabo Dungeness (Figure 49) have gone through this cycle several times. To add an interesting complication to this already complex story, the boundary between Argentina and Chile bisects the foreland, with Argentina owning the eastern part of the foreland and Chile the rest. Obviously, Chile has the better end of this arrangement, inasmuch as, through time, the foreland is essentially migrating out of Argentina.

Realizing that, I thought, "There is going to be one hell of a border dispute here some day. Maybe I can be an expert witness!"

[NOTE: On 26 January 2011, I checked on Google Earth and found that the boundary between the two countries still appears to be close to where it was in 1976, so I guess this is not a job opportunity for me after all.]

Without giving much consideration to such things as political borders when I designed this study, I had decided to measure two beach profiles on the eastern face of the foreland, which was located in Argentina. Only problem was, we did

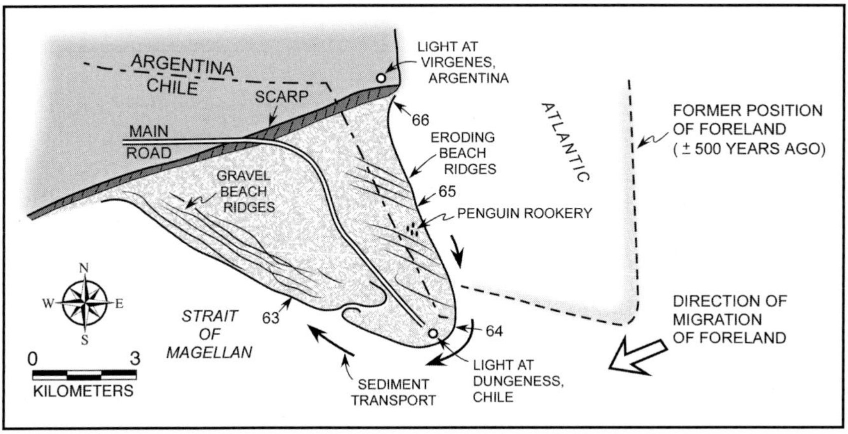

FIGURE 49. Field sketch of Cabo Dungeness, a cuspate foreland located on the north side of the east entrance to the Strait of Magellan (Figure 48B). Numbers represent locations of the beach profiles we surveyed. This is from a letter I wrote to JM on 19 February 1976.

not have permission to work in Argentina, only in Chile.

Well, regardless of this troubling detail, we drove to the lighthouse at the south end of the foreland, which is in Chile and asked the lighthouse keeper's opinion on how to solve our dilemma. He said it was against the rules for us to cross over the border unannounced, but that it would probably be okay if he went along with us.

I said, "Great idea, please join us," so he went inside the lighthouse and came back to the jeep carrying a small caliber rifle.

Then we drove bumpily over the beach ridges into Argentina, the boundary of which was marked by a rickety sheep fence. I foolishly didn't have my passport with me, so I felt a little naked being there. Our Chilean companion, carrying a rifle across his lap, was riding shotgun, so to speak, in the back seat. Finally, we reached a gravel road that crossed obliquely over the beach ridges and headed north. This road we were on was full of dead penguins, hundreds of them. Our tour guide said Argentine *turistas* killed them. I looked toward the beach and could see about 500 penguin heads peering up over the tops of some shrubs in a swale between two gravel ridges. Our guardian also told us that approximately 1.5 million penguins spend their summer vacation on Cabo Dungeness. Shortly thereafter, we drove up a steep hill to the Argentine lighthouse at Punta Virgenes and met the lighthouse keeper, who gave us permission to measure the beach profiles on his beach.

[NOTE: Looking on Google Earth again in January 2011, I can see pictures of penguins on the foreland, so they have not been completely wiped out. The area where the penguins are located is now called Preserva Natural Cabo Virgenes. Also, if you are interested, on Google Earth you can also see a photograph of the Argentine lighthouse (Faro de Cabo Virgenes-Santa Cruz), as well as the Chilean one at the end of the foreland (Faro Punta Dungeness).]

We ran the profiles over the steep gravel beaches, which had been cleaned of the oil spill long ago by the big waves, and then dropped our Chilean friend back at his lighthouse at the tip of the foreland. It turns out that he was taking the rifle back to his Argentine pal at the other lighthouse, because he had borrowed it from him to shoot rabbits. And, apparently, another reason he wanted a ride over there was so he could smuggle a carton of American cigarettes and a bottle of Argentine wine into Chile.

That night back at the oil company camp at Posesion (Figure 48), we had a super dinner in the company cafeteria composed of a huge steak with two fried eggs on top of it. That dish was called "meal for the poor" or *como pobre*, and it was considerably better than the usual sheep's tongue broth.[16]

After dinner, a Chilean Army lieutenant paid us a visit. He said, "We know what you were doing out there today, but you shouldn't cross the border again like that, because the Argentineans are 'having a little trouble' these days."

We had a multitude of adventures on that particular trip, which included me leaving my famous orange camera bag stuffed with American money, about $16,000, sitting in a bar overnight. The bartender handed it over untouched and unopened the next morning.

JM visited me in Chile toward the end of this field project, and we spent a couple of days in Bogota, Columbia on the way back. One of the toughest border crossings anywhere in those days probably was getting into Miami on a return from Colombia. JM and I were actually on our way to St. Croix, in the Virgin Islands, by way of Miami, where we would join another group of my graduate students carrying out a research project there. I was hand-carrying with me a large depth recorder that we had thought we would need in Chile, but didn't, and that was definitely needed by the crew in St. Croix. I had never bothered to study that bulky instrument, which weighed about 40 pounds, so when the customs official in Miami asked me to open it for him, I said, "Heck, I don't know how to open this thing."

This pleased that gentlemen none in the least and, as other agents converged on the scene, I started turning knobs and twisting screws, being something of a smart-ass all the while.

Finally, I did get it open for them to gaze intensely at the wires, recording graph paper, and complex electronic components inside the cover of the instrument. As soon as they were satisfied that I was carrying nothing illicit in the depth recorder, three or four of them ordered me inside their own version of the "little room" (I have been in quite a few of them in my time, my favorite being in Dammam, Saudi Arabia), and told me to strip, and bend over, and spread. But, for some reason, they never told me to take off my shoes.

Yes, Miami can be a tough border crossing at times, but New York at Kennedy Airport can be just as bad, and even surpass it sometimes, when it comes to the rudeness of the customs personnel. Once along about this same time period, I was in line at Kennedy, returning from a trip to Africa, I believe, when I noticed one of the women customs agents severely upbraiding an old lady in the line, who obviously couldn't speak English and was crying pitifully.

I went to the head of the line and said, "Hey, take it easy on that old lady, can't you see that she can't understand a word you are saying?"

"Keep out of this."

"Well, nobody should be treated like that. I want to speak to your supervisor."

"Sure thing, he's right over there."

I walked over to the supervisor and started to tell him about the abusive treatment the old lady had been put through.

His response was, "Shut up and get back in line."

9 AH KUWAIT!

A known mistake is better than
an unknown truth.

ARABIC PROVERB

We have a saying in Arabic—
Whoever farms a land,
must some day harvest what he farmed.

ABDUL HALIM AL MOMEN (FRIEND AND COLLEAGUE)

The Science

Without a doubt, 1976 was by far the most productive year I have ever had with respect to scientific output. Some of the key events, in chronological order, follow:

1) The detailed field study of the *Metula* oil spill site in the Strait of Magellan carried out in February and March.

2) Our fieldwork at the *Urquiola* oil spill site in Spain in the second half of May.

3) Field survey of Lower Cook Inlet in the month of June with JM and Jeff Brown, during which the ESI concept was formulated.

4) In late July, Chris Ruby and I took an overflight with an indomitable bush pilot named Cowboy Pierce that covered the entire Alaska shoreline between Anchorage and Prudhoe Bay, even to the Canadian border. This was a great learning experience. Many of the thousand plus or so oblique aerial photographs I took are still periodically appearing in books by myself and others. A picture I took of Point Barrow, Alaska during this flight will soon be published in Pilkey et al.'s book, *The World's Beaches*. This photograph is also included here (Plate 6).

5) Meanwhile, Tim Kana and I were preparing course notes (Hayes and Kana, 1976) for the first of our field seminars on modern terrigenous clastic depositional environments on the South Carolina coastal plain.

This first field course was sponsored by the American Association of Petroleum Geologists (AAPG). More discussion of these field seminars is given in Chapter 13 – THE ROCK RECORD.

6) The first of our scientific projects on the Arabian Peninsula, which continue even until this day (see Chapter 21 – NIGHTMARE IN JUBAIL CITY), was conducted in Kuwait in November. This work in Kuwait is the subject of this chapter.

(14 November 1976, Sunday - Kuwait-Sheraton Hotel, Kuwait City)

A significant opportunity to test the waters of the business world occurred in the fall of 1976, when my old friend from the *Suzanne* days, Paul Hague, who was working for an environmental consulting firm in New England at that time, recommended that I join him on a team of architects and engineers who were beginning a major project to reconstruct the waterfront of Kuwait City. That 15-mile-long waterfront was basically a pile of rubble created by pushing part of the old, historic town into the ocean, a common practice all along the Gulf shoreline in the early boom years of Middle East oil production. Kuwait is located on the map in Figure 50, and a modern vertical image of the Kuwait shoreline is given in Figure 51.

Some people call the water body that Kuwait is located on the Persian Gulf, but on the Arabian Peninsula, including in Kuwait, it is referred to as the Arabian Gulf (Figure 50). That is lesson number one for doing consulting work there; never refer to the Gulf as being Persian!

The Kuwait project included construction of a set of artificial beaches and other shore structures, thus someone with my experience in coastal geomorphology was required to ascertain how the natural system would react to such modifications. The proposed design of the shoreline that was the result of this project is given in Figure 52. I took this job on privately, running it through a fledgling business we called Miles O. Hayes, Associates.

Therefore, around midnight on a Sunday night in mid-November, I was lying on the bed in my fancy room in the Kuwait Sheraton Hotel, thinking about the flight into Kuwait two nights before. The six-hour flight from London had arrived in the dark. As we circled the field, thousands of lights outlined the city, which was surrounded by the burning gas flares from the oil fields. I could see the entire country in a single view, a spectacular sight. The offshore waters were full of oil tankers, and on the ground, the streets were full of Mercedes and other high-priced automobiles.

As I lay on the bed, I pulled out a pad and started scribbling "Bent over with jet lag, I landed in Kuwait City in the dark. Rode the boat in the sunset and watched the porpoises jump. Talked with Peter the Great Engineer, smoked dope and drank some Scotch. Kuwait was pretty in the sun, the part they didn't yet push into the sea. My field assistant drives an eighteen thousand dollar Mercedes Benz. I pay $93 a night for my hotel room. I've made the big time!"

The architects I was working with represented Sasaki Associates, formed by a landscape architecture professor at Harvard and located in Watertown, Massachusetts. Paul Pawlowski, a hefty, articulate dark-haired fellow in his mid thirties, was the major spokesman for the team. His lingual skills were phenomenal.

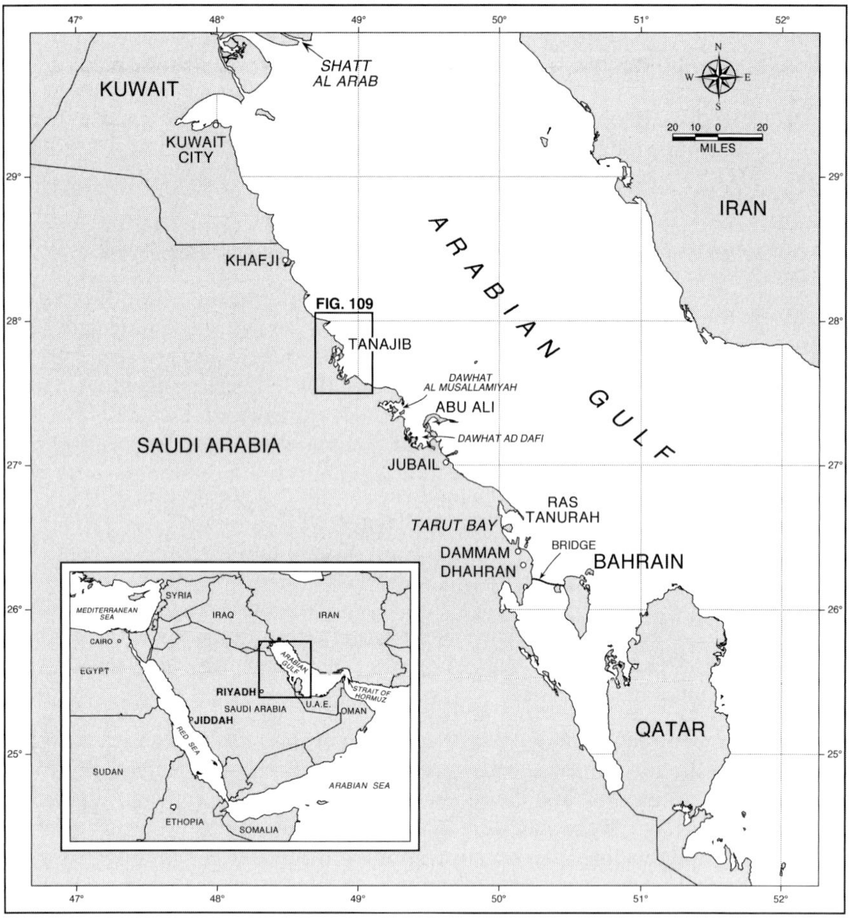

FIGURE 50. Map of the Arabian Gulf area. The Gulf War oil spill of 1991 came onshore all along the coast from Kuwait City to Abu Ali, Saudi Arabia.

One night he brought out a stack of the proposals submitted by all the competitors for the project, firms from Greece, Great Britain, Japan, Italy, and so forth, and read the first two paragraphs of the executive summaries of their proposals with perfect accents representing each country. We laughed and laughed, both at his amazing verbal skills and at the fact that our American firm, its executive summary being read by Paul with a Texas accent, had managed to beat out all of those world-class operators. Of course, the fact that Ghazi Sultan, the chief Kuwaiti architect-manager for the project, had studied under Prof. Sasaki at Harvard may have had something to do with it. The other principal architect from Sasaki was Frank James, a diminutive bearded chap about 45 years old, who was the creative genius of the group, with a captivating personality that easily won over all the Arab counterparts we worked with. I loved working with Paul and Frank, who are shown running a beach profile on the Kuwait shoreline in Figure 53, because they forced me to "think outside the box," as they say.

FIGURE 51. Vertical image of Kuwait. GeoEye-1 satellite image acquired April 2009, with permission from GeoEye, Inc.

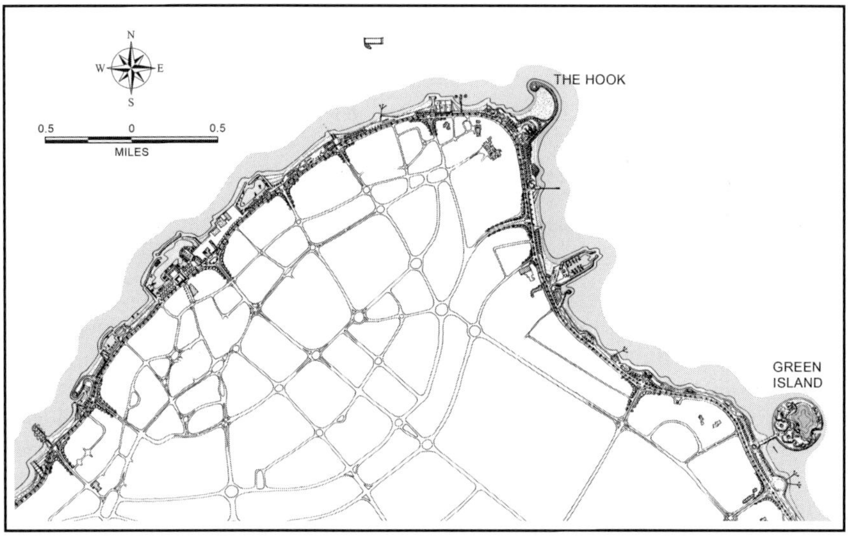

FIGURE 52. Paul and Frank's original design for the waterfront of Kuwait City, completed around 1978. It contained many new and original features, such as "The Hook" and "Green Island."

FIGURE 53. Paul Pawlowski (right) and Frank James measuring the topographic profile of a beach in Kuwait. Photograph taken at low tide in November 1976.

[NOTE: As I was reviewing an earlier version of this chapter in mid-August 2007, I thought back on an e-mail I had received a couple of days earlier. It was from Paul Pawlowski, who informed me that Ghazi had died recently. When he was 65, Frank fell off a boat in Puget Sound and drowned. This had happened several years back. Ghazi would have been 66 that year, and Paul said his 65th was coming up the following month. We reminisced that we were the only two left of that original foursome, and Paul, with whom we did some work in Abu Dhabi in November 2009, recommended that we "watch our step" for a while.]

On the following Saturday night, Ghazi invited us to a party at his house, which he shared with his dazzling wife, an Indian aristocrat from Delhi. The central room of his rectangular shaped house had a ceiling 25 feet high, with a study at one corner at the head of a spiral staircase from which he could survey his kingdom. The walls were hung with huge, hand-woven rugs, and on one side of the room, a massive fireplace was surrounded by cushions upon which the guests lounged, drinking alcohol from the best-stocked bar I had ever seen. Ghazi had a policy - "You don't come into Kuwait to work on the project without bringing me a bottle of something," which was illegal, of course, but we seemed to get away with it just fine.

The guests at the well-attended party were mostly wealthy, American-educated Kuwaiti men accompanied by their gorgeous dark-skinned, dark-eyed wives. That was the one and only time in my whole career of working on the Arabian Peninsula, which has extended from 1976 until now (2011), that I had the opportunity to have an extended conversation with an Arabian wife. I informed them that their country's investment in Kiawah Island was considered to be a smashing success in South Carolina (discussed in Chapter 5 – KIAWAH ISLAND).

I had a great time at the party, that is, until I accepted an invitation from Ghazi to indulge in one of his Havana cigars. That, on top of the Scotch I wasn't used to drinking, and a few puffs on the ever-present joint that Frank always carried with him, made me a little woozy.

That was Kuwait in the good old days. How is it now after the Iraqi wrecking crew finished their job? Definitely not the same; however, the product of the finally completed waterfront master plan, which conforms closely to Frank and Paul's original drawings, is something to behold. It isn't quite up to Dubai standards, but it is still a remarkable piece of architectural engineering. Paul and Frank's original plan is shown in Figure 52. For comparison, see the modern vertical image in Figure 51.

The tidal range at Kuwait City approaches 10 feet during spring tides, and when the tide is low, a broad intertidal platform composed of a type of limestone called "gatch" encircles the waterfront. An oblique photograph of the intertidal zone of the waterfront at low tide, which shows this intertidal platform, is given in Figure 54. On the morning following Ghazi's party, I was walking far out on that platform, feeling a little hung over, inspecting the rocks. Suddenly, a loud horn right behind me made me jump several feet off a well-worn track along which trucks hauled pieces of the rock mined from the platform in order to build a revetment along the waterfront, ostensibly to protect the shore road from wave erosion at high tide. This near miss by the truck jolted me out of my morning blahs, all right.

As I watched the truck drive away toward the shore, I thought, "What a stupid thing to do. They are removing the outer edge of the platform, a formidable barrier to wave action, to create a new barrier along the high-tide area. If they keep this up, the waves at the shoreline will get bigger and bigger all the time."

FIGURE 54. Oblique aerial view of Kuwait City waterfront at low tide in November 1978.

That afternoon I informed Paul and Ghazi about my concerns regarding the offshore rock platform, and Ghazi said, "I will make an appointment for you to meet with the City Manager tomorrow."

I met with the city officials and told them that they shouldn't be mining the platform, a natural breakwater system. And guess what? A couple of weeks later, they stopped doing it!

What? Somebody listened to me and did what I suggested? This was not academia, this was something different. I was hooked, academia would never be the same for me and, from then on, my major focus was chasing practical work of this type and building a business in which to do it.

We didn't publish as many papers on this project as we probably should have, but we did publish a few (e.g., Hayes, Michel, and Kana, 1986).

Story One

(December 1976. Watertown, Massachusetts)

In early December, I flew in a jet to Boston, where I met with Paul Hague at the airport and we drove to the Sasaki offices in Watertown to discuss the results of our first field excursion to Kuwait. The meeting went well, and we made plans for a couple more years of data gathering on the waterfront area.

Paul Pawlowski invited us to spend that Friday night with him in Watertown, as well as to join in a little celebration of some kind that he was sponsoring for several of his friends and guests at his house that evening. When we arrived, Paul proudly showed us around his mansion, which was located right on the central square of Watertown, surrounded by numerous church spires. It had either 32 or 53 rooms, depending on how you counted them.

Paul said, "Abraham Lincoln, John Quincy Adams, and many other notables have slept here."

"Now count Billy Whiskers among them," I thought.

The party went on for a long time. Frank was there with a supply of joints, as usual, and Hague and I partook up to our limit. It was a big party. I thought maybe about a third of the citizens of Watertown were there. In any event, the house was probably large enough to hold at least that many people.

Late in the evening, an earthy fly-fishing preacher came in and started in with his crap. After talking with me for a while about football in the South and fly fishing in New England, the preacher said, "There's more ego in this room than in the whole of Watertown."

A few minutes later, as Hague and I were sitting on a couch munching on the *hor d'oeuvres,* I was brought down from a fantasy walk around the edge of the high ceiling of the great room, when he nudged me.

"Hey Miles," he asked, "ever seen a Fellini movie?'

"Sure."

"I think we're in one."

Then the two space cadets laughed their silly laughs, not knowing why that was so funny, as they prepared to launch themselves up the spiral staircase and to bed.

Next morning, as I flew over the New Jersey barrier islands on an Eastern Airlines jet, I thought, "I don't really like all the emotions this type of work puts you through. I don't have to feel any of that at the University of South Carolina."

But, thinking about this a little later on, I couldn't get out of my mind the fine homes Paul and Ghazi lived in. Paul said that where a man lived spoke volumes about him. I still lived in a five-room bungalow on a side street in Columbia, South Carolina, which had seemed just fine for a university professor.

"It must be grand to live like that," I thought. "Maybe making some money out of the consulting business isn't such a bad idea after all."

Story Two

We continued working in Kuwait for many years. The following story starts as we were getting ready to conduct one of our last projects there.

(March 1997 - The neutral zone border between Saudi Arabia and Kuwait)

In March 1997, JM and I returned to Saudi Arabia in order to continue our evaluation of the long-term effects of the Gulf War oil spills on the Saudi shoreline (discussed in some detail in Chapter 21 – NIGHTMARE IN JUBAIL CITY). We finished this brief survey on schedule and were on our way to Kuwait to conduct a training seminar in oil spill response techniques. Our friend and long-term associate, Abdul Halim al Momen, drove us past Khafji and on to the Kuwait border (Figure 50), where our Kuwaiti sponsors were waiting to meet us.

However, we had a delay on the Saudi side of the border, where the customs agent insisted that JM go to the women's line so the female customs agent there could lift her veil to see if her face matched the picture on her passport.

"But, I am not wearing a veil," JM protested, "You can see my face just fine."

The customs agent continued to insist that protocol be followed, which pissed Halim off pretty badly, so he went looking for the man's senior officer. He found him, and the officer laughed, waving JM on through. This guy was probably a graduate of some school in the U.S.–Georgia State, Pepperdine, George Mason, or some school like that. I was going to ask him where he matriculated, but JM had already gone through and, after thanking Halim for his help once more, so did I.

The week-long course in Kuwait was tough, because of the language barrier. But we survived, and the students seemed to like it okay.

We split up in New York after the long ride across the Atlantic, with JM going to Providence to participate in a meeting on the *North Cape* oil spill. Back home the next morning, I decided to go trout fishing on the Lower Saluda River, the only cure I could think of for the stultifying jet lag I was suffering. The Saluda is a tailrace fishery, located below the Lake Murray Dam about five miles northwest of Columbia. When I arrived at my favorite fishing spot, they were apparently running the power generators full blast, because the river was very high and I couldn't get to the good water. That's my excuse for getting shown up by a couple of old boys who had never been trout fishing before in their lives, but caught three brown trout about 20 inches long while fishing in a jon boat out over the best pool. They caught the fish on cut-up pieces of small trout on a large size 4 hook with a sinker as big as a baseball, which is probably what you needed in that rushing torrent.

To make matters worse, I wasn't watching the water closely enough while fooling around with some dry flies, and I was trapped on an island out in mid-stream when the river suddenly rose three feet in five minutes. To get back to the shore, I would have had to swim in the cold water through a four-knot current. The three of us were the only fishermen on the river, so luckily they were able to hear me call to them to come back upstream and give me a ride over to the bank, another border crossing of sorts.

When they picked me up, I said, "This also will give me a chance to see the big trout you caught."

They had yelled to me when they caught the first one, and I had assumed that it was the only one they caught. Imagine my surprise when they opened the live well and there were three of the monsters in there.

"How did you do?" they asked, as they let me out on the shore.

"Well, I caught nine, but none over 10 inches," said I.

In other words, I had just barely caught enough fish for them to use for bait. I always released all the fish I caught in that river. They said they weren't even sure that the fish in the live box were trout and asked if they were good to eat.

"Biggest trout I've ever seen in this river and, yes, they are the best eating fish you can catch in fresh water in South Carolina. See you later. Thanks for the lift."

And then I walked back down the trail toward the car.

As I passed out of sight in the trees, one of them was heard to say, "He must not be much of a trout fisherman, if all he can catch is bait," or something like that.

If I had been my Uncle Luster, no doubt I might have thrown my rod and vest out into the river and swore never to go fishing again. But I didn't do it. Thought about it though.[17]

10 PRICE INLET

I also learned from Shepard that the best science
is done by those having the most fun,
and that we are students all of our lives.

ROBERT S. DIETZ [18]

The Science

Between the fall of 1972 and the end of 1983, the eleven years I was associated with the Geology Department at USC-East, it was common for 10-15 graduate students to be working at any given time for their degrees in the Coastal Research Division (CRD) of the Department. As the head of CRD, I was responsible for obtaining a lot of the supporting funds, and I was the designated "Director" of the thesis/dissertation committees of the students who got their graduate degrees under my supervision (listed in Appendix A). This meant that I signed the top line on the title page of the thesis/dissertation and chaired the final thesis or dissertation defense for each of the students.

During this time, eight Ph.D. projects were conducted far away from the friendly confines of South Carolina. These included the projects by Ed Owens, Patrick McLaren, and Chris Zabawa discussed in Chapter 6 – BRIDGES OVER TROUBLED WATERS, and Jon Boothroyd's study of the outwash fans on the southern Alaska coast discussed in Chapter 7 – THE MEN THAT DON'T FIT IN. There were four other dissertations completed outside of South Carolina: 1) Mohammed al Sarawi's study of the alluvial fans and shoreline in Kuwait; 2) Erich Gundlach's summary of research at two oil spills, the *Metula* in Chile and the *Urquiola* in Spain; 3) Al Hine's study of sedimentation on the Bahama Banks; and 4) Mike Stephen's studies of coastal processes at Caxambas Pass, southwest Florida. A number of M.S. theses were also conducted on areas located far away, most notably J.J. Gonsiewski's study of large-scale fluvial features on Mars. Other thesis projects located elsewhere included: 1) Anne Blount's research on the *Metula* spill; 2) Grant Wach's analysis of the tar sands in Northern Canada; 2) Craig Shipp and Mike Vogel's projects on the nearshore bars and tidal inlets in the Long Island, New York area; 3) Kenny Finklestein's study of the crenulate bays on Kodiak Island, Alaska; 4) Chris Ruby's geomorphology and sedimentation study of the shoreline of southern Alaska discussed in Chapter 7 – THE MEN THAT DON'T FIT IN; 5) Larry Ward's study of the morphology and hydrology of outwash streams in southern Iceland; and 6) Ian Fischer's investigation of the macrotidal flats in the Strait of Magellan.

[NOTE: The detailed references to the theses and dissertations referred to in this Chapter are given in Appendix A.]

All of the other projects were carried out on the coastal plain and shoreline of the Georgia Bight, most of them focusing on habitats in South Carolina. These studies fell into the following categories:

A. Coastal Plain, Flood Plains, and Santee/Pee Dee Delta

These included Ray Levey's thesis on the sandy point bars on the Congaree River, Ray Kaczarowski's dissertation on the Carolina Bays, and Tim Eckard's thesis on the stratigraphy of the Santee/Pee Dee Delta.

B. Tidal Inlets

This work was initiated by Bob Finley's dissertation on the morphology and processes of North Inlet, which produced a couple of early publications (Finley, 1975, 1978). Stan Humphries also completed a thesis on North Inlet. Duncan FitzGerald did a dissertation on Price Inlet, shown by the vertical image in Figure 55, which

FIGURE 55. Vertical image of Price Inlet, South Carolina acquired 1994 courtesy of SCDNR. Arrow points to the Magwood property, the former residence of Granny Magwood, who is the focus of Story One.

will be discussed further below. Other inlet studies included Dave Nelligan's thesis on the ebb-tidal delta at Breach Inlet, and Doug Imperato and Walter J. Sexton's theses on the Captain Sams and North Edisto Inlets. Denny Hubbard's dissertation covered tidal inlet morphology and sedimentation throughout the shoreline of the Georgia Bight. This work was elaborated upon and summarized by Hubbard, Oertel, and Nummedal (1979; see Figure 56).

C. Dynamic Coastal Processes

Tim Kana's thesis and dissertation on longshore sediment transport, which will be discussed in more detail below, were centered mostly at Price Inlet. Other such studies at Price Inlet and surroundings included Jeff Knoth's thesis, a tracer study of sand transport on the adjacent Bull Island, and Cary Fico's thesis on wave-refraction patterns. Also, Chris Reel's unique thesis examining the origin of "bubble sand" was conducted in the Price Inlet area (discussed in Chapter 21 - NIGHTMARE IN JUBAIL CITY). Other studies of coastal processes included Mike Svetlichny's thesis on beach scraping at Myrtle Beach, Martha Griffin's thesis on longshore sediment transport based on sand composition, and Kathy McCreesh's thesis on sediment transport at Hunting Island. Larry Ward's dissertation presented details on the hydrology and sedimentation patterns in the tidal creeks landward of Kiawah Island.

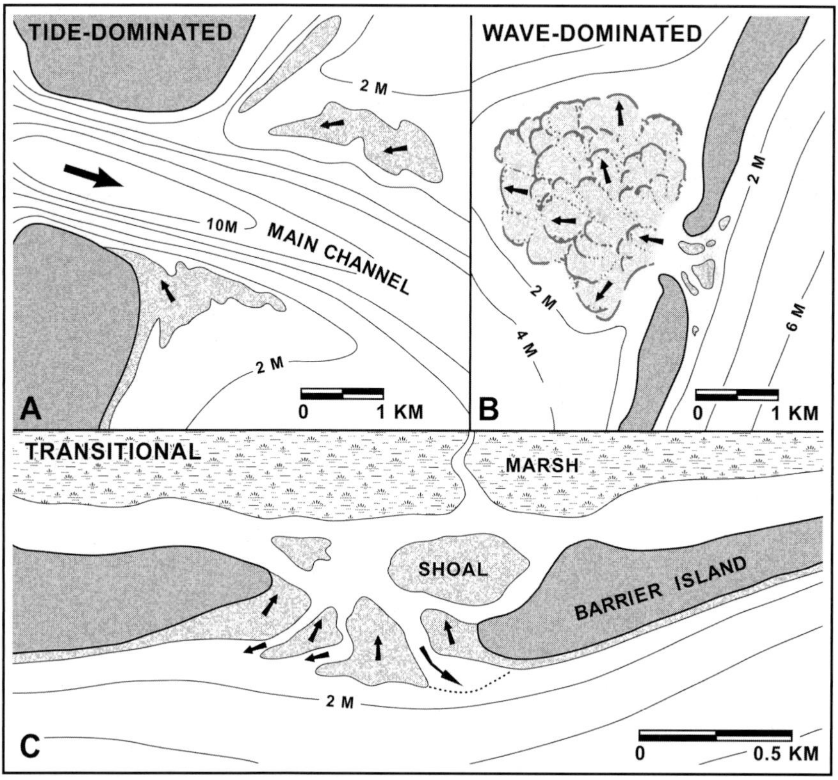

FIGURE 56. Types of tidal inlets in the Georgia Bight. Slightly modified after Hubbard, Oertel, and Nummedal (1979).

D. Barrier Island Morphology and Stratigraphy

Jeff Brown's thesis was a general treatment of the barrier islands throughout the state. Going from north to south: 1) Dan Domeracki's thesis discussed the stratigraphy and morphology of Pawleys Island; 2) Chris Ruby's dissertation dealt with the stratigraphy of a classic transgressive barrier island system at Cape Romain; 3) Bo Tye's thesis presented the stratigraphy of Capers Island, a prograding barrier island; 4) Tom Moslow's dissertation, discussed in Chapter 5 – KIAWAH ISLAND and illustrated in Figure 30, dealt with a classic, large prograding barrier island; and 5) In his dissertation, John Barwis discussed the internal structures of preserved beach ridges on Kiawah Island.

E. Tidal Channels, Tidal Point Bars/Sand Ridges, and Tidal Flats

Helen Mary Johnson's thesis dealt with the morphology and stratigraphy of the tidal river point bars in the Edisto River in the middle to upper reaches of St. Helena Sound. In his thesis, C.Y. McCants focused on the tidal flats in the southern reaches of St. Helena Sound, whereas Pete Rhinehart's thesis dealt with the stratigraphy and processes of the tidal sand ridges off the mouth of the sound. The morphology and sedimentation patterns of the tidal point bars in the tidal creeks landward of Kiawah Island was also covered in John Barwis' dissertation, and their stratigraphy was described in detail in Aileen Duc's dissertation. In his thesis, Tom Freeman, Jr., one of the "stars" of Chapter 21 – NIGHTMARE IN JUBAIL CITY, discussed the bedform dynamics and preserved bedding types in a tidal sand ridge in Port Royal Sound.

These are a lot of studies, so one could rightfully ask how I could possibly have kept up with all of them in the short eleven years available. Dag Nummedal came into the program early on as a post-doc appointment. He not only added the physics and dynamic processes expertise needed to supplement the basic geomorphology, but he also became an enthusiastic mentor for the students while they carried out their research. All of this work could not possibly have been done without Dag's input. If you scanned many of these theses and dissertations, you would see Dag's signature (as a committee member) on many of them. Dag was likewise instrumental in getting most of the students to publish their results.

Also, other full time research associates in CRD included Pulak Ray, Mike Stephen, and Gary Zarillo. Larry Ward and Tim Kana became full-time research associates as well for a while after completion of their dissertations. All of these full-time researchers were instrumental to getting the work done. Another part of the performance model was to let the new students become "apprentices" under the more experienced and more senior grad students for a time, being field assistants on their projects, and so on, until they "learned the ropes."

Obviously, I cannot cover all of these studies, so I have decided to focus briefly on the work that was done in the Price Inlet area. This is a very special area that was described as follows by Tim Kana.

"There is a sixty mile stretch of South Carolina coast which is home for countless deer, alligators, and other wildlife. From the Isle of Palms to Debidue Island near Georgetown, there is an unbroken chain of barrier islands as primitive as any in the world. The land and surrounding marsh are held in trust as a wildlife refuge by several agencies, including the U.S. Government and the South Carolina Heritage Trust."

As a result of this, Price Inlet and the neighboring islands, Bull to the north and Capers to the south, are essentially unmodified by man, the only 'impediment' being the Intracoastal Waterway which crosses the salt marshes and natural channels two miles inland of the inlet throat. Needless to say, this is a perfect place to study natural processes without the influence of jetties, roads, bridges, and so on that are so commonly associated with tidal inlets throughout the U.S.

For purposes of brevity, I will focus on the two major dissertation projects of Duncan FitzGerald and Tim Kana. Duncan finished his project pretty early (FitzGerald, 1977), so he and Bob Finley more or less set the stage for the inlet studies to follow. One of his major conclusions, which provided the basic explanation for ebb dominance at tidal inlets in South Carolina and Georgia, with their large ebb-tidal deltas and small to absent flood-tidal deltas, follows:

"Countering the general landward transport direction...(on the surface of the ebb-tidal delta in the form of landward migrating swash bars)...is the ebb dominance of the main channel. This dominance can be explained by higher inlet efficiency at low water than at high water. Consequently, bay tide phase lag is larger at high than at low water resulting in a longer flood duration. This caused higher mean ebb-tidal currents and a consequent larger potential net ebb transport of sand."

Price Inlet, with its huge ebb-tidal delta (and no flood-tidal delta) is pictured in Figures 55 and 57. The sand circulation pattern at Price Inlet as determined by Duncan and associates is shown in Figure 58.

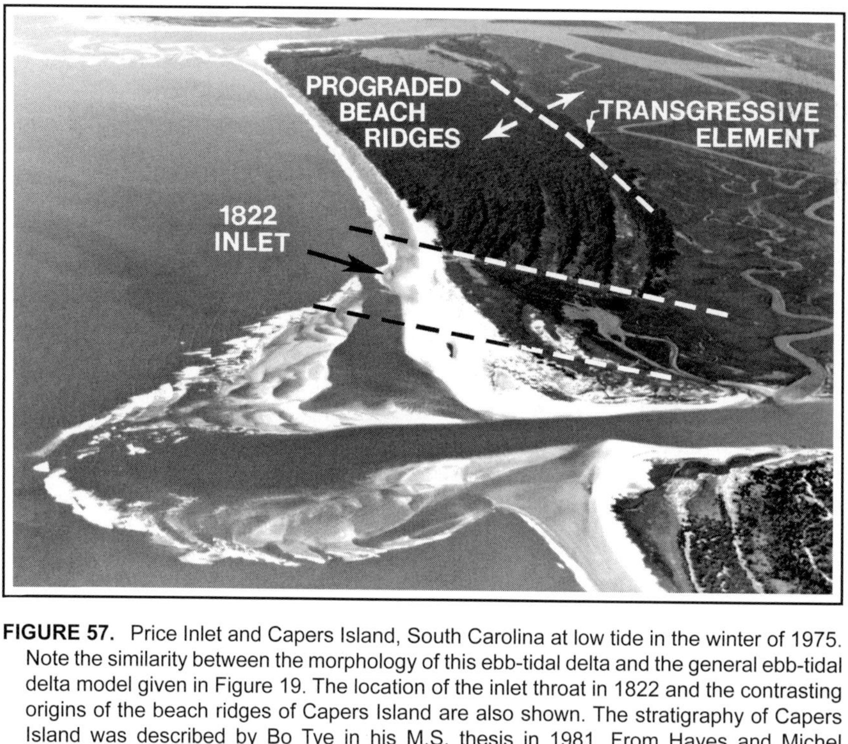

FIGURE 57. Price Inlet and Capers Island, South Carolina at low tide in the winter of 1975. Note the similarity between the morphology of this ebb-tidal delta and the general ebb-tidal delta model given in Figure 19. The location of the inlet throat in 1822 and the contrasting origins of the beach ridges of Capers Island are also shown. The stratigraphy of Capers Island was described by Bo Tye in his M.S. thesis in 1981. From Hayes and Michel (2008).

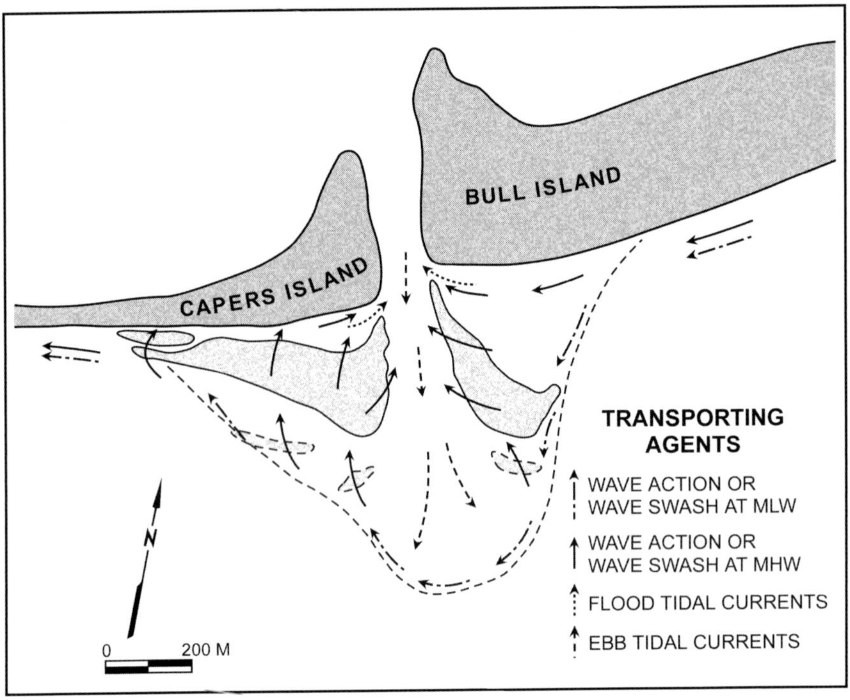

FIGURE 58. Sand circulation pattern at Price Inlet, South Carolina (compare with photo in Figure 57). Based on wave-refraction diagrams, littoral process measurements, bedform orientation, and inlet hydraulic data. From FitzGerald, Nummedal, and Kana (1976).

Another important observation made by Duncan was the effect of changes in the inlet morphology on the erosional/depositional patterns on adjacent barrier islands (FitzGerald, Hubbard, and Nummedal, 1978; FitzGerald and Hayes, 1980). Duncan has continued his work on tidal inlets over the years, becoming perhaps the world's foremost authority on their morphodynamics.

Tim Kana's dissertation (Kana, 1979) was based to some extent on sampling of suspended sediment in breaking waves on Capers Island a few hundred meters south of the Price Inlet in the summer of 1977. He invented what he called a "portable *in situ* bulk water sampler" for determining the distribution of suspended sediment in the breaker zone. Figure 59 shows the field team collecting samples in one of the 235 breaking waves that they sampled. "As many as 10 instantaneous 2-liter water volumes were obtained in each wave for a total of 1,500 samples."

Among other findings, Tim's study determined that "the principal factor controlling suspended sediment concentration at a point in the breaker zone is breaker type. Plunging waves typically entrain one order more sediment than spilling breakers." This and other projects on the South Carolina coast have allowed Tim to conclude that a typical rate of around 100-150,000 cubic yards of sand moves from northeast to southwest along the barrier beaches in the state in any given year (Kana and Knoth, 1977).

Tim's company continues to work on a number of projects in South Carolina

FIGURE 59. Field team collecting suspended sediments in a breaking wave on Capers Island in the summer of 1977. These three COASTAL HEROES from left to right are Robert Logan, Walter J. Sexton, and Dan Domeracki.

and North Carolina, illustrating this work with a colorful 2nd edition of his booklet – *Coastal Erosion and Solutions: A Primer* (Kana, 2011).

Story One

The first part of this story is excerpted from an article Tim Kana wrote for his journalism class on 8 March 1977. It was titled – *ETHEL MAGWOOD: A Gracious Woman Living an Ungracious Life*. His article was a description of a short visit he had with Ethel, a permanent resident out on the shore at Price Inlet, during which she made dinner and they went into town, first by boat to the boat landing, of course, and then on in Tim's truck, to buy some groceries.

After giving the general introduction to this part of the South Carolina coast in which he touted its pristine character, in this story Tim pointed out that, however–"There is a small island, no bigger than three acres at high tide, tucked in between Capers and Bull Islands, which is home to Ethel Magwood. In this entire section of coast, Mrs. Magwood remains the only private property owner and full time resident. But most important, this 68 year old woman keeps alive a way of life which left America fifty years ago." Her home site is located on Figure 55.

First a few facts:

1) Ethel grew up on the mainland and had not been on this island until 1927, when, at the age of 18, she married one Clarence Magwood, a widower over 40, who promptly moved her to the island and sired 11 children. These children lived a varied life, some becoming famous millionaires and fishermen, and some becoming jail birds.

2) When Tim and Duncan were doing their fieldwork in the late 1970s, Granny lived in a small house on the upland by the inlet throat that stuck up only a few feet above the marsh (Figure 55) that she had occupied since 1927. At high tide, the house appeared to be almost surrounded by water, but at low tide, a wide marsh extended away from the house for hundreds of meters.

3) When Tim's team was doing the detailed sampling, they stayed in another small shack on the "island." I should point out that Tim was ably assisted by three of my M.S. students, Dan Domeracki, Walter J. Sexton, and Ray Levey, among several other students.

4) Granny was very glad to have company, so she took it upon herself to prepare huge dinners for the guys after they had completed a hard day in the field (Figure 59). Fresh seafood, including oysters, shrimp, and fish, were on the table every evening. Also, Tim was glad to provide large supplies of other groceries as needed.

5) By the time of this particular visit, the detailed fieldwork was over, so she said, "I sure do miss you boys. You all are gentlemen." She seemed to especially favor Duncan, and she was often heard to say the following about him. "Now that boy can put his shoes under my bed anytime." She was apparently not too fond of Duncan's latest girlfriend and future wife, Stormy Attaway, alluding to her as "Salty." She referred to Stan Humphries, who apparently assisted Tim and Duncan at times, as follows. "That boy is too pretty to be a man." In fact, Stan probably was the most handsome of them all.

Tim's article was fairly long with a lot of great details, a few of which follow. Soon after arriving at the inlet, Tim helped Granny pick some oysters but, as usual, she insisted on preparing a great feast of oysters, turkey, and so on. As they carried the oysters to the house, she said–"Clarence asked me to marry him, so I did. I didn't know how to row a boat, but he taught me that–and to fish. And I couldn't cook. Clarence liked to have things cooked just right. ... I had to learn to cook oysters and clam fritters."

As the meal was cooking, she told Tim that one time when her children were young, she asked Clarence for a dollar to take them to the movies. He wouldn't give it to her, so she told him that she'd never ask him for another dollar again. She didn't either. Instead, she shucked oysters–the men would never do that kind of woman's work–and carried them to Charleston to hawk on the street. At first, she got fifty cents a quart, but by the late 1970s, she could get up to two dollars a quart. But, she also had to pick most of her oysters right alongside the men. When Dan and Andrew (her sons) oystered in the winter after the shrimping season was over, Granny still went out with them and helped bag and tag. And yet, somehow, she managed to cook three big meals a day in between the work.

Tim concluded his story with the following description of a part of their trip in Tim's boat to town to deliver some oysters and buy groceries. "We didn't try to speak over the drone of the engine, but I could tell Granny was enjoying the scene as much as I was. With the sun low, the white down of snowy egrets just jumped out at you from the deep green and blues of the marsh. A flock of geese was silhouetted under the sun as they flew along the horizon. And every now and then, a fish would be scared up from the bottom, leaving only a wake to mark its presence. We turned up the Intracoastal Waterway, and ten minutes at half throttle brought us to the landing. I put the oysters and ice chest in the back of the truck, then helped Granny into the cab."

[NOTE: As I write this in January 2011, JM is working on the *Deepwater Horizon* oil spill with Duncan FitzGerald, presently a Professor at Boston University and a consultant with a BP contractor. A few days ago, JM told Duncan about my putting Tim's story about Granny in this book, and he added one of his own that follows.]

One year, probably 1977, Duncan took a small crew, which included Jeff Knoth, to Price Inlet to conduct some studies on the weekend before Thanksgiving. Naturally, they went by to see Granny as soon as they arrived, and there she sat with her daughter Lizzie beside the dining table topped with a large roasted turkey and many other side dishes, a recently prepared feast. Duncan said, "Granny, what's going on. Thanksgiving is not until Thursday." He had not forewarned her of his visit.

She said, "Well Duncan, I prayed that you and the boys would come by this weekend and here you are. Sit down and have an early Thanksgiving dinner." Which, of course, they did.

Story Two

(November 1979; Cape Romain, South Carolina)

Unlike many of the other stories in this book, I had to do this one from memory, so if anyone reading it was there, please chalk up any misstatements of fact to my bad memory. As noted earlier, most of the other stories throughout this book are based on notes in the many journal entries I have kept over the years.

This one takes place in November 1979. By that time, I was in my fourth and final year as Chairman of the Geology Department at USC-East. On this particular Sunday in late November, we were in the field at the arrowhead point of Cape Romain, probably the most remote spot on the South Carolina coast. This spot is located about 20 miles as the eagle flies from the outlet of Price Inlet. Also, the cape is near the middle of the natural area so eloquently described by Tim Kana at the beginning of this Chapter. On this particular weekend, I was leading a field trip as part of my graduate level course in coastal geomorphology, what else? There were around 15-17 graduate students on the trip, riding aboard three separate boats.

After navigating our way through a complex network of tidal channels in the eight-mile-wide salt marsh complex in the lee of the cape (see Figure 60), we reached the outer beach sometime in mid morning. The plan was to take a long walk to the northeast along the outer beach to observe sedimentary structures in several trenches we would dig across the beach, spend some time in the dunes, observe the waves, measure the longshore currents, and so on. Three separate students were driving the boats, their leader being Craig Shipp, a graduate student studying under me, who had worked as a field logistics person for some time at the St. Croix Marine Station before joining my team. As we started on the walk down the beach, I asked Craig and his associates to secure the three boats and join us

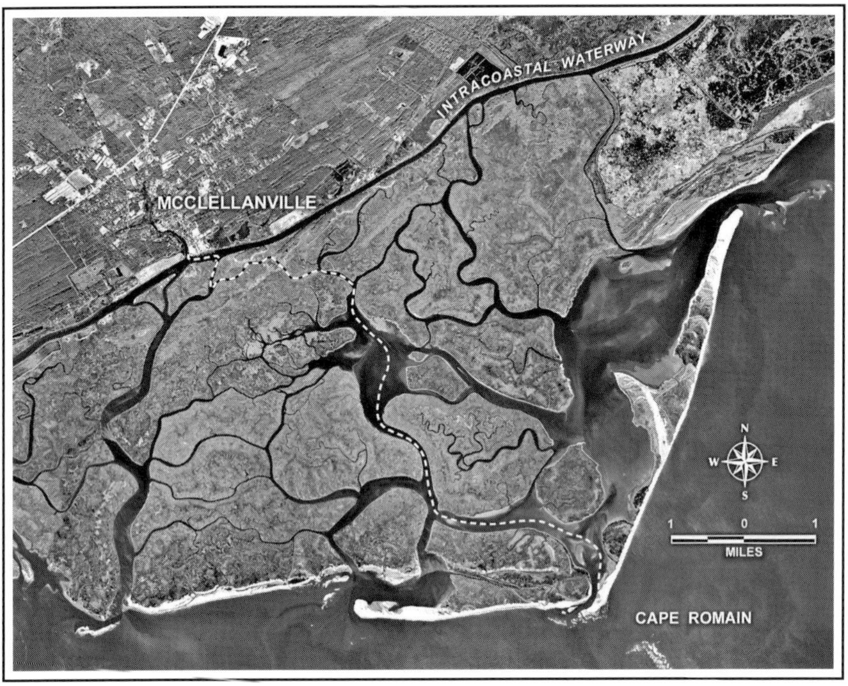

FIGURE 60. Vertical image of Cape Romain area, acquired in 2006. Courtesy of SCDNR. Dashed line marks the probable course we took in the dark following the North Star (Polaris) as a guide (discussed in Story Two).

when they finished.

This course was open to graduate students in other programs in the department, as well as in the separate Marine Science and Engineering programs. In addition to Craig, the only others of my students that I can remember being on this particular trip were two Ph.D. candidates doing research on the South Carolina coast, Aileen Duc and Tom Moslow (their dissertations, Moslow, 1980 and Duc, 1981, are cited in Appendix A), and Walter J. Sexton, who was working on a M.S. thesis.

By early in the afternoon, the tide was out and it was a spring tide by design. I always scheduled such trips to take place during especially low, mid-day spring tides. This was a clear day and we were really enjoying the outing.

We arrived back at the spot where we were supposed to pick up the boats probably around 4:30 p.m., leaving enough time to work our way back to the boat launch site in McClellanville before dark.

Oops!! The boats were lying on their side on a tidal flat about 75 yards from where we were standing on the back side of the washover terrace that composed that part of the outer beach. Hmmm. I guess the guys driving the boats had forgotten about the tide going out. It would be a while before the tide came back in, and the boats could be brought around to pick us up. According to Walter J. Sexton, he, Craig Shipp, and Tom Moslow stayed with the boats all day.

Afterwards, as we moved around and waited at the designated pick-up spot, we could not see the boats anymore, because an intervening salt marsh blocked our view, which didn't make us feel any more secure about our short-term future.

Maybe I should point out that along about that time, a miracle occurred. Numerous bottles of wine and beer suddenly appeared from the inside of almost every backpack on the trip. Needless to say, these miraculous arrivals were soon being consumed at a rapid pace as we sat there on the beach watching the sun go down.

At one point, as I looked out over the surface of the marsh, I could see this lone blond head pop up and move along the surface for a while and then it would disappear down into a tidal channel for another while before popping up again. Eventually, straight across a relatively deep channel that separated the spit we were on from the marsh, this blond head that had been moving toward us peered out through the side of the high marsh grass and greeted us, saying, "Dr. Livingston, I presume!" It was Tom Moslow, who waved his half empty wine bottle at us before heading back to the boats.

As the sun dropped out of sight, creating a beautiful sunset we were looking at straight west across the channel and the marsh, it became quite cold. One of the students on the trip, an engineer, did not have the usual long-sleeved sweatshirt and rain slicker in his backpack that the in-the-know coastal students who had been in the field with me before did. Instead, he was wearing a t-shirt and shorts, as he sat on the beach shivering violently and consuming as much of the beer as possible.

Not to be outdone by the benumbed engineer, a woman geologist (not my student) revealed the contents in her backpack–magic brownies!! By the time it became completely dark, three-fourths of the group were either slightly drunk, completely drunk, or stoned.

Sometime around 9:30 or 10:00 p.m., the boats arrived and the group staggered on board. Neither Craig Shipp nor yours truly had partaken in the extracurricular activities, so we were both stone-cold sober. Yes, both cold and both sober, that is. We took the lead boat with Craig at the wheel and me by his side occasionally taking a quick look at the nautical chart I held in my hand by flicking on it with a small, hand-held flashlight. However, it was so dark we could only use the light briefly, because we had to have our most acute eyesight possible to stay in the channel.

It was pitch black with no moon to be seen anywhere in the sky. We could just barely make out the line along the top of the marsh by the side of the channels as we moved along. Also, I knew that McClellanville, where we were hopefully headed, was located about 45 degrees to the northwest of the arrowhead point of Cape Romain. I was doing some old-fashioned navigation, keeping Polaris, the North Star, which stood out sharply in the crystal clear sky, at about a 45 degree angle to the path we were trying to take, while also hoping to stay in the middle of the major tidal channels along the way.

We had only been underway for 10 or 15 minutes, moving along at about 3 mph, when the participants in the third boat decided not to follow us and took a sharp turn to the left up another channel.

"Moslow (who was in charge of the second boat), go bring those stupid bastards back!"

"Well, they are convinced that channel is the right one to take."

"We will wait here until you get them back. I can see clearly on the chart that they are in a minor side channel. Either use your famous Moslow charm to convince them to get back in line, or, if that doesn't work, threaten to kick their ass."

"Okay. Be right back."

After a few minutes of thrashing around in the small channel, Moslow had them follow him back into the procession. Finally, we were moving along again, guided by the North Star. Now I am not saying that the Global Positioning System (GPS) had not been invented yet, but it would be quite a while before I could afford to have them on any of our boats.

"Craig, can you see the marsh line?

"I think so. Just barely."

"We probably shouldn't get too close, might overrun an oyster reef or two, which wouldn't be much fun, would it? Also, looks like we've got about 100 yards to go before the main channel takes a 90 degree diversion to the left, so let's look out for the turn."

"Roger that."

Even though boat number three did that same trick of wandering off into a side channel two more times, we finally, about two hours later, broke out into the open and could see the lights along the Intracoastal Waterway. The probable route we took is shown on Figure 60.

It was after midnight when we finally pulled the last boat out of the water at the town ramp in McClellanville, and about 3:30 a.m., or so, when we drove into the parking area at USC-East in Columbia. Needless to say, Craig and I, as well as Moslow and Duc taking turns (they were still sober as well), did the driving of the motorcars back home.

A few days later, we had our weekly science meeting of the Coastal Research Division, with 20 or so members in attendance (made up of graduate students, technicians, post-docs, etc. working in my program). As the meeting started, Aileen Duc interrupted to give new T-shirts she had just had made to Craig Shipp and me. On the back of the shirts, COASTAL HERO was printed in honor of our navigating the group back to McClellanville using the North Star as our guide under those conditions of extremely limited visibility.[19]

As I took the shirt, put it on, and looked around the room, I was remembering the long days the people in that group had spent in the sun and the cold, and how so much time had been spent wading through the mud and fighting off the mosquitoes, and so forth and so on. I was also thinking about the long workdays, with them commonly being in their offices at 11:00 p.m. on even Saturday and Sunday nights, as they tried to understand that complex zone where the land meets the sea. And I concluded to myself, "Everybody in this room is a COASTAL HERO."

The name COASTAL HERO caught on after that, usually when someone was alluding to the grad students and researchers on our team. It was regularly used to refer to them as they did the logistics for the 150 or so weeklong field seminars that we ran on the South Carolina coast for the American Association of Petroleum Geologists (AAPG) and several companies in the oil-and-gas business between 1976 and 2005 (discussed in Chapter 12 - THE ROCK RECORD). We still commonly use that sobriquet, which is why I have chosen it for the title of this book.

11 A RELOCATED TIDAL INLET

If along the high road
I caught hold of your sleeve,
do not hate me;
old ways take time to overcome.

THE BOOK OF SONGS

The Science

In Chapter 5 – KIAWAH ISLAND, I discussed the work we started on Kiawah Island in the summer of 1974, which resulted in the establishment of construction set-back lines and allowed us to start gaining a better understanding of the geomorphology and stratigraphy of the barrier islands in South Carolina. Over the next few years, we continued to monitor the beach profiles we had established earlier on Kiawah, as well as a couple on the neighboring barrier island to the south, Seabrook Island (Figure 28).

Like Kiawah, Seabrook is now a gated community, which is a very pleasant place to stay, if you are so inclined. Our continuing studies showed that the geomorphology, stratigraphy, and so on of this island are similar to Kiawah. It has had some erosion problems, but no set-back line approach was ever attempted. Like Kiawah, the development of this island for residences began in the early 1970s. In 1978, our group, the company RPI in this case, was contacted by the developers regarding their shoreline erosion problems, of which there were three:

1) The original Beach Club and marketing center, a good-sized two story building, was built on the side of the main channel of the North Edisto Inlet, which attains depths of over 90 feet in the inlet throat. As a result of this proximity to the deep inlet throat, the beach became erosional almost as soon as the Beach Club was built. We were not involved with the solution to this problem, which was to build a revetment composed of large riprap along the side of the channel, which preserved the building, but some problems with erosion and loss of the original sand beach continue. Tim Kana's company, Coastal Science and Engineering, Inc., has done three beach nourishment projects in the area to deal with this problem.

2) As shown by the oblique aerial photographs in Figure 61, an eroding arc in the beach in the middle of the island was threatening the golf course,

FIGURE 61. Beach erosion at the golf course on Seabrook Island. (A) Photograph (originally infrared) was taken at low tide on 17 October 1978. Note the erosional arc that was developing in front of the golf course (arrow). (B) Photograph taken on 28 January 1983. Arrow A points to sandbag revetment in front of the golf course and arrow B points to riprap revetment in front of houses in the lower right. From Hayes and Michel (2008).

as well as houses nearby. This was erosion caused by waves approaching the shore from the open ocean.

3) Captain Sams Inlet was migrating into Seabrook Island, with a loss of many acres per year of highly developable land as a result of this erosion. In 1982, we advised the developers that the inlet would be at the golf course in 14 years. This inlet migrates rather consistently to the south at a rate of around 200 feet/year.

A review of historic charts, maps, and aerial photos showed that the outlet of Captain Sams Inlet (Figures 61 and 62) had migrated south and eroded large parts of Seabrook Island several times. We also noticed that the neck of the spit was last cut through a fairly short time before the 1949 vertical aerial photograph was taken. The cut may have happened during a hurricane a year or two earlier, but we have not been able to determine which one, if indeed a hurricane was responsible. It is possible that the migrating meander bend of the large tidal creek behind the spit eroded into the foredunes to such an extent that waves generated by a relatively minor storm could have eroded through the spit during an unusually high spring tide. We think the odds favor a hurricane, but we just don't have enough data to say which one. No major hurricane crossed the South Carolina coast during that time, but the eyes of a couple did cross the Georgia coast, which means there should have been a significant storm surge along this part of the South Carolina coast during these storms.

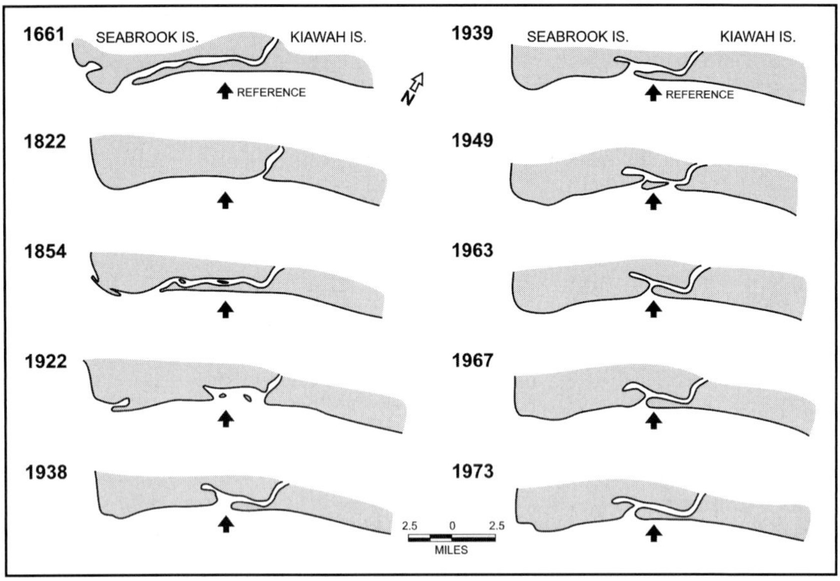

FIGURE 62. Historic changes Captain Sams Inlet between 1661 and 1973, based on a combination of historical charts and aerial photographs. These maps and photos show three periods of breaching of the spit: 1822, 1922, and 1949. Breaches occurred at numerous other times, of course, at intervals we estimate to be about every 40-50 years. From Hayes et al. (1976b).

Based on studies of the aerial photos taken soon after the new inlet had cut through at the neck of the spit in the middle to late 1940s, we observed that the shoreline of Seabrook Island built out dramatically within a few years afterwards. At the time of that 1940s cut, the inlet throat was near the spot where it was located in the early 1980s. As you can tell from the oblique aerial photograph taken in 1978 (Figure 61A), a modest-sized ebb-tidal delta was located off the inlet's outlet in the late 1970s (and no doubt was always there). Therefore, it stood to reason that the sand that accreted to Seabrook Island in the early 1950s was sand that originally composed the ebb-tidal delta of Captain Sams Inlet in the late 1940s! Once the 1940s inlet was abandoned, shoals of the ebb-tidal delta no longer had the strong ebb currents that formerly exited the inlet throat to maintain them. Therefore, wave-generated currents eventually moved the sand on shore, resulting in the major outgrowth of the shoreline evident on the old aerial photographs.

With that observation in hand, we had a solution for two of the erosion problems plaguing the Seabrook Island development (Hayes, Kana, and Barwis, 1980), namely, beach erosion near the golf course and the migration of the inlet into the front third or so of the island. The solution was to go back and artificially cut the neck of the spit, which would:

1) Allow the ebb-tidal delta to weld onto the beach in the area of the golf course, solving the erosion there; and

2) Eliminate the threat of the inlet moving into Seabrook Island for many years, depending on how far up the spit the cut was made.

Once the plan was approved, we made the cut for the new inlet and closed the old Inlet on 4 March 1983 (see vertical image of the new inlet in Figure 63), using land-based equipment, at the cost of about $300,000. However, for reasons given later, we could only go about a third of the way up the spit to make the cut.

The on-scene management of the project was carried out by Tim Kana, who has published several papers on the project since (e.g., Kana and Mason, 1988; Kana, 1989; Kana and McKee, 2003). Tim's company, Coastal Science and Engineering, Inc., continues to monitor and deal with this issue (Kana, 2011).

To make a long story short, almost a thousand feet of beach was created in front of the old eroded scarp in front of the golf course and, needless to say, the migrating inlet was no longer an immediate threat to Seabrook Island (see Figure 64). The progressive welding of the old ebb-tidal delta sand onto the beach in front of the golf course is illustrated by the photographs in Figure 64. The inlet was relocated again in 1996, and sand continues to be added to the beaches on Seabrook Island. The width of the added sand in 2006 is illustrated in Figure 64B. By that time, the beach in front of the golf course has built out almost 1,000 feet. Kana (1989) estimated that over one million cubic yards of sand from the abandoned ebb-tidal delta had accreted to the shore of Seabrook Island after the inlet was relocated in 1983. Those kind of numbers make this inlet relocation project one of the most effective and inexpensive techniques of beach nourishment known to man. If you don't think so, check out the costs for some of the other nourishment projects that take place on the South Carolina coast about once a year (these two relocations combined cost a total of considerably less than one million dollars). As far as I know, this was the first time such a nourishment technique had ever been tried.

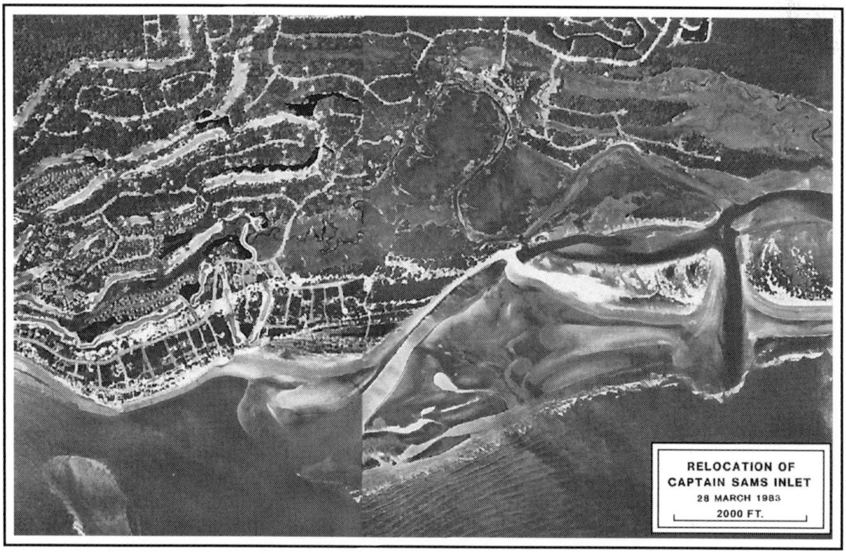

FIGURE 63. Vertical image of the relocated Captain Sams Inlet on 28 March 1983. Relocation of the new inlet was completed 24 days earlier, on 4 March 1983. From Hayes and Michel (2008).

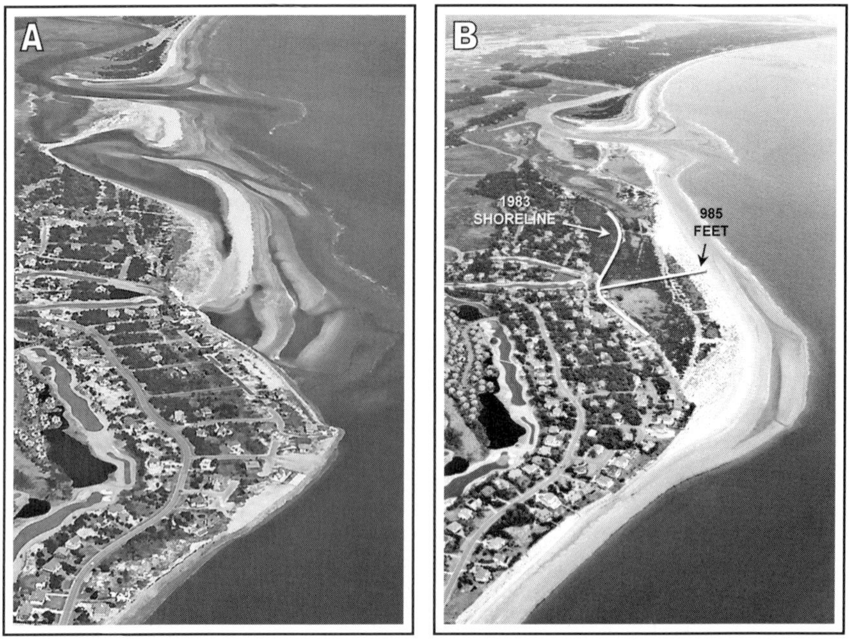

FIGURE 64. Sand added to Seabrook Island as result of relocating Captain Sams Inlet in 1983. (A) Photograph of Seabrook Island by Tim Kana taken at low tide on 25 February 1986, two years and eleven months after Captain Sams Inlet was relocated. Note the welding and landward migration of swash bars composed of sand released from the abandoned ebb-tidal delta after the inlet was relocated. (B) Photograph by Tim Kana taken on 10 February 2006. The beach in front of the golf course had built out 985 feet in the 23 years since the inlet was originally relocated. It was relocated again in the spring of 1996. Note the position of the 1983 shoreline (compare with photograph in Figure 61B). From Hayes and Michel (2008).

The Story

(Throughout 1980, 1981, 1982 and on into 1983, Seabrook Island, South Carolina)

Presumably at least partly because of our designation of a construction setback line that excluded the development (construction of vacation beach cottages, etc.) on the recurved spit at the south end of Kiawah Island (for obvious reasons; see Figure 62), the managers of the Kiawah Island development project apparently decided to donate a portion of the spit to Charleston County for the construction of a public beach access, which is now called Beachwalker Park. This is a popular park, inasmuch as it offers the only public access to the many miles of splendid sand beach on Kiawah Island, one of the most beautiful on the east coast. This natural and picturesque spit is shown in Figure 61A, and its history of formation and erosion is illustrated in Figure 62. Because of the consistent migrations and cut-offs of the spit over time, some questions were raised about whether it belonged to the owners of Kiawah or Seabrook Islands, inasmuch as a case could be made that the owners of both islands had a claim to it. Because of the timing of some of the early land grants, the spit was a part of Kiawah on some of the old charts and a part of Seabrook on some of the others (see Figure 62). Apparently, this did not cause an unsolvable problem with regard to the establishment of the new park.

As early as 1980, I was making speeches telling anyone who would listen about our idea to cut a new outlet channel through the spit (Hayes, Kana, and Barwis, 1980). We noted that if the new channel were excavated through the spit where it cuts naturally during hurricanes, it would be fifty years before the Seabrook Island owners would have to worry about the inlet migrating through their property again. Also, after the new inlet were constructed, much of the sand in the abandoned ebb-tidal delta would come on shore, ending the erosion problems at the Seabrook golf course.

Once when I made such a presentation, one of the listeners was heard to respond: "Are you out of your mind Hayes? The place where the natural cut occurs is now Beachwalker County Park. Can you imagine how many kids would drown in the strong currents in the new inlet throat?"

Okay, so we couldn't make the cut at the neck of the spit. How about going up the spit a shorter distance, say several thousand feet or so? Anyway, our group (RPI by then) drew up a plan to cut the spit about a third of the way up, and I presented that plan numerous times during trips from Columbia to the coast. Part of this plan was that the cut would have to be repeated about every 14 years, because of the limited distance up the beach we would be able to relocate the inlet. And, lo and behold, the owners of the two islands agreed with the plan! Also, compared to other alternatives, such as permanent jetties and other "hard" solutions, this was a very inexpensive solution, even if it had to be repeated upon occasion.

Oops! Minor detail. Little old lady in tennis shoes sued the management of Seabrook for even contemplating such an insult to the environment, which slowed up the process for another year. Meanwhile, during this additional year of delay, the island lost many more of its valuable acres.

As the ensuing legal deliberations unfolded, there was a hearing on Seabrook Island at which the two sides of the issue were heard from. Coastal geologists Orrin Pilkey of Duke University, and Bill Cleary of UNC-Wilmington, were opposed to the plan, so they represented our technical opponents. At that time, Pilkey tended to disagree with many of the proposed modifications of the natural coastal system, such as sea walls and groins.[20]

In any event, the plaintiff's lawyer shoved a map into my chest after asking me some simple-minded question. Luckily, I was not my usual volatile self that day; therefore, we "won the day," and the project was finally approved. On completion day, 4 March 1983, Tim Kana and I and a group of other interested spectators celebrated as we watched the new inlet being cut by the rising tide (see photo in Figure 65).

However, I immediately made plans so that such a scene (i.e., having a lawyer piss me off like that in such a setting) would not be repeated again anytime soon. How? By getting out of that line of business! So I did stop doing these types of projects in South Carolina, but we did do a few overseas, as the following chapters will show.

Also, I was heavily involved at that time in the oil-and-gas business with RPI companies in Colorado, Texas, California, and western Canada which, for a

FIGURE 65. The group of us involved in the creation of the new inlet through the Kiawah Island recurved spit celebrating on 4 March 1983 as the rising tide does the job (initiates the new channel, which is shown in Figure 63). I am the guy in the middle in the dark vest and, hopefully, I was not as drunk as I appear to be in this picture. The guy on my right waving the champagne bottle is Tim Kana.

while, was considerably more profitable than working with coastal developers (for coastal geologists; coastal engineers did okay). After the major drop in oil prices in 1986, this bonanza out west came to a halt. But, that is another story not in any way related to coastal heroism.[21]

The first relocation project was initiated and carried out while Tim Kana was still our partner in the RPI company in Columbia. Not long after that, he formed his new company, Coastal Science and Engineering, Inc. (CSE), which is also located in Columbia and, as noted earlier, all of the major future projects involving Captain Sams Inlet were carried out by his company (Kana, 2011).

12 THE TRADE SHOW

Through want of enterprise and faith,
men are where they are, buying and selling,
and spending their lives like servants.

HENRY DAVID THOREAU

The Science

March 1977 was when we incorporated our original company, Research Planning Institute, Inc. (RPI), and later on that year, we bought an old motor cycle shop on the back streets of Columbia in which we set up housekeeping for the business. This original company had three additional primary partners, my wife, Jacqueline Michel (JM), and two other Ph.D. graduates of the USC-East Geology Department, Tim Kana and Erich Gundlach. During the years 1976-1979, we established ourselves as oil spill experts and began responding to oil spills for NOAA and creating oil spill sensitivity maps (discussed in Chapter 19 – ENVIRONMENTAL SENSITIVITY INDEX), commonly under NOAA sponsorship, along different parts of the coastline of the U.S.

Our experience at so many major oil spills allowed us to offer oil spill training courses related to oil spill response, coastal clean-up strategies, and sensitivity mapping for many groups around the country, especially the U.S. Coast Guard and individual state response groups (e.g., Texas, California, Florida, and New Jersey). Many private companies, such as Shell, Chevron, and Amoco, also sent participants to these courses.

By the time of the *Exxon Valdez* oil spill in Alaska in March 1989, we had been working under contract with NOAA for eleven years, providing scientific support to the U.S. Coast Guard, during which time we assisted in the response to most of the largest oil spills of that era (e.g., *Amoco Cadiz* in France and *Ixtoc I* in Mexico and South Texas, as discussed in Chapter 8 – OIL SPILLS). As also already noted, JM and I, and our other partners, had formed RPI by that time, to which the NOAA contract was assigned. I had resigned from USC-East and worked full time for the company, and JM, who would spend many months of field work at the *Exxon Valdez* oil spill site, was in charge of the oil spill response work in the company. We conducted follow-up studies of the *Exxon Valdez* spill site for many years afterward. During that time frame, we conducted coastline sensitivity mapping projects for all of the Alaska coast. During one of the mapping projects in Southeast Alaska in 1990, I barely escaped death in an airplane crash off Mitkof Island south of Petersburg (discussed in Chapter 15 – BY FOOT, BY CAR, BY BOAT, BY PLANE, and in Chapter 16 – DOOMNESS).

At the time of this writing (November 2010), JM has just returned from a two-week stint at the *Deepwater Horizon* spill in the Gulf of Mexico, having been working on it in the field, two weeks on/two weeks off, since April. In total, our company still has 18 people working on that spill, which resulted from an explosion on an offshore drilling rig on 20 April 2010.

We also kept working on beach development projects, such as the ones described previously for Kiawah Island, Seabrook Island, and Kuwait, which ceased for projects in the U.S. soon after we completed the inlet relocation project, as noted in the previous chapter. We continued, however, doing overseas projects in that field, one of the most notable ones being dealing with erosion near Keta Lagoon, Ghana (Nairn and Hayes, 1997; Hayes, Montello, and Nairn, 1997; discussed in detail in Chapter 13 – WEST AFRICA). The oil-and-gas projects are another issue, which are discussed briefly in footnote 21.

Story One

To keep a business such as this going, you have to have an aggressive and active marketing program. One of the standard ways of doing this is to attend trade shows where you rent space for a booth in which you display your wares to potential clients and customers for a period usually lasting from three to five days. Over the years, we have made several attempts to determine if going to such shows does indeed increase sales. Thus far, we have no conclusive answer to this question, but we keep doing it, because, if we didn't show up, people might conclude we've gone out of business or something. In fact, as I was writing an earlier draft of this chapter in March 2001, we were busily planning to exhibit RPI's products and services at the International Oil Spill Conference in Tampa, Florida in a couple of weeks. Our group presented nine papers at that conference. I even had my name on a couple of them. Our graphics department had prepared a blow-up of the sensitivity map we had made the previous year of the Golfo de Fonseca, on the Pacific shorelines of El Salvador and Honduras. This multicolored extravaganza decorated the back wall of our booth in hopes of garnering interest from potential clients in Latin America, and Linos Cotsapas, our multi-lingual biologist and a senior partner in the company, manned the booth to explain our work to potential clients in their native tongue. We felt pretty good about our performance, and possibly even landed a couple of jobs because of it.

Sometimes you go to these trade shows just to show off. The first time I did this was in the fall of 1973, during my first semester as Head of the Geology Department at USC-East. The Geological Society of America (GSA) met in Dallas, Texas that fall. Of course, we didn't call the meeting a trade show then, no, it was a professional scientific conference. Anyway, some of us decided we would make a splash at the meeting and put the new wave of coastal geology at USC-East on the map, as it were. Therefore, Bjorn Kjerve (BJ for short), of Billy Whiskers fame, who had just arrived as a new professor of physical oceanography in the department, and I designed a booth that would catch everyone's attention. We enlisted the services of BJ's wife, Lucy, who was a creative interior designer, to help with the colors and design. The back wall of the homemade booth was colored bright yellow and green with a large, blue breaking wave painted on it. I decided we needed a beach in the scene to go with the wave, after all, we were supposed to be experts on beach erosion. I also concluded that nothing but natural beach sand would do the job. Therefore, I had a couple of my graduate students drive a pick-up truck by the beach near Pensacola, Florida, which is composed of beautiful, pure white quartz sand, to get a load of that sand to take to Dallas so we could build a small artificial "beach" in the back corner of the booth for the wave to break on.

As we were loading our disassembled booth in a van for the long drive to Dallas, BJ said, "We are the Gamecocks, are we not? We need a rooster for the booth!"

"Great idea, let's rent one for the week."

My administrative assistant, a local fellow named Francis Marion Reeves, III, who knew about such things, quickly found for us a *bona fide,* oversized fighting gamecock rooster named Killer, who had done just that to numerous opponents, or so I was told. On the ride to Dallas, Killer was happily crowing away in the back of the van.

As it turned out, our space was exactly in the center of the exhibit hall. You could see it plainly from every entrance. In addition to the bright-colored background and the breaking wave on the sparkling white sand and the strutting, brilliant red, purple, and black Killer walking about on his tether crowing loudly every ten minutes or so, we had two speakers mounted on the roof of the booth. We played continuously and loudly a tape of rock music that included Otis Reading singing *Sitting By the Dock of the Bay*, Bob Dylan's *Nashville Skyline Rag*, and other hits. Every once in a while, Killer would walk over to the white sand on the "beach" and take a dump, scratching the sand wildly afterwards, much to the amusement of the spectators, and to the dismay of the graduate student, or BJ, or me, who was minding the booth at the time. After cleaning up the accident and sweeping the sand back on the "beach," things were back to normal, and Killer would crow out his approval of our efforts.

Well, the booth definitely aroused people's attention. I'm sure some of the uptight scientists didn't approve of the show, but you can't please everybody.

One afternoon, as we were standing around the booth, Clyde Moore, a fellow grad student at UT, who was then a professor at LSU, said to me, "Well babe, we all thought you would do okay at South Carolina, but we didn't expect you to own the place."

Therefore, I guess the booth achieved at least a part of what we wanted it to.

With such experience in our resume, it made sense for us to get into the trade show mode in a big way once we started the private business. One of our very first attempts was at the Offshore Technology Conference held annually in Houston, Texas. This was a giant trade show for the offshore oil-exploration industry, which drew up to 50,000 people for the week. It was so huge, in fact, that it was staged in the Astrodome.

We had a local carpenter build us a nice-looking, stained wooden booth, and Dan Domeracki and I loaded it in the back of a pick-up truck and headed off to

Houston for one of the biggest trade shows in the world. I believe this was in either 1979 or 1980. We had opted for one of the cheaper booth spaces, which happened to be way back in a dark corner under the bleachers. Thus, while the big-time operators had offshore drilling platforms erected on the playing field, replete with dancing girls in bikinis, we were well off the mainstream of activity back under the cheap seats. However, with such a large attendance, some people were bound to wander by our booth, at which time Dan and I would gallantly try to sell our expertise in oil spill response, beach erosion, and the like.

But it was kind of a wearisome job, and it soon became apparent to both of us that our potential clients were either: a) Not at the conference; or b) More interested in what was happening on the playing field than anything that might be hiding under the bleachers. One dreary late afternoon, a skinny, nervous looking fellow wearing a Colombo-style raincoat walked up. He was carrying three shopping bags stuffed with goods he had collected from the different booths.

After helping himself to the goodies in our booth, he waved us up close to him and said in a confidential tone, "They're everywhere!"

"Who?" Dan asked.

"They are taking bodies back with them, back to God knows where. Maybe another galaxy. You can't trust anyone. Be careful."

Then he quickly went on down the row to the next booth, where he continued filling his shopping bags.

Late in the morning of the last day, a young man in an expensive business suit walked up and started gingerly fingering one of our brochures. A few years later, we would have called him a classic yuppie.

I walked over, extended my hand toward him, and said, "I'm Miles Hayes, President of our company, could I tell you about our services?"

He raised his left arm a bit, exposing his expensive watch at which he gave a long glance, and then said, "I'm in a hurry, make it short."

I leaned over to him, cupped my hand over his ear and said in a loud voice, "Fuck you! That short enough?"

When Dan told this story back at the office, it was decided that MOH should not man the marketing booths at trade shows anymore.

Story Two

During that same period of time, I developed a friendship with Dr. Ram Saxena, an LSU graduate, who quickly became a millionaire oilman while working out of New Orleans. Ram was also a *bona fide* sedimentary geologist with numerous published papers and a reputation for being a good lecturer; hence, he was in demand at many schools, especially since he did not always require a speaker's fee. One day he called and asked if I would fill in for him at a lecture he had committed to deliver at LSU. Something else had come up unexpectedly that he had to take care of in Baton Rouge.

"I can pay you your standard lecture fee," he said, as if I had one.

"No, no Ram. I'll be glad to do it. I have lots of academic connections at LSU. It will be nice to see some of my old students again, as well as my buddies Jim Coleman, Clyde Moore, and Harry Roberts. Uh, but you will cover my travel expenses, though, won't you?"

"Certainly I will. We'll make reservations for you in the finest hotel in Baton Rouge. My chauffeur will meet you at the airport in my limousine, and he will take you wherever you wish to go during your stay."

I showed up at the Baton Rouge Airport, still not dressed too well, a holdover from my university days. I was wearing a ratty old, tan corduroy jacket, no tie, and my scuffed up one pair of shoes that I wore interchangeably on field excursions and business trips.

I was surprised that nobody met me as Ram had promised. I did see that a guy dressed in some kind of uniform was hustling around looking over all the passengers, but he passed me up with little more than a quick glance. I picked up my bag and walked out to the street. Soon, I was the only person left who had arrived on the flight from Atlanta that hadn't gone on about their business. It was just me and the guy in the uniform, who still was pacing around in an agitated manner. A long limousine waited a couple of parking spaces down from him.

I finally concluded that this was Ram's chauffeur, but he couldn't seem to bring himself to accept me as one of Ram's friends and associates.

The words from one of The Band's songs kept running through my mind, something about being judged by my shoes.

I finally walked over to him and said, "I think you are looking for me."

"Are you Dr. Hayes?" he asked incredulously.

Well, the lecture went okay. I was glad to visit with my pals at the Geology Department and Coastal Studies Institute at LSU.

Sometime after that visit to LSU, I went to the GSA Annual Meeting in New Orleans. The ever-in-demand Ram had asked three of us to help him teach a one-day seminar on the importance of understanding how the rocks in the Gulf Coast Province and elsewhere were made. This seminar was sponsored by the GSA and was advertised as a part of the meeting. The other speakers, besides Ram himself and me, were John Ferm, my colleague at USC-East and Ram's dissertation advisor when he was a grad student at LSU, and John Barwis, one of my ex-graduate students, who was on his way up in Shell Oil Company by then.

During the evening of that same day we conducted the seminar, Ram invited the speakers to a very expensive dinner at one of the better restaurants in New Orleans. During the dinner, we discussed why we bothered to do things like these one-day, stand-up seminars.

Ram said, "I do it for the money."

Ferm said, "I do it for Ram."

I said, "I do it for my ego."

Barwis said, "I do it for science."

We all laughed.

Story Three

About mid summer that same year I got another call from Ram who told me that, unfortunately, he had gotten himself into another speaking engagement and could I help him out, again. However, this one would be a little more complicated, because he had been signed up for a five-week tour in the following fall that involved giving twenty talks at various and sundry universities and oil-and-gas societies throughout the central U.S.

"I've just recently found out that I will have three wells coming in at that time and I really can't afford to be away then. I had some other people in mind to do the lectures, but the only substitute the organizers of the tour would accept was the talk you gave at the recent GSA meeting. I told the organizers the title was the *Role of Tides in Determining the Defining Characteristics of Reservoir Quality Sand Deposits.* Was that right?"

"Close enough, but Ram."

"Look, I can make it worth your while, you would get all of the rather generous speakers fees I occasionally charge. Also, all other expenses would be covered."

"I'm afraid that really would be necessary this time, I guess."

"You could also do some marketing at the same time, now that you and John are going into the business. Also, you can have my limousine and chauffeur again. You will be able to travel in comfort."

"Oh, I don't know about that."

"I understand that you had some trouble with my last chauffeur, but I will assign someone else this time. His name is Alvin Green. A great guy. Very smart. He is a former member of Special Forces. He even sometimes acts as my bodyguard. That should give you another level of comfort."

As you have probably guessed by now, I agreed to do the tour and showed up in New Orleans in early October. Soon after that, Alvin and I headed off to the University of Arkansas, then to St. Louis and then on to Lawrence, Kansas, where I lectured to a packed house at the University of Kansas, home of a famous and historic Geology Department.

Meanwhile, Alvin and I were becoming good friends. We both grew up in the South and both played baseball for many years. The World Series was beginning and we were making one-dollar bets on who would get the first hit, strike out first, and so on. Riding along in luxurious comfort listening to the big games on the radio was a lot of fun.

Let's see, where else did we go? There was a memorable visit to the Colorado School of Mines, where I was able to get together with some of the RPI/Colorado folks. Then we went to Tulsa and on to the University of Oklahoma, and to what seemed like every possible gathering of oil industry people in the great state of Texas.

Lecture number 18 was on the Rice University campus in Houston, and number 19 was on the campus of the University of Louisiana at Lafayette (oo-la-

la).

As we drove away from Lafayette that night on our way to a hotel in New Orleans, I said, "Alvin, I can't do this anymore. If I have to get up on another stage and tell the same old stale jokes one more time for another hour, I will go nuts."

"Maybe after a good night's rest in NOLA, you will change your mind. After all, we only have one more to do, tomorrow night at good old Southern Mississippi University. By the way, do you know any of those people over there in Hattiesburg?"

"No, I don't think so."

"Hmmm."

At mid morning the next day as we rode along through the pretty swampland between New Orleans and Hattiesburg, Mississippi, Alvin said, "I've been thinking. You said you do not believe you can do one more of these. Why don't we switch roles. We are about the same size. We can stop somewhere along the way and we can swap clothes."

"Oh come on Alvin."

"No, I'm serious. Remember the slide that shows the general model for shoreline embayments?"

"Okay, sure."

Then he quoted word for word what I usually said when that particular slide was on the screen. He told me that while listening to me do the talk 19 times, he had memorized it, just to amuse himself, I guess.

Upon hearing Alvin's recitation about the embayment slide, I said, "Good. Let's do it."

After meeting all the proper people, making sure the projectionist had things in order, and so on, we usually walked into the auditorium together, with Alvin, who was dressed in his chauffeur's uniform, always sitting at the isle side of one of the middle rows so as to be available if needed. I always walked on up to the front.

This is exactly what we did in the auditorium in Hattiesburg, with me dressed in Alvin's chauffeur's uniform taking a seat at the end of the middle row and Alvin, dressed in a typical university professor's attire, walking up front where he waited to be introduced.

I thought his lecture went great. He didn't forget a thing. Actually, I thought he got more laughs for my jokes than I usually did.

At the end of the talk, he received a long and loud round of applause.

Then came the questions. He handled the first two just fine.

The third question was asked by a somewhat grizzled middle-aged guy, who I guessed was an oil-company geologist.

He said, "Well Dr. Hayes, I have been reading some of your papers on tidal inlets, so you will no doubt be able to answer my question. I think maybe one of the units we are working on in the Tertiary rocks not too far from here, which will remain unnamed for obvious reasons, contains some tidal inlet deposits. However, I have been having a little problem telling whether certain cores are penetrating the ebb-tidal delta or the flood-tidal delta. How do you distinguish them?"

To which Alvin replied, "Boy oh boy, did you sleep through my whole lecture? This is the twentieth time I have given this lecture on this tour and you have just won first prize for asking the stupidest question I have been asked yet!"

"I will tell you what. Just to prove to you how stupid that question was, I am going to ask my chauffeur, who is sitting out there in the middle of the audience, to answer it."

Then he looked out at me as I stood up and moved into the aisle. Then he said, "Okay, Alvin would you please answer this man's question?"

"Sure. Just study the cores carefully and if you find one that contains a number of thin layers of mud, it is probably from the flood-tidal delta. However, if the core contains little mud but has fairly large-scale crossbedding, and even some plane beds, it will have been taken from the ebb-tidal delta."

"Thanks. That is just what I wanted to know. Now you have solved my problem."

Nobody had the nerve to ask any more questions. Therefore, the meeting was over and the tour was over.

Once we had stopped laughing as we rode along on the way back to New Orleans, I told Alvin, "Please tell Ram that I am not available to do any more talks for him, even if this one was great fun and the pay was adequate. Also, wish him good luck on his new wells for me."

[NOTE: Unlike all the other stories in this book, this one is not based on fact. I just made it up. I got the idea from Dan Domeracki, who recommended that I listen to the comedian Jerry Clower's 1970 album *Yazoo City, Mississippi Talkin'*. The title of this particular piece in his album was "The Chauffeur and the Professor." I just recently scanned Google and learned that Einstein possibly once pulled a trick just like this himself, donning a chauffeur's costume at a scheduled speech. I have told this story many times, and have been surprised how many people fell for it, especially family groups.]

PLATES

PLATE 1. Low-tide aerial view of sand waves on the sand flat in the lower portion of the Parker River Estuary, Massachusetts (the southeast side of the flood-tidal delta, also known as Middle Ground; see map of the area in Figure 12). View looks offshore (east). Sand waves in right portion of the image area are flood oriented. Photograph taken at low tide in September 1966.

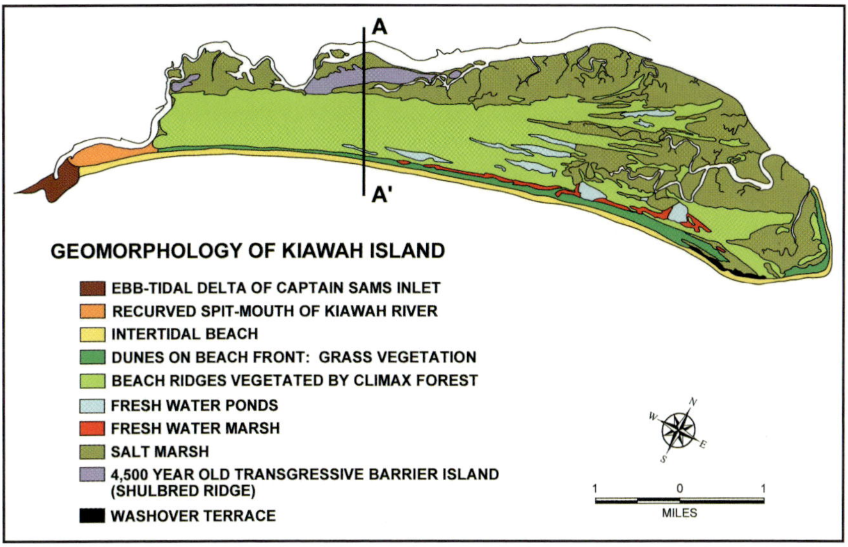

PLATE 2. Map of the geomorphology of Kiawah Island using a 1971 aerial photograph as a base. Constructed based on fieldwork completed in 1974. The line labeled A-A' gives the approximate location of the stratigraphic cross-section shown in Figure 30.

PLATE 3. Oblique view of the north end of Kiawah Island, showing a large swash bar welding to the beach. This is the main mechanism by which such prograding barrier islands attain their drumstick configuration, which is illustrated in Figure 29. Infrared photograph taken on 29 May 1980.

PLATE 4. Heavily oiled, fine-grained sand beach during low tide at the *Urquiola* oil spill site near La Caruña, Spain. Photograph taken in May 1976.

PLATE 5. Cuspate foreland at Punta Catalina, Chile, which is on the south side of the east entrance to the Strait of Magellan (see Figure 48B). View looks north. These recurved beach ridges, which indicated longshore sediment transport from north to south, were not formed by open Atlantic Ocean waves, but rather by waves created by strong westerly winds blowing across the "sheltered" interior of the strait during high tides. Photograph taken during at low tide in late February 1976. These beach ridges are composed of gravel (including abundant cobbles).

PLATE 6. Point Barrow, Alaska, a cuspate spit. Photograph taken in late July 1976, when the ice still covered much of the surface of the Arctic Ocean in that particular area.

PLATE 7. MOH lecturing on the beach near Lomé, Togo in late January or early February 1979. Arrow A points to the great American/French sedimentologist George Allen, and arrow B points to Roy Lee, the UN representative and overseer of the project. Photograph taken by Tim Kana.

PLATE 8. Oblique aerial view of the southern end of Kayak Island, Alaska (Figure 22) in the summer of 1970, one of the featured photographs in the tone poem, *Suzanne's Lament*.

PLATE 9. Maylo Murday (right) and JM (on his right) during a conference in the northern city of Kaduna, Nigeria sometime in the early 1980s.

PLATE 10. Classic example of a transgressive barrier island on the South Carolina coast. This is a view of the southwest end of Cape Romain (Sandy Point) a few days after Hurricane *Hugo* (11 October 1989). The white band of sediment, a mixture of sand and shell, is a washover terrace that advanced landward several tens of feet during the storm. The dark layer seaward of the washover terrace is composed of exposed backbarrier muddy sediments. As a result of the landward migration of this barrier island, a new tidal inlet was created where it intersected the large tidal channel in the foreground. From Hayes and Michel (2008).

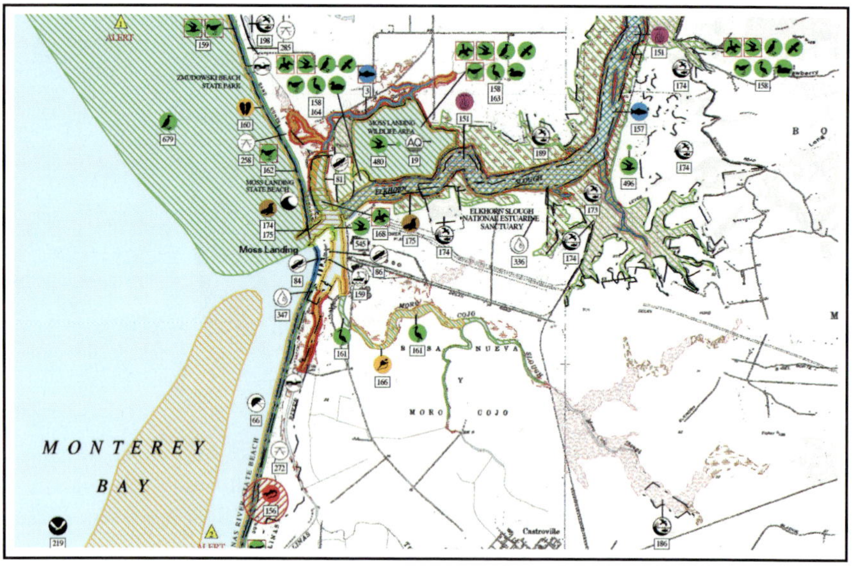

PLATE 11. Environmental Sensitivity (ESI) Atlas of the Elkhorn Slough area in Monterey County, California (NOAA/RPI, 2006). This area has a high density of birds. Note all of the green icons (showing specific bird species involved), and the green hatch patterns showing concentration areas. The blue color on the outer beach represents sand (ESI = 4), and the red color inside the Slough represents marsh (ESI = 10). From Hayes and Michel (2010).

PLATE 12. Low-tide photograph of a heavily oiled intertidal flat, which is several hundred meters wide, located on the Gulf Coast of Saudi Arabia. This photograph of oil spilled during the first Gulf War was taken by JM in May 1991.

13 WEST AFRICA

Die religion ... ist das opium des Volkes.

KARL MARX[22]

The early part of 1979 was a busy time for us. In January and February, Tim Kana and I spent two weeks in Togo and Benin in West Africa on a UNEP-sponsored mission to recommend solutions for beach erosion problems in those two countries. It was the first of many trips that we would eventually take to that unique part of the world, the east-west oriented West African coast between the Ivory Coast and Cameroon (see Figure 66). Over time, with my last trip being in 2003, we carried out numerous projects in Togo/Benin, Nigeria, and Ghana. These three areas will now be discussed separately.

TOGO/BENIN

The Science

The erosion and deposition along this coast in the early to middle 1900s was variable, but since about 1960, the beaches in both countries showed some dramatic erosion in places. This accelerated erosion was so acute that the two countries requested that the Environmental Programme of the United Nations (UNEP) initiate a program to study the problem. The UNEP-sponsored workshop that Tim Kana and I attended was held in the two countries between 29 January

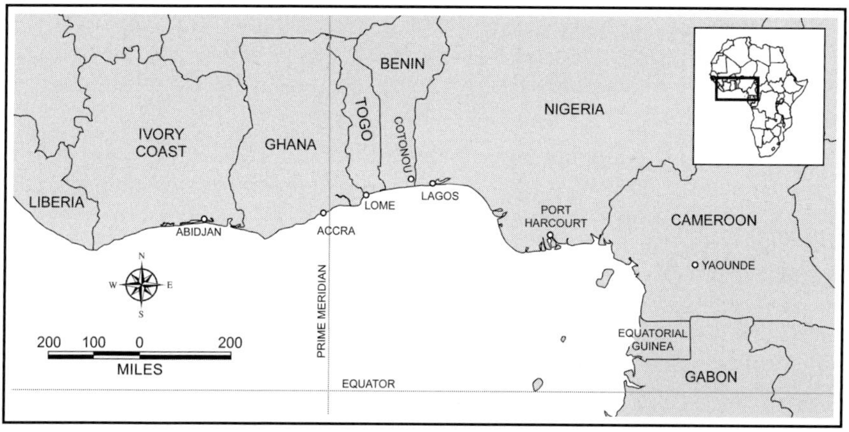

FIGURE 66. Location map for West Africa.

and 9 February 1979. The working technical group consisted of 33 expert engineers and technicians from Togo and Benin, some of whom are pictured in Plate 7, as well as experts from four other countries. In addition to Tim and I, there were three engineers from Holland, two geologists from France (George Allen, the great sedimentologist, and Laurent D'Ozouville, who worked with us on the *Amoco Cadiz* oil spill), and one engineer each from Mexico and India. The U.N. representative was Roy Lee, a very congenial guy who was fun to work with. I was eventually made the convener of the visiting technical group, which published a final report (U.N. Technical Group, 1979). This report is the basis for the remainder of this scientific discussion. It was also summarized by Hayes (1985a).

The state of erosion and deposition on the shoreline of the two countries in 1979 is shown in Figure 67A, and some of the recommended solutions suggested by the visiting UNEP team are given in Figure 67B. The key cause of this increased erosion was the construction of deepwater ports at Lomé, Togo in 1966, and at Cotonou, Benin in 1963. The jetties of these ports interrupt the longshore transport of sand, one of the strongest in the world (greater than 1,000,000 cubic meters per year; NEDECO, 1978). Transport is from west to east almost every day of the year due to the influence of waves arriving at an oblique angle to the shore from the southwest. By 1979, no sand-bypass systems had been installed at either harbor entrance, so erosion became a serious problem downdrift of both harbors.

The jetties at Lomé provided a classic case study of man's impact on beaches. The shoreline updrift (west) of the jetties was building out in excess of 50 meters per year, whereas downdrift (east) of the port, the beaches were receding at rates of 8 m/year. The Tropicana Hotel, located in this downdrift area, was threatened. However, later information implies that the beachrock exposed in the intertidal zone by the original erosion event has slowed down the erosion some in that area. In this same zone, many fishing villages had been abandoned as a result of the erosion.

The erosional/depositional patterns at the port of Cotonou, Benin mimic those described for Lomé. West of the port in 1979, the beach had built out 700 meters since the construction of the jetties. The area just east of the jetties is complicated, because of the presence of a dam and some groins, but a highly erosional crenulate bay had formed east of the last groin by 1979. A maximum of 250 meters of erosion had occurred in the middle of that crenulate bay, which is located approximately 1,000 meters east of the jetties (Figure 67A). Many villages had been moved and an industrial zone was threatened. The rest of the coast of Benin was relatively stable.

[NOTE: The term **crenulate bay** appears in several places in this book, thus a brief definition of these very common features is in order. They are numerous on rocky coasts like Kodiak Island, Alaska (Finkelstein, 1979). These bays typically have a log-spiral, or fishhook shape, with the shank of the hook pointing in the downcurrent direction in response to the dominant approach of the larger waves and swell along the shoreline. The sandy beaches in the crenulate bays are most commonly located between two segments of rocky shoreline. The shape of the bay is the result of refraction/diffraction of the dominant waves around a rock headland area located at the updrift end of the bay. The waves arriving at the shoreline refract and diffract into the **shadow zone** in the lee of the up-coast bedrock headland, creating longshore currents that remove sediment from that zone and transport

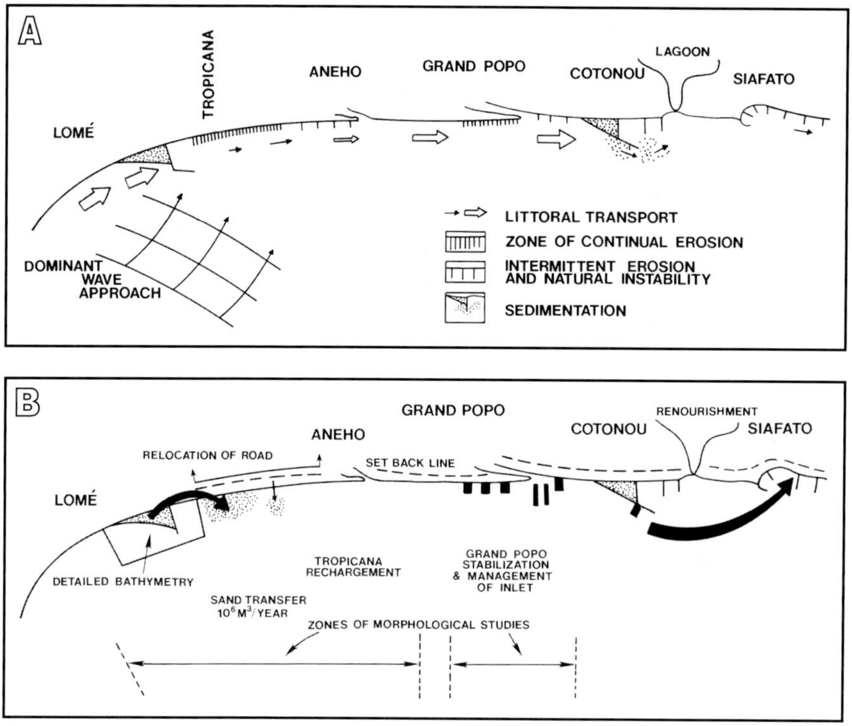

FIGURE 67. (A) Conditions on the coasts of Togo and Benin in January/February 1979. Note the major role played by the two ports in causing beach erosion downdrift (east) of them. From U.N. Technical Group (1979). (B) Some suggestions for dealing with erosion on the coasts of Togo and Benin. The large black arrows indicate sand that should be moved past the jetties to their downdrift sides by means of sand-bypass systems. From U.N. Technical Group (1979).

it toward the opposite end of the bay (except for some reversals, minor in some bays and major in others, in which the finer-grained sand moves back toward the updrift headland), eventually producing a stable equilibrium beach with a crenulate, or log-spiral shape.]

After ten days of evaluating the problem of shoreline erosion in Togo/Benin, the group of experts, relying heavily on data supplied by NEDECO (1978), formulated several options for dealing with the specific erosion problems in the two countries. Some of these recommendations are outlined on the map in Figure 67B.

The most obvious solution to the major problem in both areas, that is, erosion downdrift of the jetties, would be to bypass sand from that accumulated on the updrift (west) side to the downdrift (east) side at both harbors. The volume required would be on the order of one million cubic meters per year at each harbor. In view of the high cost involved, many millions of dollars for each project, some alternative solutions were also proposed by the experts, including:

1) Establish set-back lines in critical areas.

2) Install some new groins to slow erosion in places.

3) Selectively pump sand onto the beaches where maximum erosion is occurring.

Also, try to prohibit mining of sand from the beaches, which was a very common practice.

The Story – TOGO/BENIN

(*On the beach at Lomé, Togo on 30 January 1979)*

Tim Kana and I arrived in Lomé on the evening of 29 January, a Monday. The first three days were spent in Togo getting acquainted with the Togo coast and each other. With slightly over 40 people involved, sometimes the field trips were a little crowded. The meetings always started late, and lunches drug out to three hours length. In case you haven't figured it out yet, I like for things to run on time, so this was a little tough for me to adjust to.

Tim and I did some teaching on the beach (see photograph in Plate 7), and Fritz Hoekfeldt (of Haskoning Engineering in Holland) presented the results of the Nedeco-Haskoning half a million dollar study of the erosion problem in Togo. In the field, it was hot and my body dehydrated. We were never given anything to drink or consulted on the schedule. So it went.

On Friday morning, we went across the border into Benin, my first visit to a Communist country. We spent most of that day studying the Benin beaches.

About 3:30 p.m. Saturday, we arrived at the Cotonou Airport for our flight for an aerial inspection of the coastlines of the two countries. We finally went up in a boiling hot, ancient Russian biplane (1930s) with small port holes looking out between the wings. These were marginal conditions for taking pictures, or even viewing the coast, as far as that goes. The plane was piloted by a pair of Russians, who couldn't speak either French or English. We were turned back at the Togo border, resulting in no aerial views of Togo. Also, we were not allowed to take pictures of the port in Cotonou. Not a very informative airplane ride.

The next morning was a Sunday and we took one of the most exotic tours I had ever been on. We went by boat to visit the Ganvié stilt village on Lac Nokoué near Cotonou. It was a gray day over the lake, but the reflections off the water made for perfect composition for photographs of the fishermen and fishing boats on the water (see Figure 68). Floating into the village itself was like a trip to another planet. The place was primitive still, but it made me sad, because the kids begged for coins, and the women covered their heads with cloths when our bunch of white intruders snapped their pictures.

That night we drove back to Lomé.

Next morning, I discovered that I had a urinary infection, probably because of the heat and lack of fluids to drink. One of the Togo representatives was a scientist named Maylo Murday, native of Mauritius, whom I befriended. Maylo took me to the local Peace Corps doctor (an American), who had the right pills, so the infection cleared up within a few days.

[NOTE: Not long after this visit, Maylo came to the U.S. and joined our team. You will hear a lot more about him in the discussion of Nigeria that follows. Also see

FIGURE 68. Two views of the "Village On Stilts" (Ganvie) near Cotonou, Benin. Photographs taken on 3 February 1979.

the discussion in the DEDICATION. A photograph of him is given in Plate 9.]

After some confusion, I emerged as the spokesman for the foreign experts. We had our work cut out for us. The Africans liked to talk things over slowly and at length. We (the foreign experts) wanted to meet separately and get organized, which pissed the Africans off no end, because they thought we were cooking something up without consulting them. Actually, I just wanted to avoid a huge open confrontation between Tim and I and NEDECO, because we obviously had different ideas on how to solve the problem. Basically, we had a fairly simple sand bypassing technique in mind, whereas they had a more complex system involving major engineering models and projects. I guess our ideas didn't cost enough to suit them, who were in the drivers seat to get any future projects that might have become available.[23]

After getting over that rough spot, we spent some long hours writing sections on "causes of erosion" and "solutions," about 50 handwritten pages in all. George Allen, the French/American, whom I always thought was the best sedimentologist I had ever met, except for the great Robert L., perhaps, was indispensable, churning out text prodigiously.

During the open meetings that followed, I acted as spokesman for the experts. I also wrote the conclusions and recommendation on Thursday night, although the Africans were supposed to. By early Friday morning, the report was completed.

We went to lunch on Friday at the U.S. Embassy (by invitation of the Ambassador herself). The food was lousy, but the house was nice, and we had fun posing for pictures in the Ambassador's huge black Chrysler.

I read the conclusions and recommendations, which were approved unanimously, at the closing meeting of the workshop at 4:00 p.m. on Friday. Tim and I left Lomé for Paris at 5:30 p.m. that same night.

NIGERIA

The Science

In the time interval from July 1981 until on into 1985, we pursued projects in Nigeria, with our initial contacts there being aided by Maylo Murday who, as noted earlier, had been stationed in Togo for some time and knew the area and the region's scientists and engineers well. In July 1981, Maylo and I were in Nigeria to meet with two of the senior technical people, Dr. Ben Akpati, of the Nigerian Institute of Oceanography (NIOMAR), and Dr. Jerry Nwankwo, of the Nigerian National Petroleum Corporation (NNPC).

The meeting with Ben involved a relatively small project with the goal of training some of the Nigerian students in techniques of conducting coastal geology studies, and to make some conclusions about the erosional/depositional trends along the coast. One of the activities carried out for this project was a two-week field trip that I led with Walter J. Sexton for Ben and five students in his Institute. A photograph of this group is given in Figure 69. A fairly lengthy discussion of this most interesting field trip is given in the following Story Two.

The NNPC project, which was funded for several million dollars, started in March 1982 and lasted for nearly five years. During that time, we established a separate company, RPI/Nigeria, with an office in Port Harcourt, which was also the headquarters for the big project – *Environmental Baseline Studies for the Establishment of Control Criteria and Standards Against Petroleum Related Pollution in Nigeria.*

FIGURE 69. The field team on the shoreline survey in Nigeria in September/October 1981. The person in the middle is Ben Akpati, and the second one from the left is Walter J. Sexton. The lady on the left is Christi and the one on the right is Iyoba.

The reason for the interest on the part of NNPC in having such a project conducted at that time was that a rash of major oil spills had recently occurred in Nigeria. Some of these spills were massive, and they killed large areas of mangroves, several of which were eventually mapped by our team. These spills also resulted in a variety of other damages to the coastal and riverine ecosystems.

Although I was listed as the Project Manager in the final report for the project (Hayes, Michel, and Murday, 1985), which was a mere 1,013 pages in length, it was actually managed by JM as the U.S. Project Coordinator and Maylo as the in-country Project Coordinator. During this time period, I was heavily involved with our oil-and-gas service companies in Colorado, Canada, and elsewhere.

Managing this project was not without its difficulties. Maylo, who was a full-time resident in Port Harcourt throughout the project, was attacked one evening by bandits seeking the payroll cash that had just arrived. He was beaten over the head with a lead pipe and just barely survived. This was further evidence for the widely held notion that Nigeria is one of the most dangerous places in the world to work.

Another problem was getting paid on time. If we asked the client when an overdue check would be available, the standard answer was "any time from now." Despite the discussion in footnote number 23, we had decided to take the risk in this case. There was a period later on in the project when several unpaid receivables from Nigeria totaling several hundred thousand dollars were late by months. As a result of these delays, the managers of our local bank in South Carolina went ballistic, causing us a few problems. However, although many skeptics told us we would never be paid in full, this did not turn out to be the case; something of an exception to the general rule, we were informed.

One of the major objectives of the NNPC project was to provide a comprehensive report on the existing ecological conditions in the oil-producing regions of Nigeria. This work was carried out with a somewhat minor Nigerian partner, Rivers State University. In addition to Maylo, we hired several scientists and other collaborators who lived in country. The studies covered a wide diversity of disciplines, including a chemistry program that analyzed over 7,000 samples of sediments and tar balls. The biological parameters studied included terrestrial and aquatic vegetation (including vegetation and land-use maps), general (nektonic) macrofauna, benthic macrofauna, phytoplankton, fisheries, and several other topics. Also, socioeconomic studies were carried out to ascertain how the Nigerian communities in the oil-producing regions viewed the country's oil industry, among other things. A total of 61 communities were visited and over a thousand questionnaires administered.

The studies most relevant to this discussion had to do with the coastal geomorphology of the area, which were managed by Maylo and Walter J. Sexton. The first field visit to do this was in March 1982, and later field studies were carried out as follows:

May-June 1983 (early wet season)

July-August 1984 (wet season eastern region)

October-November 1984 (wet season western region)

February 1985 (dry season)

With regard to the coastal studies, the entire shoreline of Nigeria was surveyed, with emphasis on the Niger Delta. The team divided the coast into five morphological zones: 1) Strand Coast (on the east side); 2) Delta Flank (east and west sides of the delta); 3) Arcuate Delta (middle portion of the delta); 4) Transgressive Mud Coast; and 5) Barrier-Lagoon Coast (on the west side). As shown in Figure 70, 60 beach stations were established within the five morphological zones, where beach profiles were surveyed and sediment samples were collected at each station. An example of the beach profile of one of these stations in the Barrier-Lagoon Coast (zone 5) is given in Figure 71. The geomorphic team also produced an Environmental Sensitivity Index (ESI) map of the entire shoreline, which is presented in Figure 72. These types of maps will be discussed in more detail in Chapter 20 - ENVIRONMENTAL SENSITIVITY INDEX (ESI).

The different river systems in the Niger Delta were classified into two categories, white-water rivers, those deriving most of their water from the uplands, and black-water rivers, those deriving most of their water from adjacent swampland. This classification is very similar to the one used for coastal plain (black-water) and piedmont (red-water) rivers on the South Carolina coastal plain (Hayes and Michel, 2008). This is a very important distinction with respect to the ecology and sediment sources of the two river types in both areas.

Story One - NIGERIA

(23 July 1981; Thursday - On the beach in Lagos, Nigeria)

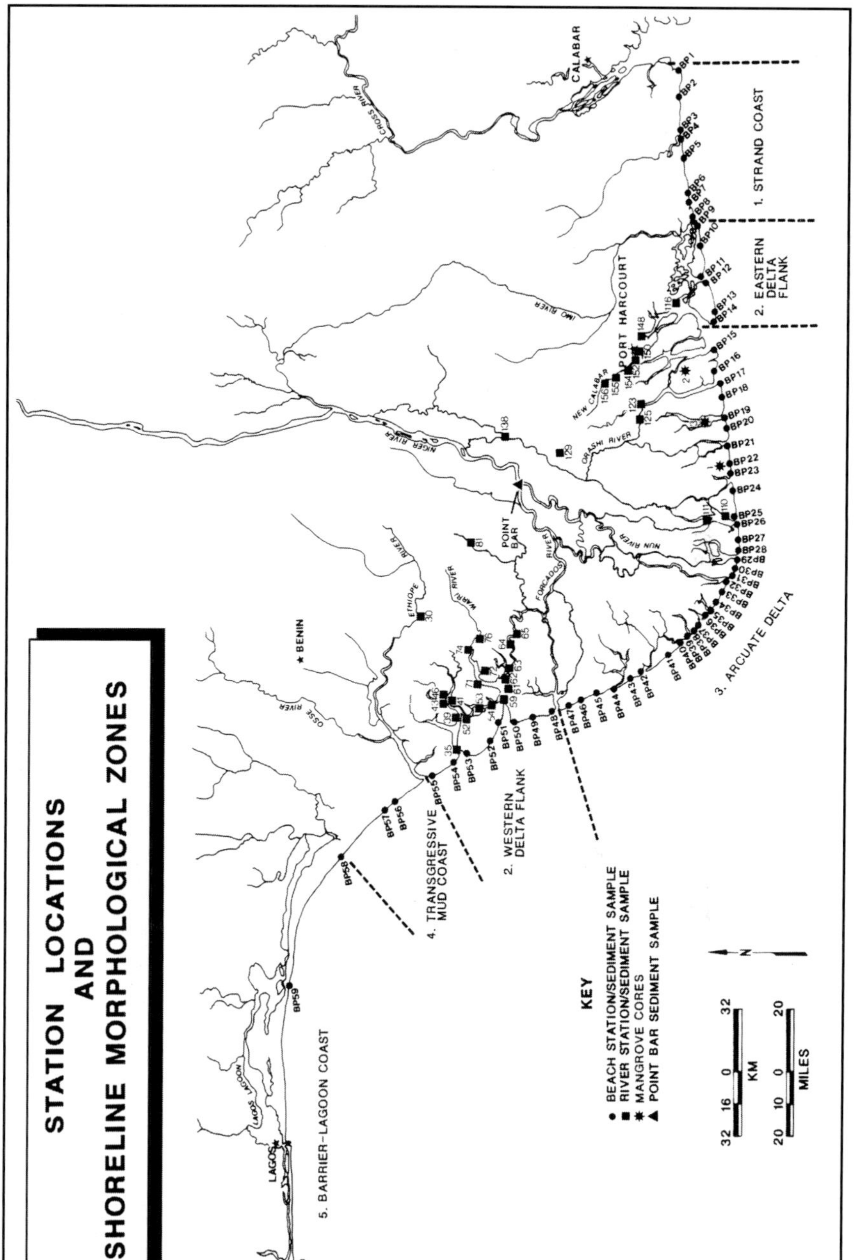

FIGURE 70. Location of the 60 beach stations surveyed and the five basic morphological coastal zones delineated during the study of the Nigerian coast carried out by RPI in 1982-1985. Topographic beach profiles were surveyed and sediment samples were collected at each station. From Hayes, Michel, and Murday (1985).

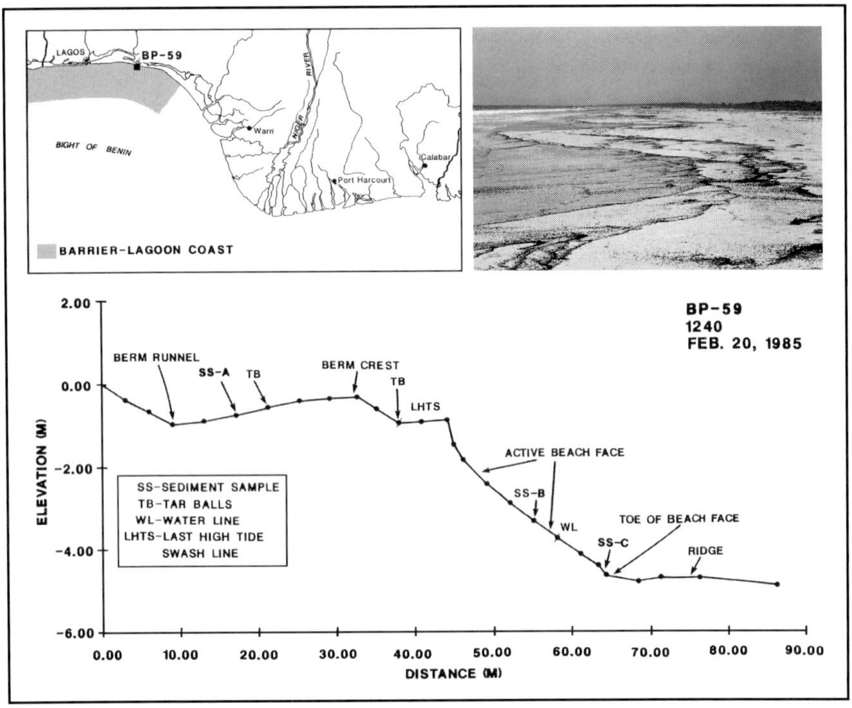

FIGURE 71. Location and beach profile at station BP-59 along the western shoreline of Nigeria (in the Barrier-Lagoon Coast, see Figure 70). This is a steep beach relative to most of the beaches in the delta region to the east. The inset on the right shows multiple oil swash lines observed on 15 February 1985 at Station BP-58, located about 40 miles east of Station BP-59. These swash lines were composed mainly of small, very soft tarballs. From Hayes, Michel, and Murday (1985).

Two days earlier, I had set out from Columbia to meet Maylo in Lagos so we could try to wrap up the marketing of a couple of projects, one a relatively small one mentioned earlier with Ben Akpati's program at the NIOMAR, and a much larger one with NNPC. Both of these marketing efforts must have gone pretty well, inasmuch as we eventually were awarded both projects.

One thing I wanted to do while there was get a better idea of what the Nigerian beaches were like, so on Thursday, 31 July 1981, I went out onto the beach on the spit offshore of Lagos with Ben and Maylo. The following notes are taken directly from my journal.

As we were driving along the beach there, I noticed a black man with his head (face, actually) buried in the sand. Ben didn't laugh when I suggested that we could use dead bodies to measure the longshore sediment transport direction and rate. This man had been dead for some time. Probably he was washed onto the beach at high tide and was left stranded up on the beachface when the tide receded.

A little ways past this body, we got out of the car and walked along the beach for a while. I was appalled by the broken down shacks, the garbage, and the stench of shit.

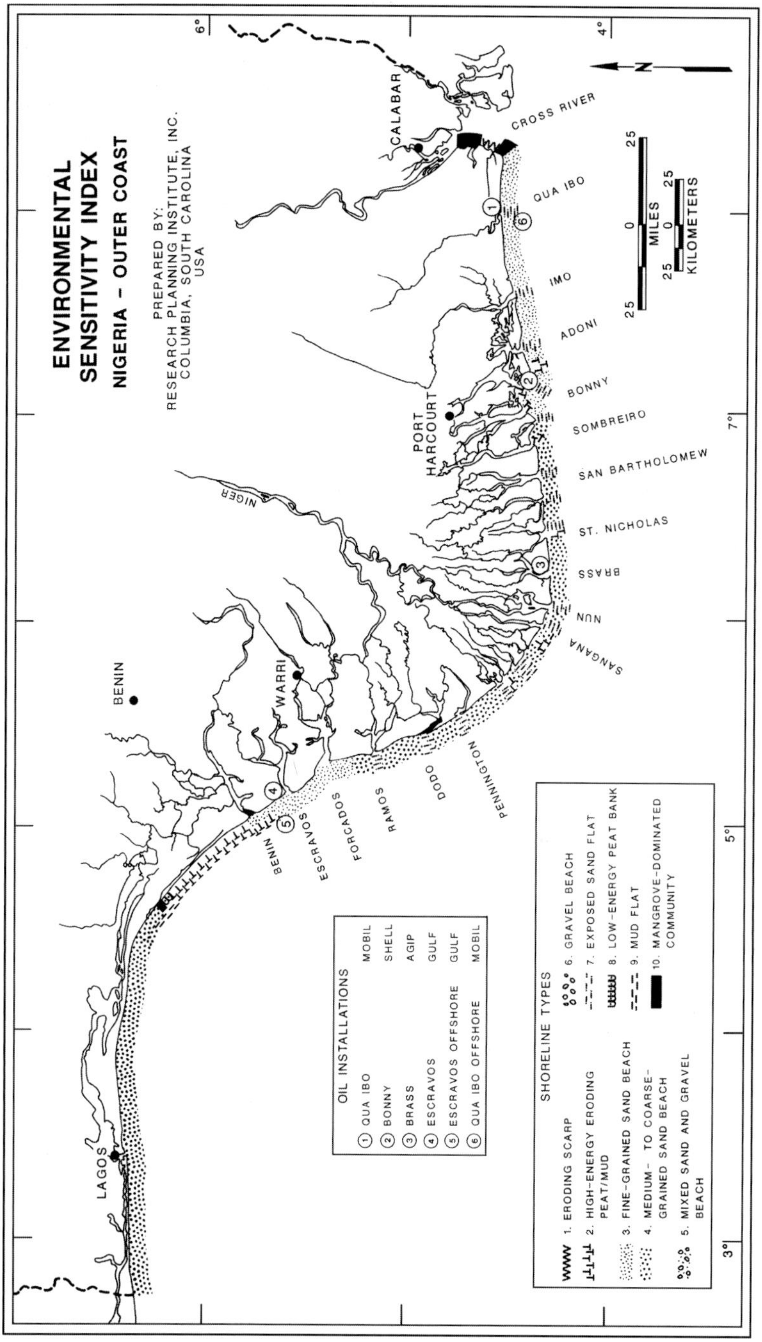

FIGURE 72. Environmental Sensitivity Index (ESI) map of the outer coast of Nigeria completed by RPI in 1982-1985. This map also shows the area covered during the field trip with Ben Akpati and his NIOMAR students in September/October 1981 (discussed in Story Two – NIGERIA). From Hayes, Michel, and Murday (1985).

I kept saying over and over to myself, "*nobody should be that poor*!" which made me think of a Cat Stevens song. Then I reminded myself that Cat was performing in some bar in Boulder, Colorado that very week. And I thought, "There aren't any dead bodies decomposing on the beach in Boulder, Colorado, where everybody has blonde hair and is rich and is less than 22 years old and is from Cleveland, Ohio." I knew Boulder fairly well by that time, because we were in the process of setting up RPI/Colorado there (discussed briefly in next chapter).

As we walked further along, we passed close by the sign shown in Figure 73, which warned people not to shit on that particular holy ground. And as we passed just beyond the sign, I looked over at the crest of the berm, where two dark rear ends hung, trying to squeeze out some turds.

NOBODY SHOULD BE THAT POOR!!

Later that evening, Maylo and I walked into a candle-lit Chinese restaurant and sat down in the midst of Japanese and European businessmen there after the quick buck, just like us. As we waited for our food, we talked about world order, Ghandi, building from the grass roots, and what to do about the starving poor. I said God and religion were bullshit. Maylo, a serious member of the Baha'i faith, said religion was necessary to give us moral standards. However, nothing we could say would give the thousands of wretched blacks out there on the beach a bed to sleep on or a decent Chinese dinner to go with some of that outstanding Star beer like we were eating and drinking.

In addition to the beach walk already described, I had spent some time the following afternoon with Ben and his students studying maps and aerial photographs of the Nigerian coastline. I guess I was playing Dr. Science again. This coast is interesting, for sure, much like South Carolina's.[24]

Then on 25 July, a Saturday, I was on a Nigeria Airways flight from Lagos to New York. That morning Maylo was concerned that I might have some problems making the flight, so he rode with me through the dark and over the bumps and in between the throngs of hundreds of colorfully attired natives who cluttered the road.

Getting checked out of the country was a crunch, a melee of arms and legs and pushing and shoving and yelling. Tempers flared. I laughed a lot. I was mentally prepared. I had been to Saudi Arabia. Actually it only took about an hour to go through the whole thing and, as a bonus, I didn't have to strip.

As we neared the end of the flight, I wrote, "Now we are going down. We took off five and a half hours ago, a pretty long flight. Nigeria wasn't nearly as bad as I was expecting with respect to living quarters (for us), ability to get the job done, and so on. Conditions appear to be similar to those in the Middle East in many respects."

"I guess I expect big things to develop in Nigeria, but there is ALWAYS an element of DOUBT!"

Story Two – NIGERIA

(Fall of 1981)

In this story, I will recount the two-week adventure Walter J. Sexton and I

FIGURE 73. Sign on beach near Lagos on 23 July 1981.

had in late September and early October 1981, as we led a field trip to explore the coastline of Nigeria (see Figure 72). The goal of the trip was to train Ben Akpati's students at the Nigerian Institute for Oceanography and Marine Research on how to conduct field studies of the coastal geomorphology of the shoreline, with emphasis on beach erosional/depositional patterns and tidal inlet geomorphology. The fact that the shoreline was so similar to the South Carolina coast, where Walter J. and I ran over five, week-long field seminars a year (discussed in the following chapter), was so similar to the Nigerian shore made the scientific part of our job rather easy. Managing the people on the trip was another matter. First, I will introduce the team.

Dr. Ben Akpati – About a 40-year old native Nigerian too good to be true. He had come by USC-East earlier to give some lectures, during which I recognized that he was one of the most dynamic speakers I had ever listened to. He had spent some years as a professor in a university in California, but was now back home in another professorial role to take advantage of all the money available due the oil boom in his native land. This was the first field trip he had been on since coming back to Nigeria months ago.

Me – A 46-year old, long-haired hippie professor with iron-rimmed glasses. In pretty good physical shape, even exercising routinely during this trip. Not yet recognized as the "Father of Coastal Geology," but working on it. A little preoccupied with the startup of a couple of new companies, but glad to get back into the field for a good look at the Nigerian shoreline.

Walter J. Sexton – Early to mid-thirties native South Carolinian. A recipient of an M.S. degree that same year (1981) under my supervision based on a study of a couple of tidal inlets on the South Carolina coast. He eventually became a full-time RPI employee as a coastal geologist. A very skillful field man and boat handler, he was often in the field on a variety of projects. Participated for many years in the instruction of the field seminars conducted on the South Carolina coast for oil-and-gas geologists and started to run them on his own after I quit doing it in 2005.

Austin – Very serious and very reliable. When I yelled at him, he said, "I didn't know!"

Iyoba Johnson – A classic chubby Negro woman. Always complaining about the food. Educated in England.

Victor – Likes to visit the girls in town. Young and strong.

Christi – A real cutie, but she doesn't shave her legs. Wants to go to Carolina.

Anti – Supposedly the smartest one. Doesn't say much. Smiles a lot.

Okay, so this is the team (see photograph in Figure 69). We set out to see the Nigerian shoreline in two separate vehicles. I rode in the back seat in conveyance number one with Ben, who sat beside me and right behind the driver Theodo. Now Theodo was probably the worst driver I have ever ridden with. Each time he did something stupid, which was about every five minutes, Ben would caution him with the conciliatory words - "My friend! My friend!"

The following is what I wrote in my journal as we scouted our way across Nigeria.

(29 September 1981 – Tuesday; Benin City)

We finished a long and bumpy ride down the "freeway" to Benin City about 9:00 p.m. Two hours in the perilous darkness with Theodo at the wheel. There were check points every five miles or so, and the potholes in the road were big enough to bury a Volkswagen in. Naturally, we had a clogged fuel filter.

We were supposed to leave at 9:00 a.m. "sharp," but instead we left at 4:00 p.m. "dull." Anyway, I was trying to relax and enjoy it, doing my sit ups when possible, getting a little sun, and checking out some new birds. So far, I had spotted 13 new species.

So, at 10:20 p.m., I was sitting, sweaty and exhausted, in my room in the Emoten Hotel in Benin City, Nigeria, a Federal hotel similar to the Federal Palace in Lagos, which means that there was no water in the bathroom, and the air conditioner wasn't working too well. As I reviewed the situation, I thought, "There are probably bugs in the bed and I can't walk barefoot on the floor and I certainly couldn't drink the water, even if there was any."

(30 September 1981 – Wednesday; Warri)

We stopped on this morning at the museum in Benin City. The history of this area changed dramatically after several British diplomats were killed in 1896 while on a mission to see the Oba, or king of the Edo people of the Benin Kingdom, which lasted from 1180 to 1897. Presumably in response to the murder of the diplomats, a British punitive expedition sacked Benin City in 1897 and exiled Oba Obonramwen, taking control of the area and establishing the British colony of Nigeria. Apparently before that, the Oba made a lot of money selling slaves to the white slave traders.

Most of the original art assembled during the centuries that the Benin Kingdom existed was somewhere else. Ben said there is a great collection in a museum in Philadelphia. According to the Wikipedia Encyclopedia, the Benin royal art was auctioned off by the British to pay for the expedition that put a sad ending to the long-lived Benin Kingdom.[25]

We arrived in Warri about noon, too late to take a boat to the beach. Plans were made to go over there the following day.

In the afternoon, we moved into rooms in a fine motel, surrounded by a sea of some of the worst filth and poverty I had ever seen. Walter J. and I walked the dirt streets for a while, dodging the traffic and puddles and slopping through the heaps of garbage all along the side of the road. They have a severe drainage problem in Warri, and it had rained very hard that morning.

As we walked along in our "running" shorts, Walter J., with a camera about his neck, and me, with a pair of binoculars around mine, the children asked us for money and chanted "white man" mixed with some other words in their native tongue. The women flirted a little bit, and the men stared back in either a semi-resentful or an absent fashion.

After that, I spent a nice afternoon checking out the birds around the motel. Somewhere out there was the famous Niger Delta.

(1 October 1981 – Thursday; Warri)

We left about 9:00 a.m. to go to the boat. After a little wait for the engine to be repaired, we crammed into a small speedboat for a one-hour ride down to the ocean's edge at Forcados (see Figure 72). This boat ride down the estuary was a trip back in time through an original African jungle, after a fashion. The area reminded me somewhat of the Santee/Pee Dee Delta in South Carolina, but it is much bigger!

On the beach, I gave a little speech about flow regime, and beach profiles, and washovers, but nobody listened. I soon gave up. We walked around to the outer beach where trees were exposed in the eroding surf. Then we ran a topographic profile across a wide protuberance. It took our erstwhile crew well over an hour to run the profile, but they had fun at the beach!

About 3 o'clock, we were caught in a terrific downpour of rain. We hurried back to the boat, which was practically swamped. My whole body, and my camera and binocs, were soaked through and through. For a while, I rode outside in the wind and driving rain. The boat was flying along through a dense overcast. Finally, I had to go inside, because I was freezing! Freezing to death in a tropical rain forest! We all huddled inside, clutching our life preservers, as we bumped our way back to Warri for another 45 minutes.

At dinner, it was pointed out to me that this day was Nigeria's 21st birthday. I thought, "You have to give these Nigerians credit, they are taking a holiday weekend to make this trip."

That night after dinner, we had a science meeting. Some comments made at the meeting:

Ben said, "It was a start."

I said, "Motivation is the key."

Walter J., in an aside to me, said, "They were a lot slower today." We had earlier done some practice beach profiling near Lagos.

Austin said, "I didn't know."

Christi said, "They really don't smoke at RPI?"

Anti said, "What is RPI?"

Victor said, "Do you have full-time employees at RPI?"

And Iyoba closed out the meeting by saying, "I had gunpowder for breakfast."

(2 October 1981 – Friday; Port Harcourt)

On this day, we rode for a long time around the edge of the jungle called the Niger Delta. We stopped at, examined, and had general discussions about several of the geomorphic features we encountered along the way. We even crossed the wide Niger River, noting for the students it's beautifully developed sandy point bars, which were later studied in some detail by our team during the NNPC project.

The towns were relatively clean. The fields were open and untended. We saw a few scars of the Nigeria/Biafra War, but the jungle heals quickly. They had been fighting and dying in these woods only ten years earlier. And as I write this chapter in February 2011, the Christians are killing the Muslims and the Muslims are killing the Christians on a regular basis further north in Nigeria (according to the reports on CNN and in the New York Times).

We arrived in Port Harcourt (Figure 72) about 3:00 p.m. After checking into our hotel, The Presidential, Walter J. and I went looking for Dr. Obu, our potential future partner in the city, walking through the stacks of humanity and garbage. All the roads in the city were paralleled by a high stack of garbage that extended throughout their whole length. Amazing. I had never seen anything like that before. Some of the roads in the smaller towns were lined with garbage, all right, but they were nothing like this. We saw a young girl covered with flies pawing through one of the garbage piles.

We finally found Dr. Obu, a pretty chunky guy with a white wife. He mumbled so badly that we could hardly understand him (just like me!). I think the first thing he told us was that he had a small cash-flow problem; the government owed him one million naira. Wow! Now that's a real deficit. Therefore, he was noncommittal on whether JM and associates should come over in November. Anyway, I delivered to him a long communiqué from Maylo that outlined our plans for establishing RPI/Nigeria and locating its headquarters in Port Harcourt, which we eventually did.

This might be a good time to check out the photograph of JM and Maylo in Plate 9, which was taken during a conference in the northern city of Kaduna.

Back at the hotel, we observed a few birds, and then I did my usual 25 minutes of heavy exercise. Covered with sweat, I went to take a shower, but there was no water, as usual.

That night we had an excellent, spicy dinner (fried prawns and rice). During the meal, we laughed about the perils of driving through Nigeria, and then I played the "What did you learn today?" game with the troops, and it went great! No doubt the three Star beers each, which came in oversized bottles, helped.

I said, "Well Christi, what did you learn today, or yesterday either, if you so choose?"

Christi said, "I learned how the water (wah' tur) moves the sand."

Walter J., in an aside to me, said, "Not to worry. Salvation should still exist."

Ben said, also in an aside to me, "I can work with you, because we think alike. We want the same things" (i.e., we both hate garbage).

Iyoba said, "This is not a double room!" Then she concluded the meeting by saying, "I had sawdust for breakfast."

Nevertheless, as we split up, I thought, "I really like being a professor, that's for sure."

But I couldn't help wondering who is going to clean up all of this garbage.

Later that night, Ben came to my room and we talked for a long time. There we were, two bright boys born under questionable circumstances, who were awarded scholarships to attend college for free, so we could get our Ph.D.s and become university professors. But now, as we sat there pondering our future, we concluded that we both wanted to be POOR NO MORE. I had changed my mind on this subject dramatically since Hadley and I had that same conversation about 20 years earlier in Port Aransas, Texas (see Chapter 1 – THE HURRICANE).

(3 October 1981 – Saturday; Calabar)

The troops rallied at 7:00 a.m. sharp, and we were on the road by 8:30, it only taking 30 minutes to check out this time. After a long and bumpy ride, we finally made it to the shore about 12:30 p.m. No lunch again this day.

We hired a dugout canoe and rode it over to the east side of a relatively small inlet located about 20 miles to the west of the entrance to the large Cross River Estuary at the border between Nigeria and Cameroon (Figure 72). This inlet was a dead ringer for Price Inlet on the South Carolina coast, which is discussed in some detail in Chapter 10 – PRICE INLET. It also has a beautiful ebb-tidal delta very similar to the one at Price Inlet (Figures 55 and 57). One difference was that the beach sediment didn't contain an abundance of shells like it does at Price Inlet. When we were there, a well-formed tombolo was present in the lee of a large welding swash bar. I tried to make an intelligible lecture on the geomorphology, tidal current patterns, and wave dynamics of ebb-tidal deltas. I even drew the general ebb-tidal delta model (Figure 18) in the sand. This was a topic a little too advanced for the students at that stage, I'm sorry to say.

We took a walk and inspected the intertidal bars. I got some sun and went for a swim, all the time thinking, "This inlet would make an excellent topic for an M.S.

thesis. It should be studied on a seasonal basis, wet season versus dry season." This never happened, not by any of my students anyway.

At 3:30 p.m., we began another perilous drive, this time to the city of Calabar, a drive of about 40 miles. Calabar is located only about 25 miles west of the Cameroon border, so on this trip, we had covered most of the width of the country of Nigeria (see Figure 72).

As we went east, the scenery improved and, in the towns, the garbage had been cleaned up pretty well, a pleasant surprise for Ben and me. I liked that part of Nigeria!

In Calabar, we checked into what appeared to be a pretty nice hotel, the Metropolitan, where we had a good dinner. This minor feast was followed by a weak "What did you learn today?" during which, for some reason, only two Star beers each were consumed. After that, I went to my "hot as hell" room in which the air conditioner was broken.

(4 October 1981 – Sunday; Lagos)

I do not have any journal notes written on this day, but it would appear that we drove all the way back to Lagos from Calabar, a distance of at least 400 miles, if I read Google Earth properly. I guess the drive was so bad that I couldn't bring myself to write about it. Also, when we finally arrived in Lagos, I was probably feeling guilty, because, in my mind, I would have murdered Theodo at least 15 or 20 times during the drive. On the other hand, maybe "My friend, My friend" worked this time.

(5 October 1981 – Monday; Lagos)

At 9:00 a.m., Walter J. and I went to the Oceanographic Institute. We were supposed to take an airplane ride with our team of students to get an overview of the coast in the Lagos area. However, the flight was canceled for a variety of reasons, one of which was that the Flying Club hadn't paid its bills, so nobody would sell them any gas. We decided to try again the next day, but the flight never happened. I am telling you all this in order to give you one small example of the difficulties involved in conducting research in a recovering country like Nigeria.

We agreed to do some lectures the next day, but Walter J. and I were at a bit of a loss for something to do. We concluded that he should head back to South Carolina now that the field work was over, which he did that evening. Later in the morning, we visited a park by the lagoon, where I saw some more exotic birds, upping my bird list to 35.

In the late afternoon, Ben and I went over to NNPC to meet with Jerry Nwankwo, who was just back from a 3-week trip to Europe. He informed us that the baseline study had been approved by his management team. This convinced me, so I became a believer at last. The big project was truly about to start! After that, we walked along Bar Beach in a pretty good mood as we watched the sun set.

(6 October 1981 – Tuesday; Lagos)

I showed up at the Institute for my 8:30 a.m. lecture, but getting the room arranged properly took almost an hour. They had trouble understanding my particular English accent, which is not surprising. I had a lot of trouble understanding them sometimes.

I was a bit discouraged at first, because they couldn't answer some of my relatively simple questions. However, I thought that the bedform lecture and some of the beach morphology lecture went okay. The tidal inlet lecture was hopeless.

They went to lunch at 10:30 am, during which I guess gunpowder and sawdust were served.

Ben showed up dressed like a million bucks. He told me he doesn't want to be an administrator, and that our technology exchange was going great. Actually, I thought he would be an excellent administrator, a dynamic guy who cares.

(7 October 1981 – Wednesday; Lagos)

Only two of the students on the field trip, Austin and Victor, showed up for my 9:00 a.m. lecture. I talked about deltas and a beach inventory system we had started using in South Carolina. After concluding the planned lectures, I put on my shorts and walked/jogged out to the base of the eastern groin on Bar Beach. The Institute is located near there. Toward the end of the beach, some locals had occupied an old house at the end of a track. They had painted crosses all over the side of the building as well as on the large riprap nearby. In one place, they painted in large letters HOLY GROUNDS. There was also a hand-lettered sign with a lot of mumbo-jumbo on it, with the largest letters being–OFF YOUR SHOES.

This sign blocked my way to the end of the groin, where I wanted to go to check out the sea gulls. Also blocking my way was a white-clad "priest," who was engaged in earnest conversation with one of his disciples, I presumed.

I finally interrupted them to say, "May I walk through?"

"Yes, if you off your shoes," was the priest's reply.

I considered taking off my shoes.

"And your socks too," he said, anticipating my thoughts.

I was not about to walk through that shit in my bare feet, so I said, "I'll go around, over the rocks."

"Okay, but you might fall down," he laughed.

I didn't fall down, but I surely was pissed!

From the end of the groin, I looked back at the city and down the length of Bar Beach, thinking about how it might have originated. I also noticed a Japanese man fishing from the groin about half way down. That explained the cooler I had noticed on top of the riprap as I went by.

On the way back, I observed a local "chopping" away on the Japanese fisherman's food in the cooler (to chop = West African term for to eat). The fisherman could not see him, because he was on the other side of the riprap. I thought, "Everything is up for grabs in Nigeria."

Later that evening, I had dinner with Robert Lee at the Eko Hotel. During these visits to Nigeria, I had become friends with Robert, a native of Trinidad and an employee of the United Nations. Robert sort of took it upon himself to look after me a little, inasmuch as he had been in Lagos quite a while and knew all the ropes.

As soon as Robert and I had finished dinner, Ben came by to see me, but he was running late, arriving just as we were leaving for the airport. He apologized for not inviting me to his home for dinner, because it wasn't finished yet.

He said, "When Jacqui comes, we'll do it up right!"

I said, "Next time we'll do better. I enjoyed Nigeria."

My final bird list totaled 38.

Thank goodness, Robert Lee had volunteered to accompany me to the airport. We left the hotel dining area at 8:00 p.m. riding with a driver named Wahswai, who was pretty good at it. A good break from Theodo.

I was scheduled to fly to Rome on Alitalia. The first hassle was the fact that Alitalia had overbooked the flight, so my ticket had to be reconfirmed for the second time in three days. So poor Robert, with his UN passport and my ticket held high above his head, thrashed his way through an untold number of gyrating bodies up to counter to reconfirm the ticket. After that victory, it was time to get a boarding pass and check in my luggage. This was an even bigger rugby scrum than the first one. I just kept getting pushed further and further back from the ticket counter, with at least two or three dozen people between Robert and me. I could just barely see his hand holding his passport and my ticket waving above the crowd.

After a while, he yelled, "Hey Miles, throw me your bags."

So I gave a couple of mighty heaves, throwing my two bags over the heads of the multitude between Robert and me. He finally completed my check-in at 10:00 p.m.

Robert and I had a quick beer, Star beer, of course, and I said goodbye. Robert had been very kind. We got together every time we could when I went back. He was a great guy. One of a kind.

It took another hour to get through customs. I was beginning to hate the loutish, pushing and shoving, undisciplined ignoramuses.

They called my flight and then there was another flail, because Alitalia had not assigned seats. But, for five naira, you could get an assigned seat, as well as be escorted up to near the head of the melee. Unfortunately, I only had four naira left. Small panic; I thought I had lost my seat. But, a 20-dollar bill American solved that problem, and I rode to Rome in first class, as well. In those days, we flew in "toilet class," but that ended for me years ago.

The airplane I was on landed in Milan, instead of Rome, because of the fog. However, it soon cleared up and we arrived in Rome early enough for me to make my connection to New York.

On these two flights, I finished reading Jean Paul Sartre's *The Reprieve*, a complicated but explicit story of people preparing for war. A quote from this book follows:

"It isn't a question of numbers," said Mathieu. "Humanity replenishes itself, none is missed, and none awaited. Humanity will continue on its futile journey, the usual people will ask the usual questions and wreck their lives in the usual way."

GHANA

The Science

Compared to Nigeria, by the late 1900s, Ghana was a wonderful place to work, although conditions were pretty rough for several years before that. A history lesson about Ghana is not the goal of this chapter, but it appears that the modern era started on 15 May 1979, when a group of junior officers led by Flight Lieutenant Jerry J. Rawlings attempted a coup, and the leaders of this group were arrested and held for court martial. However, on June 4, 19 days after his arrest, some of Rawlings supporters in the military finished the job (Wikipedia Encyclopedia).

The story of the rule of Jerry Rawlings, who was born on 22 June 1947, is long and colorful. He ruled the country as a military dictator from 1981 to 1992, and then as the first elected President of the Fourth Republic from 1993 to 2001. He retired from political office in 2001. An important element of this story is that his mother, Victoria Agbotui, was a native of a coastal village called Keta. His father was a Scottish chemist.

During the 1980s, the country was in bad economic decline, but it had apparently recovered pretty well when we started working there in the late 1990s. As you will notice if you read Story One, we didn't have much fun in Ghana in July 1983, but all those problems aside, we kept on working in West Africa and, on April Fool's Day in 1996, I was off to Ghana again.

In Zurich, I met up with Rob Nairn, an engineer and modeler who works with Bill Baird at Baird & Associates (with headquarters in Canada), and Wayne Valley, a construction engineer and our primary contact for our client on this project, Great Lakes Dredge and Dock (with headquarters in Illinois). Our assignment was to investigate a shoreline erosion problem that had been going on since 1907.

This field study was the opening round of a major multi-million dollar project funded by the Export-Import Bank to save the town of Keta (Figure 74), which had lost its road and perhaps over half of its buildings, including residences for tens of thousands of people, to the "inexorable advance of the sea," as they say. A massive structure, called Fort Prinzenstein, a major holding center for human cargo during the days of the slave trade, had been cut exactly in half by the waves. A photograph of a couple of the eroded houses in Keta, taken in 1996, is given in Figure 75.

This project, managed in country by The Ministry of Works and Housing located in the capital, Accra, had four main objectives:

1) Sea defense. During the project, the berm crest along the study area was surveyed in January 2000. When this line was plotted on a couple of 1960 vertical aerial photographs, as illustrated in Figure 76, it showed that the shoreline eroded on an average of around 300 meters during those 40 years. As the aerial photos show, many houses were located within that

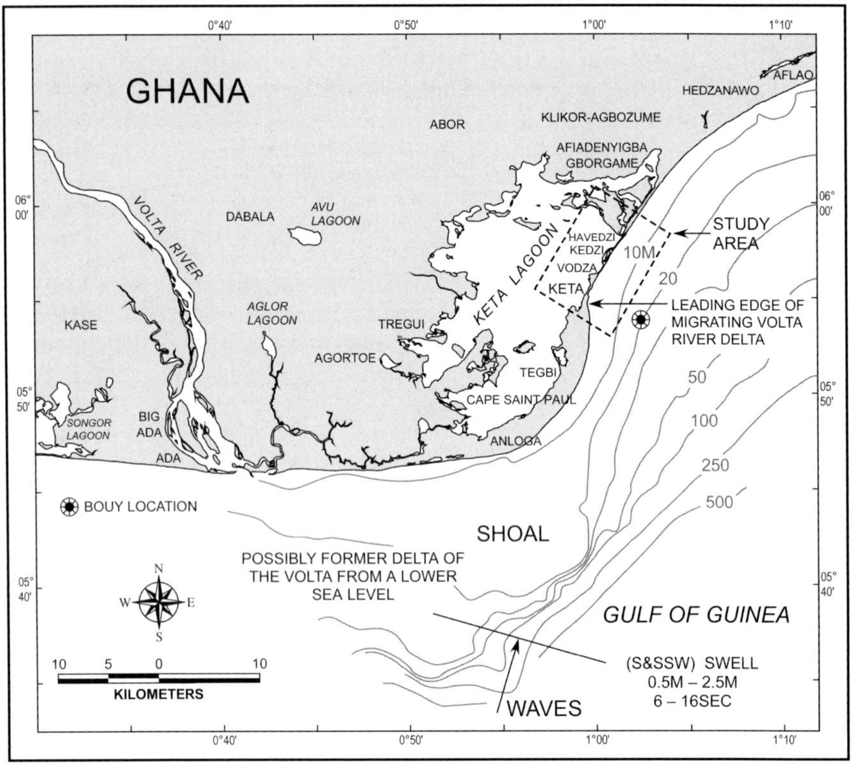

FIGURE 74. Map of the Keta/Keta Lagoon study site. Courtesy of Baird and Associates.

zone of lost land. Therefore, a major goal was to stop the ongoing erosion, thereby saving the town of Keta.

2) Road Link. In earlier years, a coastal road connected Keta with Lomé, Togo, a distance of only about 20 miles. With the old road gone, however, one had to drive all the way around Keta Lagoon to get to Togo, an added distance of about 40 miles. The goal, then, was to build up enough land behind the beach so that a permanent road could be re-established.

3) Flood Relief. During the wet season, at times the Keta Lagoon became so full of water that the town and the houses that remained on the narrow spit to the east of town were flooded. Raising the land behind the beach and establishing better control on the inlet into the lagoon would improve this situation.

4) Land Reclamation. A lot of people had lost their homes, so adding some land located safely above sea level would allow some of these homes to be replaced, as well as let the town expand. This would be accomplished by dredging sediment from the lagoon and adding new land behind the beach.

Now I will continue my description of my first trip to work on this project.

FIGURE 75. Eroding houses in Keta in 1996.

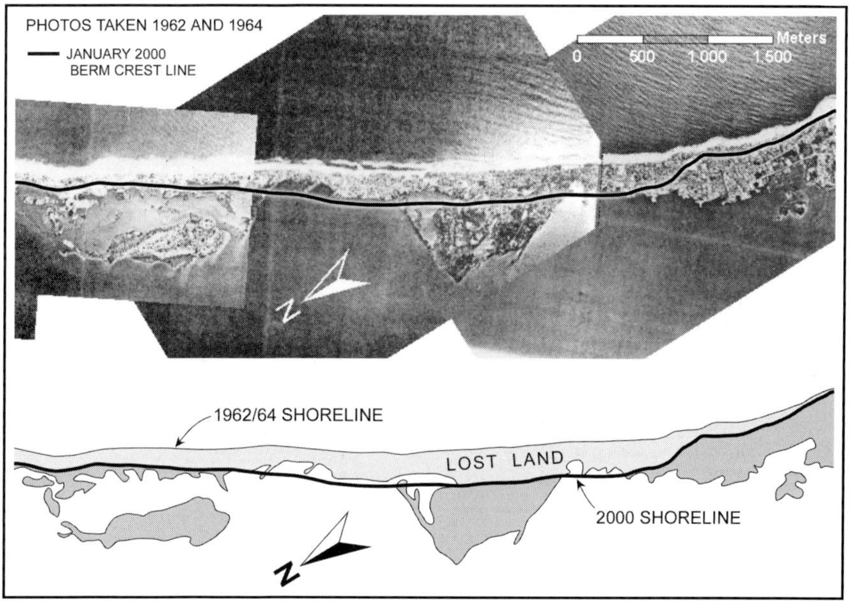

FIGURE 76. The berm crest line mapped by the Baird/Great Lakes team plotted on 1960 aerial photos showing the tremendous amount of loss of land and houses on the spit immediately to the northeast of Keta, Ghana (Figure 74).

Three nights with hardly any sleep and two rushed days of meetings and fool's errands later, we settled in at the Paradise Beach Hotel at Ada, at the mouth of the Volta River. After that, we embarked on four days of intense fieldwork, scheduled to coincide with the four-day Easter holiday.

"The more people you have on the field team, the less work you get done," is how I had voiced my philosophy on that subject. So, the field team consisted of only the three of us and a Ghanaian driver name Hope, which is what you needed an abundance of as you ricocheted along the pot-holed roads through the throngs of chickens, goats, pigs, and people. Hope was a good driver, though, and in the five total days we spent at the field site, he only killed one chicken and one goat, both on Easter Sunday, so they might well have been destined for the table anyway.

A very shallow, twelve-mile long, water body called Keta Lagoon separates the narrow beach zone from the mainland. We surmised that the lagoon is so shallow because at least once before a distributary of the Volta Delta emptied into it. The lagoon loses much of its water during the dry season.

As our research progressed, Rob determined that the dominant waves are long-period swell ranging from 0.5 to 2.5 meters in height with periods ranging from 6 to 16 seconds. Pretty big waves!! These waves are relatively unidirectional, approaching the coast from the south-southwest. At Togo/Benin, it had been determined that the longshore sediment transport rate was around 1,000,000 cubic meters per year (discussed earlier in this chapter). Two factors made for smaller longshore sediment transport rates in this area: a) Its somewhat sheltered location in the lee of the bulge of the Volta River Delta; and b) The presence of an old, submerged delta lobe offshore of the Volta River mouth (Figure 74). Rob's group determined that about 300,000 cubic meters per year of sand was making its way around the deltaic protuberance east of the river mouth to the main erosion zone, but in that area of greatest erosion, the rate was about 500,000 cubic meters per year. Therefore the erosion zone was losing 200,000 cubic meters per year (one football stadium full; Hayes and Michel, 2008).

My early conclusion was that probably the major long-term cause of this erosion, which had been a problem since 1907, was the usual shifting of the mouth of the Volta River, a large river over a thousand miles long. This is not an unusual occurrence for major river deltas, and the Volta is no exception. To add to this problem, however, a major dam, called the Akosombo, was built on the river in 1965 about 70 miles upstream from the coast, forming the largest man-made lake in the world (by surface area). It is over 200 miles long. This power dam supplies 80% of the electric power for the country (Barry et al., 2005). We could not determine exactly how much, but this loss of part of the natural sand supply (i.e., the sand deposited in the lake and not carried to the shore by the river) no doubt contributed to some extent to the major erosion rate that has occurred in more recent times (e.g., since 1960; see Figure 76).

During this quick trip, Rob and I were also getting some ideas on how to deal with the problem. It was clear that: a) More sand needed to be added along the beach in the erosion zone, which could only be done by a major beach nourishment project; and b) The rate the sand was leaving the erosion zone should be slowed down, with structures such as groins or offshore breakwaters.

FIGURE 77. Oblique view of the eroding spit east of Keta, Ghana during a low tide (looking northeast) that I took from President Rawlings helicopter during a low tide in 1996.

In July 1996, I was back in Ghana for a much longer stay this time. One of the first things we did during this visit was go on an overflight in President Rawling's personal helicopter. We flew the shoreline from the Volta River mouth to east of the erosion zone. Rob and I got the prime seats, and I was able to get some very good oblique aerials, such as the one shown in Figure 77.

We landed several times, and large crowds surrounded the helicopter hoping to see the President, only to be severely disappointed when two white guys got out of the plane. Actually, we were accompanied by three or four Ghanaians, as well, but no President, of course.

As we were flying over the Keta Lagoon, it occurred to me that although it was nearly full of mud at that time, earlier in its history, it had been a functional, fairly deep lagoon. As a result of it being in this microtidal setting (tidal range about one meter) with such big waves, there should have been a good-sized flood-tidal delta inside the lagoonal entrance during that earlier time before the lagoon became filled with mud. The old flood-tidal delta, which should contain a significant volume of sand, would now be buried in mud, of course. Because of the west to east longshore sediment transport direction, the inlet for the lagoon and its flood-tidal delta would be located at the eastern end of the lagoon. Therefore, that end of the Lagoon (on the beach side) would be the location to look for the buried sand.

There had already been considerable effort made to find a source of sand for the beach nourishment project, with one area being on the offshore shelf, an extremely dynamic and difficult area to work. If, indeed, they could obtain a significant volume of sand from inside the quiet lagoon, millions of dollars would be saved.

Therefore, I was feeling pretty good about this new theory when the helicopter landed on the tarmac at the airport in Accra. We landed about 50 yards shy of the hangar and got out. As we were exiting the plane, I looked toward the hanger and saw a lone individual approaching us. As he got closer, I noticed that all of the Ghanaians with us were in a subservient crouch.[26] Rob wasn't in such a crouch, but he did look a little awed. Yes, it was the President himself coming to greet us. Maybe I should point out again that Keta was a special place for President Rawlings, because it was the original residence of his mother and, presumably, many of his Ghanaian relatives on his mother's side.

Rawlings was a tall guy, about six feet four, who looked to be in very good physical condition. Very handsome. He walked right up to the group and, as I remember, there was no handshake with anyone. I guess because I was in the middle of the group and standing upright, he walked directly up to me, stopped only a few feet away, looked straight at me, and asked, "Can you solve the erosion problem at Keta?"

Without any hesitation, I said, "Yes. We know why the beach is eroding, and we know how to deal with it. It will just take some time."

He nodded and said, "Good."

Then he turned to one of the head men in the Ministry, who was on the flight with us, and said, "Well, in that case, it is time to take somebody's (he gave the name of one of the local engineers who had been working on the project for years) plate right off the table. He has been eating at our expense too long."

What else could the Ministry man say, but "Yes sir." I did, however, upon further reflection, wonder what actually happened to that poor guy, who was now out of work.

I guess it was a good thing I didn't give a negative answer, but I was feeling pretty confident anyway, with the flood-tidal delta solution to the sand source problem in mind. He was the first President of a country I had met personally, and he was really an impressive man. Rob and I both were a little amazed at his forthrightness.

We couldn't just take for granted my guess that an old flood-tidal delta in the lagoon would provide us with near the volume of sand we would need, so we were awarded a small project to take vibracores of the sediments in that part of Keta Lagoon. This project was done fairly soon thereafter by Mike Bise, Todd Montello, and couple of terrific Ghanaian field assistants. I helped Mike and Todd describe the cores, which showed that sand was there, as expected.

Another thing I noticed during the ride in President Rawlings's helicopter was that the spit to the northeast of the town of Keta was a classic landward-migrating (transgressive) spit or barrier island (illustrated in Figure 77). The coring project showed the details of the internal stratigraphy of the spit, which is illustrated in Figure 78.

After completion of the coring project and submitting our results in a report

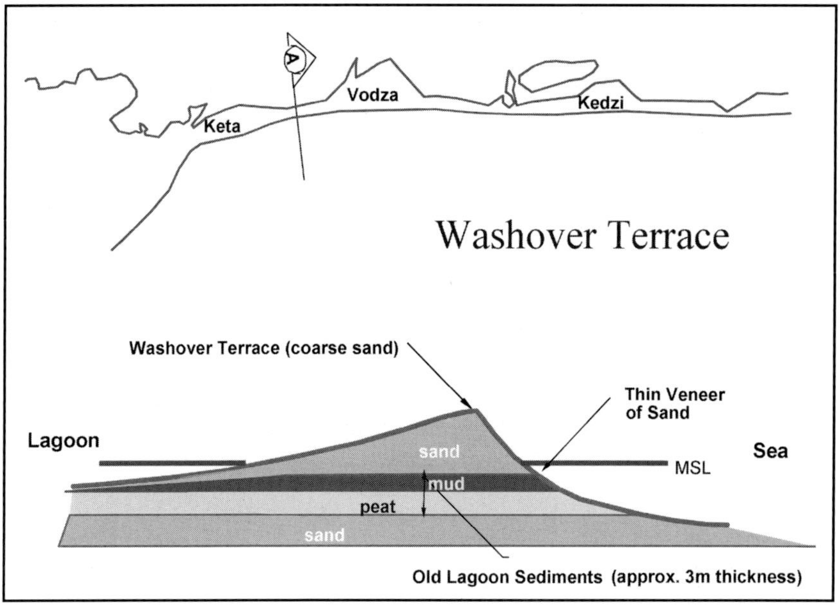

FIGURE 78. Stratigraphic and geomorphic cross-section of the spit east of Keta, a classic migrating washover terrace. This spit is shown in the photograph in Figure 77. Courtesy of Baird and Associates.

to Rob, my work was pretty much over for this project. One of the final products, which are shown in Figure 79, included building several groins between which the nourished sand was placed. Dredging in the lagoon also produced some sediment for a major landfill project behind the beach, which now has a number of houses on it. Also, RPI, under the direction of Linos Cotsapas, carried out a five-year environmental monitoring program for the project, with the help of a couple of hardworking Ghanaian biologists.

Story One - GHANA

(26 July 1983; Tuesday - Continental Hotel, Accra, Ghana)

It was 9:00 p.m. and I was sitting in my room in the dark, in the broken nation of Ghana. Similar to the lights, the air conditioning didn't work either, of course, so it was very hot and stuffy in the room. We had just arrived from the airport after flying in from Lagos, and I was finally beginning to calm down a little after a very disappointing day in Nigeria, waiting for a contract for a job we had spent a small fortune getting off the ground, only to learn that we would get paid "any time from now." I kept saying over and over to myself, "I will never do business in West Africa again."

John Horne, my senior partner in RPI/Colorado, had arrived to take the lead in selling a major geological study to the Geological Survey of Ghana we had heard about. Maylo Murday, who was then the head man in our office in Nigeria, was also there to help out with the logistics. In preparation for the border crossing into Ghana, Maylo had warned me that all American money that went in had to also

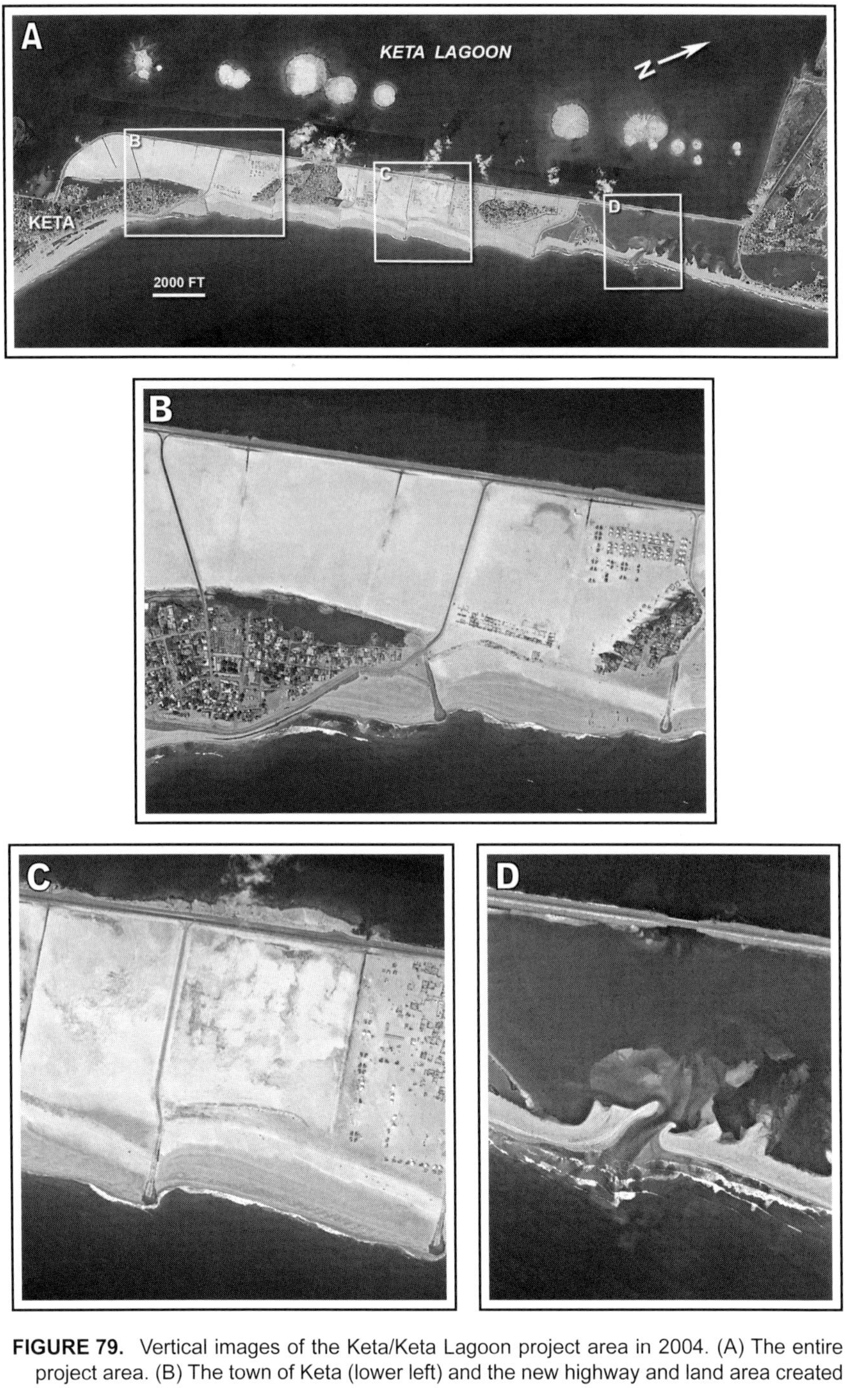

FIGURE 79. Vertical images of the Keta/Keta Lagoon project area in 2004. (A) The entire project area. (B) The town of Keta (lower left) and the new highway and land area created along side of the lagoon. (C) Two of the new groins. Note the nourished sand beaches. (D) The active tidal inlet on the far east side of the lagoon. QuickBird image © 2004 DigitalGlobe, Inc. All rights reserved.

go back out. We were carrying some Nigerian money with us to spend during our brief stay. Therefore, I confidently wrote $700 U.S. on the form the customs agent gave me to fill out.

I slept with the covers pulled up around my chin, because there were mosquitoes in the room. There were several large blotches of blood on the wall marking the spots where various mosquitoes had met their demise. The air conditioning finally started working around midnight.

Breakfast was very meager in our 5-star hotel and restaurant. There was no bread in Ghana, but we ate better than 99.9% of the rest of the people there. JM always made a little care package for me to take on such trips which, in this case, consisted of several granola bars and a bag of nuts. Refraining from indulging myself with those favors, I guiltily gave them to the maid in the hotel before we left.

On the way to the Survey, our taxi driver pointed out the lines of cars that had waited one week for 5 gallons of petrol. A long drought and abandonment by English interests after the revolution had just about put that country out of business.

The driver asked, "The United States, is it a nice place?"

"We have no complaints. It's been very good to us," I replied.

After spending a day at the Survey, John and I concluded that we had better peddle our business elsewhere.

On the way to the airport late that afternoon to catch a plane back to Lagos, Maylo said, "Now count your dollars to make sure you have the same as what you came in with."

"I have $700 and you have paid for everything here," I assured him.

"Well, if you don't mind, count it again."

I did, and guess what, I only had $625. I had forgotten that during a brief visit with my daughter, Joy, in New York on the way over, I had given her the missing $75 for spending money.

"Oh boy, this is not good," Maylo lamented, "We may be detained for questioning."

By some miracle, the customs officer that checked me through did not miss the wayward $75 dollars, but that was just the beginning of the fun. As we were sitting in the gate area, suddenly everyone, except we three, ran like mad down to the ramp. I couldn't understand why, because there was only one plane at the airport, and it was a West African Airways plane, not the Nigerian Airways flight we were waiting for. But, soon our plane landed and everyone ran out onto the tarmac, pushing and shoving so that they could be first in line to get on the always overbooked flight. John, a brawny ex-rugby player, was doing okay in the scrum at the loading ramp, but Maylo and I were being pushed slowly to the back of the line. Luckily, we were able to get the very last two seats left on the plane when we finally pushed our way inside the door. During the flight from Accra to Lagos, I was reading Charles Darwin's *Origin of the Species*, thinking that somehow the process of natural selection had passed this place by.

Because of some canceled flights, Maylo and I would have to spend another night in Lagos, but I had no entry visa for my overnight there. At the counter, the customs agent, who was wearing a military uniform, listened impatiently to our sad story. He whispered aside to me, "You have to understand the system, a tip is required."

Finally getting the picture, I rummaged through my briefcase for some cash. Ten U.S. dollars bought me a four-day visa in Nigeria.

Story Two - GHANA

(Keta, Ghana – July 1996)

I was in Ghana working on the Keta erosion project in July 1996. Wayne Valley, manager of the project at that time for Great Lakes Dredge and Dock, was also there, worrying about the logistical details concerning bringing some of his dredging equipment into the Keta area. Thus, he decided to take a quick trip to check out the harbor in Lomé, Togo, and invited Rob and I to go along to view the phenomenal sand deposition on the beach on the upcurrent side of the jetties at Lomé, which is discussed in the section in this chapter on Togo/Benin. We went in one of the project's vehicles driven by our redoubtable driver, Hope. During the drive to Lomé, as we were in the unpopulated section that separates the two countries, we saw some road signs warning us that a small bridge ahead was out. As we drove off the road and down by the bridge across the dry stream bed, we observed that the driver of a car that had preceded us had apparently ignored the sign and had plunged into the gap where the bridge used to be, with the car still left there pinned up in the air at a 45 degree angle between the two bridge abutments. That sobered us up for a few minutes.

"Hope, let's keep this spot in mind as we head back this way," I suggested.

At the Togo border, Wayne–to say he was a little tight with his available funds would be a major understatement–was quibbling with the customs agent. We didn't have a visa, but Hope had assured us that the border was open to Ghanaians and, presumably, Ghanaian drivers like him. This debate between Wayne and the customs agent lasted for a long time.

"For God's sake, Wayne, dash (West African term for bribe) him ten bucks so we can get on about our business," I suggested with Rob's approval. "You know the border closes at six, and I want to spend the night at Paradise Beach, not in some cheap hotel of your choosing over here in Togo."

I slipped a ten-spot American to Hope, and in the local dialect, he passed the word on to the agent, who let us through. I assume Hope went to the border later, as we looked around the port, and paid the agent.

Despite our urging, Wayne was late getting back to the car and we arrived at the border after it had closed. We barely managed to dash our way back across the border into Ghana, but the price this time was considerably higher than before. After we had been driving in the dark for twenty or thirty minutes along that remote section of the road, we all three called out almost simultaneously, "Hope, don't forget the hole!" referring to the missing bridge.

14 THE ROCK RECORD

ODE TO THE GEOLOGIST

And some ryn up hill and down
dale
napping the chucky stanes
to pieces wi' hammers,
like sae many roadmakers
run daft.
They say it is to see
how the world was made.

ALFRED, LORD TENNYSON

The present is the key to the past.

JAMES HUTTON

The past is the present to the key.

ALAN J. SCOTT

The Science

By 1982, with the assistance of a number of graduate students and RPI employees, I was leading as many as eight seminars a year on the South Carolina coast aimed at helping oil-company geologists interpret potential oil-producing rock units by using the coastal sedimentary deposits of South Carolina as modern analogs for those rocks. Such an understanding was aimed at improving their batting average in predicting which of the sandstones buried thousands of feet below the Earth's surface that they were exploring would contain economically viable oil and gas reserves. These courses started in September 1976, the first one being sponsored by the American Association of Petroleum Geologists (AAPG). Eventually totaling about 150, these field courses (many sponsored by private companies, as well as the AAPG) were carried out for 30 years. Probably around 4,000 geologists attended these courses, from such diverse areas as Western Europe, the Middle East, Japan, a large number from Louisiana (under sponsorship of Schlumberger Well Services), the Appalachian region, and even a few from Texas. However, probably nearly half of the students worked with oil companies located in Calgary, Alberta. Apparently, a significant number of the Cretaceous formations in western Canada were deposited along shorelines that closely resembled the mesotidal, depositional coast of South Carolina. At least, the South Carolina coast was a better model for those particular rocks than other areas where field seminars were taught, such as on the Mississippi River Delta.

As the years passed by, I heard several success stories from the participants based on using the South Carolina coast as an exploration model, so the field course remained popular for three decades. I finally decided the seminar had run its course, for me anyway, and led my last one in September 2005. Walter J. Sexton, my partner on these courses for many years, continued this tradition, running them through his company, Athena Technologies, Inc.

The courses included visits to a sandy point bar on the Congaree River a few miles south of Columbia and a full-day inspection of the Santee/Pee Dee River Delta. One of the special features of the course was an overflight of the coast between the Santee/Pee Dee Delta and St. Helena Sound (see map in Figure 28). This typically involved eight or nine fixed-wing (high-wing) Cessna aircraft that usually flew in single file at around a thousand feet altitude. In the early days, several of the planes were piloted by my graduate students.

Such a complex exercise required a course coordinator, a position first held by Tim Kana, the co-author of the course syllabus published in 1976 (titled *Terrigenous Clastic Depositional Environments*). Others included Aileen Duc, Leslie Sautter, Patrice Cunningham, Patty Tate, Becky Cox, Wendy Early, and Christy Suggs, all serving admirably and keeping things running on time. Why so many? Remember, we did this around five times per year for 30 years.

With respect to part of the coastal zone exclusive of the fluvial and deltaic components, there were three basic habitats that received the most attention on the field trip, as well as the most interest on the part of the participants.

1) <u>Morphology and Stratigraphy of the Barrier Islands</u>

There are two primary types of barrier islands on the South Carolina coast, landward migrating and prograding, which are illustrated in Figure 80. As this diagram shows, landward-migrating (transgressive) barrier islands are composed of coalescing washover fans, or a washover terrace, that is overtopped at high spring tides, usually several times a year. In the process of migration, the entire washover terrace complex moves landward, leaving an eroded nearshore zone in its wake (see example in Plate 10). As a result of this type of migration, in three dimensions the entire complex consists of a relatively thin (<6-12 feet) wedge of sand and shell of the washover terrace that overlies muddy sediment originally deposited in the estuaries or wetlands behind the islands (see cross-section in upper half of Figure 80). The islands on the outer margin of Cape Romain, the subject of a dissertation study by Chris Ruby (1981), are of this type. The landward-migrating barrier islands in South Carolina are relatively short, 1-5 miles on the average, because new inlets are created where the landward-migrating islands intersect large tidal channels,

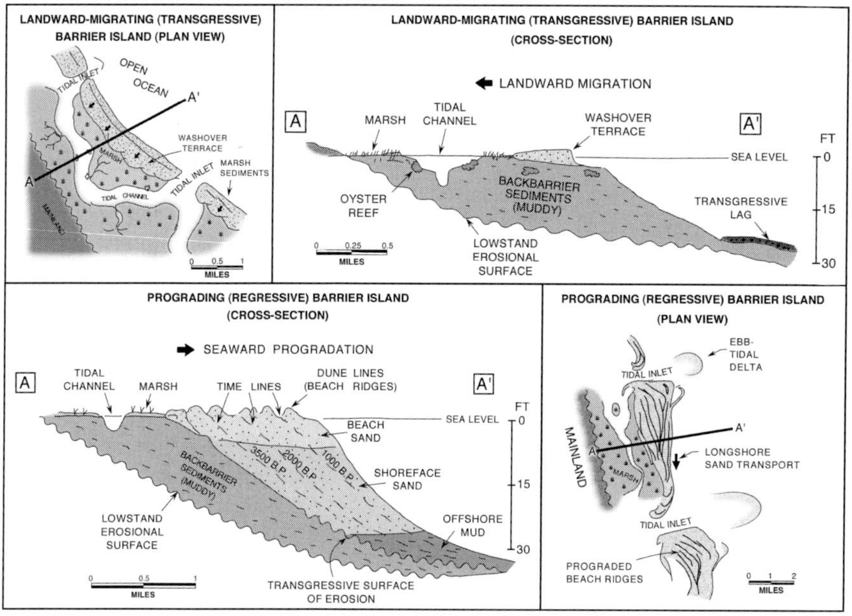

FIGURE 80. Morphology and subsurface three-dimensional configuration (stratigraphy) of prograding (regressive) barrier islands in the bottom row; and landward migrating (transgressive) barrier islands in the top row. The cross-sections (A-A') are located in each of the plan view maps. From Hayes and Michel (2008).

Prograding (regressive) barrier islands (lower diagram in Figure 80) are typically composed of multiple, relatively parallel linear ridges of sand topped by vegetated sand dunes that originally formed as front-line dunes (foredunes) on the backbeach. The most notable changes on these types of islands occur where adjacent tidal inlets migrate into them or when the inlets expand dramatically during hurricane storm surges. As a result of this type of seaward growth, in three dimensions they consist of a wedge of sand 25-30 feet thick that has built seaward over offshore mud (see cross-section in the lower half of Figure 80 and Tom Moslow's stratigraphic section of Kiawah Island in Figure 30). Most of the major developed barrier islands in the state, which commonly are greater than 10 miles long, are of this type (e.g., Isle of Palms, Kiawah Island, and Hilton Head Island). The morphology of the prograding barrier islands longer than about 6 miles takes on a characteristic drumstick appearance, as shown in Figure 29. Stratigraphic studies of barrier islands carried out in the area visited during these field seminars were conducted on Kiawah Island in Tom Moslow's dissertation study, on Cape Romain in Chris Ruby's dissertation study, and on Capers Island by Bo Tye's thesis project (see Appendix A for references). I should point out that a number of cores through the barrier islands, as well as other habitats, were taken as part of the field seminars, and core labs during which the students examined these cores were also held during some of them.

2) Tidal Inlet Deposits

One of perhaps the most surprising discoveries by the numerous thesis and dissertation projects was the overwhelming dominance of tidal inlet deposition along the central and southern South Carolina coast, primarily in the ebb-tidal delta systems, for reasons discussed in Chapter 10 – PRICE INLET. This was determined by a detailed study of Holocene sand volumes between Bull Bay and Fripp Island, which was summarized by Hayes, Sexton, and Sippel (1984), Sexton (1987), Hayes and Sexton (1989), and Sexton and Hayes (1996). This survey, which was based on data derived from several hundred vibracores, plus detailed analysis of bathymetric data, showed that 77% of the clean, Holocene sand (<15% mud) in the area was stored in ebb-tidal deltas. The relative abundance of shoreline sand deposits in the different depositional environments is shown by the bar graph in Figure 81.

It is probably worth elaborating a little further on the fact that on mesotidal, more tidally influenced coasts like South Carolina's, ebb-tidal deltas tend to be much larger than flood-tidal deltas (Hayes, 1991). In fact, flood-tidal deltas are almost non-existent on both the South Carolina and Georgia coasts. This strange occurrence has been studied in some detail, as is pointed out with respect to Duncan FitzGerald's dissertation project in Chapter 10 – PRICE INLET. In the simplest terms, because of the complex nature of the back-barrier regions of the barrier islands in South Carolina (extensive marshes and tidal flats cut by numerous tidal channels), water "running uphill" into these areas during the flood portion of the tidal cycle takes longer to fill the backbarrier region with water at high tide than it takes that region to empty when the tide falls. This means that the flood phase is considerably longer than the ebb phase. With essentially the same amount of water going in as comes out, much stronger currents are generated during the

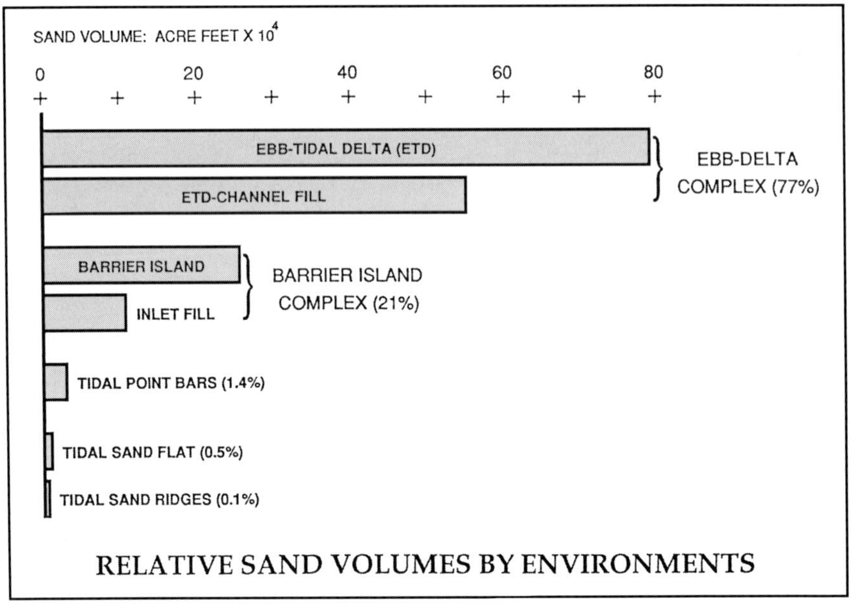

FIGURE 81. Relative volumes of clean sand occurring in the Holocene depositional environments of the central South Carolina coast. Note the dominance of tidal-inlet deposits. From Hayes and Sexton (1989).

falling tide. These strong ebb currents are able to sweep sand brought into the inlet by the wave-generated and flood-tidal currents out of the inlet and deposit it on the ebb-tidal delta.

These findings on the tidal inlets in South Carolina were of keen interest to the geologists doing oil exploration in the Cretaceous units of western Canada, and several of them informed us that they had drilled a number of successful wells that produced from tidal-inlet deposits of this type. It was common in those days to try to find the ancient shoreline trends during their subsurface exploration, so wells could be located along that trend, producing from the long axes of barrier islands, for example. A key point we made was that tidal inlet deposits are commonly oriented perpendicular to the shoreline trend, giving the explorationists a different direction in which to drill (i.e., perpendicular to the ancient shoreline trend) should they believe they had discovered a tidal inlet in the barrier island system they were drilling into.

3) Beach Morphology and Structures

During these training courses, the guys assisting with the course dug hundreds of long and deep trenches across the beaches and tidal flats. Mike Bise, who did this for at least two decades, generally in a supervisory capacity, although he probably shoveled more sand than anybody else, should be especially acknowledged for his hard work. Description of these trenches allowed us to create the general cross-sections presented in Figure 82, which have been published in a number of articles and books (e.g., Hayes, 1994; Hayes and Michel, 2008).

4) Role of Tides; Shoreline Embayments

The spring tidal range in South Carolina varies from about 5.5 feet at the North Carolina border to 8.2 feet at the Georgia border. This places much of the South Carolina coastline in the mesotidal class, for which we developed a depositional model, emphasizing tidal inlets (Hayes and Kana, 1976; see Figure 2B).

Whether the average tidal range in an area is large or small is dependent upon several factors, with the largest tides occurring where: 1) The water body has on open connection with the world ocean; 2) The continental shelf is wide; and, most importantly, 3) Along shoreline embayments (Hayes, 2005b).

Other factors, such as the hydrodynamic forcing of oceanic currents, position within an amphidromic system, and presence of resonant standing oscillation play a role in some areas, but none of these are as relevant as the first three factors with respect to the tides on the South Carolina coast. The largest tidal range along the southeastern U.S. coastline occurs at the head of the Georgia Bight, which is why the tides are so large at the South Carolina/Georgia border. With the exception of the Bay of Fundy, an area where resonance plays a role, most, if not all, of the larger tidal ranges around the world occur at the head of shoreline embayments such as the Georgia Bight. Komar (1976) gave the following verbal explanation for this: "As the tidal front approaches a narrowing indention of the coastline,...the enveloping shores constrict its movement and wedge the water together."

In Figure 83, the tides and wave conditions of the Georgia Bight are compared with two other shoreline embayments, the German Bight and the west coast of Florida (from Hayes, 1979).

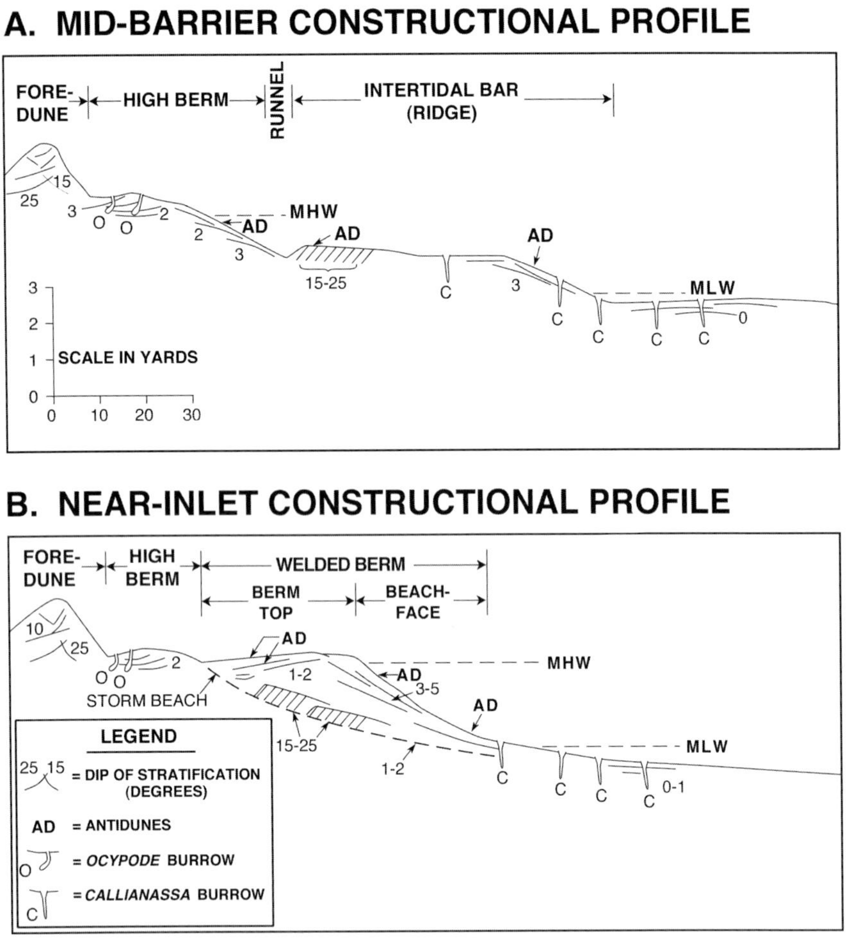

FIGURE 82. Representative sedimentary structures occurring on the beaches of the prograding, mixed-energy barrier islands of South Carolina and Georgia. (A) Structures found on a typical mid-barrier beach in a constructional mode. Modified slightly after Hoyt and Weimer (1963). (B) Structures found on beaches near tidal inlets in a late constructional mode. These two diagrams are based on inspection of hundreds of trenches dug on the beaches of the South Carolina coast. *Ocypode* = ghost crab; *Callianassa* = ghost shrimp. From Hayes and Michel (2008).

One of the topics I repeatedly championed during the short courses was the variation of depositional systems around shoreline embayments, as is illustrated in Figure 84. John Horne and I presented this idea at a national AAPG meeting, but it ended up being one of those papers that should have been published but wasn't.

All this work with oil-and-gas company explorationists caused us to think about getting into that business ourselves. On 18 January 1980, John Horne, who was contemplating leaving the academic world (he was a research scientist in John Ferm's coal program at USC-East) and moving to Colorado to concentrate on business activities in the energy sector, came by the motor cycle shop (i.e., the original RPI office) to talk a deal. He said we could make a lot of money working

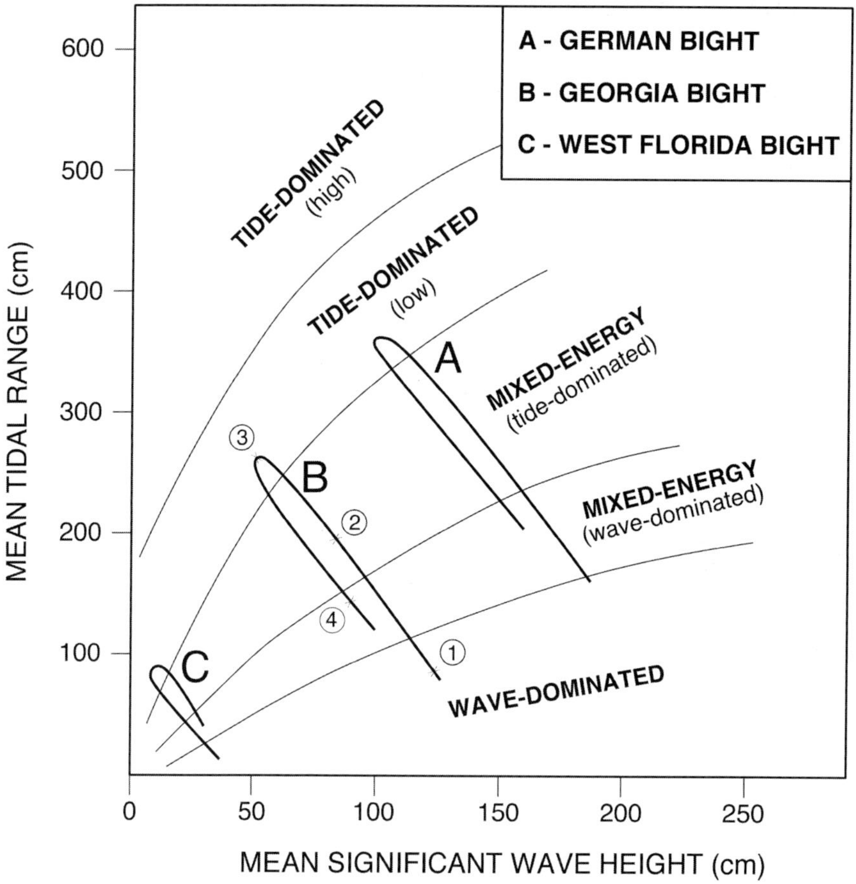

FIGURE 83. Comparison of the wave and tidal regimes along the shoreline of the Georgia Bight (curve B) with that of the German Bight (curve A) and the embayment of the western coast of Florida (curve C). The stations located along the Georgia Bight curve represent: 1) The wave-dominated barrier islands of the Outer Banks of North Carolina; 2) The mixed-energy barrier islands and tidally influenced inlets near Charleston, South Carolina; 3) The tide-dominated estuarine complexes at the South Carolina/Georgia border; and 4) The mixed-energy barrier islands and wave-influenced inlets in the vicinity of Matansas Inlet, Florida. After Hayes (1979).

jointly with certain coal and oil interests that we both were acquainted with. We agreed to think it over.

During one of the AAPG field courses that same year, John and I convinced some managers from Dome Petroleum in Calgary, Alberta that we could help them with their rock interpretations, thus a new arm of the company was born. This eventually led to the formation of RPI/Colorado, RPI/Canada, RPI/Texas, and RPI/Pacific. John Horne remained a partner in all these operations, which finally came to a close in the early 1990s, but that is a story, except for the short account given in footnote number 21 and the following NOTE, that I will not elaborate upon here.

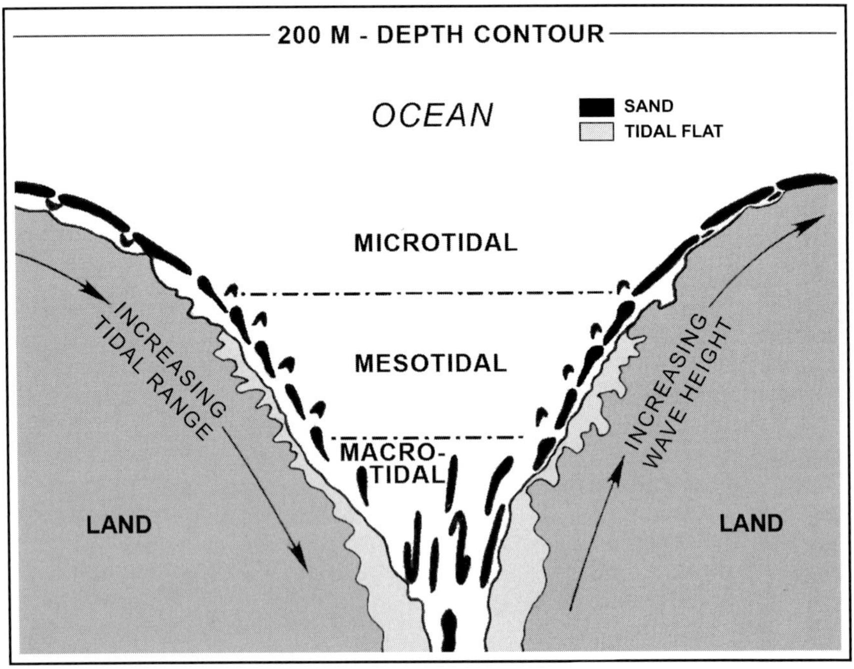

FIGURE 84. General model of shoreline depositional systems within a shoreline embayment on a coastal plain coast without a major river delta system. These systems include: a) Long, wave-dominated barrier islands along the outer shores; b) Mixed-energy barrier islands with large ebb-tidal deltas along the flanks; and c) Long, linear tidal sand ridges at the head of the embayment.

[NOTE: On 4 December 1981, we had an open house for the new Boulder office, home of RPI/Colorado. I wrote the following in my journal about that momentous event: "Probably 50 or 60 people showed up, mostly Tom Ryer's friends and groupies, fellow Buddhists, coal brothers, and local merchants."

Tom Ryer, a Yale graduate who had just left the U.S. Geological Survey and was an expert on the oil-bearing rocks of the Rocky Mountain region, and John Horne, who, as noted earlier, had suggested that we start such a business in the first place and was a world-renowned coal geologist, were the two senior scientists in the Boulder office, which was managed by Dan Domeracki. Several of the employees, most notably Valerie Williams, who eventually became Marketing Director, were practicing Buddhists.

I spent that weekend in Boulder so John and I could go to Calgary, Alberta Sunday night on another of our many marketing trips. On Saturday, I hiked up to the base of the flat irons above Boulder and watched some birds.

While walking around in the town on my way to the hiking trail, I thought: "Everybody smiles at you in Boulder. Everybody is happy in Boulder. Boulder just isn't normal; it's a strange Brave New World."]

In any event, at the beginning of all this in 1980, John Horne thought that if I were going to go back to working on rocks, I should go see some, namely on a field trip that he was leading to eastern Kentucky. So, about noon on 6 April 1980, JM and I left town in our "new" used white Mercedes, which my accountant had found for us at the tidy sum of $10K, and drove to Caryville, Tennessee,[27] where we slept in an offbeat motel in the shadow of a mountain. In order to get to Caryville, we took a haunting trip through the Smoky Mountains. Driving along in the dusk through the hills of East Tennessee brought back memories of an earlier time, one with fewer cares, when the mysteries of the natural world were less understood, but better appreciated by a younger soul not yet jaded by the motivation to become The King of Marvin Gardens.[28]

After looking at some rocks in East Tennessee the next day, we drove to Berea College, where I showed JM the Science Building and the baseball diamond built by the hands of Monarchy Wyatt, my old baseball coach, himself. I didn't see the baseball team, but the rocks that I saw and photographed on that field trip led by John Horne were featured in some of the lectures that I did during the many field seminars we conducted on the coast of South Carolina, the primary subject of this chapter.

Story One

In the first few years of the field course, we started at the coast and finished the trip in Columbia. The last night we were together, we had a big party at our house in downtown Columbia. The course ended the next day with a half-day session on the Congaree River area so the participants could catch a plane back to Calgary, or wherever, in the afternoon, with more than a few of them suffering from a bad hangover.

[NOTE: In February 1978, JM and I bought an elegant English Tudor style house built in 1910 and located on Gregg Street, about three blocks from the University. I became interested in interior design and we completed the job the previous owner had begun in creating a handsome residence, decorated with photographs of the coastlines we had studied all over the world, that served as an excellent site for parties, class-discussion sessions, and business meetings. We moved from Gregg Street in December 1985 to a new house we had built on a large tract of land located on the west bluff of the Congaree River Valley about 20 miles south of town. Even today, however, JM and I often refer to our stay on Gregg Street as "our eight greatest years."]

One of the main attractions after dinner at that last night's party, during which grilled delicacies of various kinds were served, we gave out "awards" to the participants. A few examples of these awards included:

1) Miles O. Hayes Geohorny Award – This was the final and most hotly contested award.

2) Rip Van Winkle Award – For the participant who slept the most during the lectures. Also hotly contested.

3) Worst Southern Accent Award – Vied for by almost everybody on the course, except the guys from Louisiana.

4) Can't Leave Home Without It Award – For the guys who brought their wives along. I don't remember any woman, and there were always several on the course, ever winning this award.

And I wouldn't want to leave out the loud music and wild dancing that usually followed the awards ceremony. I remember one occasion when the entire group joined hands, formed a continuous line, and circulated in a loop throughout the downstairs rooms of the house to the tune – UP AGAINST THE WALL, YOU REDNECK MOTHER. The house was rocking that night.

Story Two

A geological consulting firm called McCallum and Stewart of Calgary sent most of their employees on our coastal field seminar during the early years. They thought that the Cretaceous sediments they were working on, the Athabasca Tar Sands of Northern Canada, were formed in tidal estuaries similar to St. Helena Sound, South Carolina when the Cretaceous Seaway covered that country around 70 million years ago. Those tar sands, which occur just beneath the earth's surface, contain some of the largest petroleum reserves in the world. In the early 1980s, one company, Syncrude, exported about 200,000 barrels of crude a day from their plant, which distilled the oil from the sand and sent it south via pipeline. According to data supplied by Syncrude, a consortium owned by eight private companies and two government entities, there are 300 billion barrels of recoverable bitumen[29] left in Alberta's oil sands using today's mining methods. This total amount is at least nearly as much as the known reserves in Saudi Arabia.

Gordon Stewart, one of the senior partners in the consulting firm, invited me to visit the tar sands with him and two other geologists. Thus on Thursday morning, 10 July 1980, I woke up in a modern hotel in Ft. McMurray, Alberta, which is located 440 kilometers northeast of Edmonton in "aboriginal" country. That day we flew in a helicopter out to the Athabasca River, which cuts a canyon through the tar sands, and the pilot landed it on a gravel bar in the river near one of the steeply sloping canyon walls. We debarked from the chopper and climbed up the precarious slope, with Gordon, who was in his early fifties, making like a mountain goat. I experienced vertigo a couple of times when I looked down the crumbling slope 30 meters to the surging standing waves in the muddy river. The sediments we were inspecting had been deposited in a riverbed similar to the one I was trying to avoid falling into.

After that perilous climb, we returned to the helicopter and flew over the strip mines themselves, where huge red machines, which cost $200M each, were gouging enormous holes in the earth–black gashes hundreds of meters long. Monster lakes had been built where tailings from the mines were dumped, creating massive "oil spills" on the surfaces of the artificial ponds. I had two different and distinct emotions as we circled that colossal wonder of the world there in the far-north wilderness: (1) The eternal, however muffled, cry of the naturalist, thinking of the huge scars that would result and the desolation of the landscape; and (2) A feeling of pride that man can master his planet, like the Fremen producing water for Dune.[30] It was a look into the future, another modern rush for the black gold.

During a return visit to the Syncrude mine to teach a two-day seminar in May 1996, my host proudly showed me a reclamation site of one of the old strip mines, where a herd of wood bison now roam. There were 48 bison in the herd at the end of 1994, a year when 212 hectares of land were reclaimed.[31] This restoration project was sponsored jointly by Syncrude and the Fort McKay First Nation of the native peoples.

After circling the mine for a while in the helicopter, we left there and landed in a small river canyon, which was lovely, except for the stench of the oil. We defied gravity again as we climbed the slopes to look at oil-saturated sands that we all concluded had been formed as sand bars along the bends of tidal creeks much like those found in the tidal estuaries of South Carolina and Georgia, except maybe the tidal range was a little larger. This interpretation is important, because the mining process requires careful mapping and interpretation of the sand deposits so as to predict wall failure–wouldn't want that $200M machine to fall into a hole–and to make accurate estimates of the amount of productive source rock expected to occur in the next swath cut across the mine wall. A team of geologists mapped the wall of the Syncrude mine each day, one of the most carefully studied sediment bodies anywhere in the world.

Story Three

In early October 1980, JM and I bought an historic ferry landing called Wide Awake, located on the Intracoastal Waterway south of Charleston. Among other noteworthy historic events, it was the site of a battle during the First Civil War. The house, which had been remodeled when we bought it, was originally built in 1790. It sat on a low bluff by the waterway overlooking several thousand acres of salt marsh across the main channel. The property contained seven acres on the waterway itself, including four acres of lawn with marvelous flowering plants, such as azaleas and camellias. Wide Awake became our weekend escape during the hectic business-building years of 1980-1983. It also became the place where the final night party for the field seminar was held for a year or two.

[NOTE: A year and a half later, in April 1982, we had a family gathering at Wide Awake. My mother and father drove to Columbia from the mountains and rode with us to the coast, and my daughters, Joy and Mya, flew into Charleston from Massachusetts. My mother wanted to see the old Cooper River Bridge in Charleston one last time, as she put it, so we drove over it before going on out to Wide Awake.

When my dad saw the house by the waterway, he said, "My goodness, son, my whole house would fit inside this living room."

It was a beautiful springtime reunion in one of the garden spots of the world.

In his book *Future Shock*, Alvin Toffler said (on page 379): "The problem is not, therefore, to suppress change, which cannot be done, but to manage it. If we opt for rapid change in certain sectors of life, we can consciously attempt to build stability zones elsewhere."

I do not believe we did it consciously to escape future shock, but we did retreat momentarily from time to time to our 1790s house at the coast with it's huge live oak trees in an 18th century garden. Was that a buffer against rocketing through the skies to the high-rise office complexes in Houston, Denver, Calgary, Anchorage, and Chicago, and to the chaos of burgeoning Lagos, Nigeria? Maybe.]

Toward the end of that first month after we bought Wide Awake (October 1980), we invited a number of high-level managers from the Calgary energy companies to take the field seminar on the South Carolina coast, so we could show them some science and, hopefully, market our services. And probably even show off the Wide Awake property during one of the boat rides.

It rained every day, but the trip went well.

Thinking back on those days in the field as I rode on an airplane headed for Anchorage a few days later, I wrote the following in my journal:

Sunday night, the party (at our Gregg Street house in Columbia) was very quiet, only the participants in the seminar came. I had made friends earlier with one of the people on the trip. His name was Frank Jamine, a Jewish geologist at Dome Petroleum in Calgary, one of the original partners with the company. He is about twelve years my senior. He told me that he was enjoying the trip immensely. Why is it that most of my good friends are either Jewish or Polish?

I was showing Frank around the house when he stopped me and said, "You are a business man, but your heart is not really in it. You are an artist. You have such a keen sense of form, of design."

I said, "Yeah, sometimes I just want to create something special, that maybe would last, and get off this wild ride."

Then he walked away, shaking his head while looking at my sparse collection of paperbacks, wondering why I didn't have more of the works of Hesse.

"You have enough money," he said, as he waved out toward the dining area, "you don't have to do this anymore!"

You don't have to do this anymore? Maybe not, but I kept at it anyway. However, I did read Hesse. By the way, here is one of Hesse's aphorisms: *Wisdom is incommunicable. Wisdom that a wise man tries to communicate always sounds like foolishness.* Okay, so we were not trying to communicate wisdom on these field courses, only how the coast might have been made. Here's another one: *The higher humor begins when a person no longer takes himself seriously.* Got me again. What was old Frank trying to tell me?

Story Four

By the mid-1980s, we changed the order of how we ran the course, starting in Columbia with an evening lecture and snacks the first evening, and then next morning going to the Congaree River point bar and Congaree National Monument. At the Monument, now a National Park, we walked across a flood plain in its natural state, with a large virgin stand of timber as a bonus. Later that afternoon, we drove on down to Charleston.

In the introductory lectures the first night, I gave a talk called "Wildlife and History of the South Carolina Coast." In this presentation, I discussed certain aspects of the wildlife they would see during the course, especially the birds. Since my early years, I had always had a keen interest in birds and, after buying Wide Awake, JM and I became ardent birders, which we continue to do, even to this day.

Also, I took many opportunities to discuss the human history of the area, in particular the happenings during the First Civil War (aka The Revolutionary War). One of the local heroes in that war was a South Carolinian named Francis Marion, The Swamp Fox. We rarely missed an opportunity to take everyone on a walk through the mud back into one of the swamps on the upper delta plain of the Santee/Pee Dee Delta. Once everyone gathered around me under the heavy forest canopy, I went through a series of quotes recounting the adventures of the great Swamp Fox himself, some of which took place in the very swamp in which we were standing. There amongst the mosquitoes and water moccasins was the perfect place to do that, and it always went over well.

Another personal change occurred when, in 1985, as mentioned briefly earlier, JM and I hired a Hilton Head contractor to build us a small, but modern, house on the river bluff at the edge of the Congaree River flood plain overlooking the 266 acres surrounding the house site, which was also a part of our property. This was a beautiful natural setting, with abundant large trees. About half of the property was swampland on the flood plain that was full of all kinds of wildlife, especially birds. We called the place simply–The Bluff.

In April of 1988, Tenneco Oil Company contracted us to conduct the one-week field seminar for them. The first night the 25 or so participants showed up, we took them to The Bluff for the introductory lectures and some catered food. This trip was also a little different, because I wanted to show them what a Carolina Bay near Georgetown (Figure 28) was like. It was tough to walk through, that's what it was like.

Anyway, this was a great group, one of the most harmonious seminars of any we had done, and, lo and behold, on the last night in Charleston (18 April 1988), after the awards to the participants were given out, one of them recited the following poem the group had written themselves about the course. This masterpiece was called –

ODE TO THE SWAMP FOX

We came from far and wide
to learn the local lores
of the marsh and of the swamp
and the first of Civil Wars.

Our guru was a wise man
though some think him insane.
He lives deep in the jungle
with the birds and with his Jane.

He'd wake us up at dawn
on the tides of a new moon
in search of nature's mysteries
the ubiquitous antidune.

He solves the riddle of the swamp
the beach and all the wood
and the Carolina Bays
though the exposure's not that good.

He knows the song of birds
what the trout likes for bait
and the intricate procedure
of how the horseshoes mate.

He led us through the bog
at times it did get rough
and if you fell behind
you were labeled a creampuff!

Gators do not scare him
the ebb he never fears
*cause he's got a bunch of heroes**
with wrack between the ears.

So now we've learned the local lore
in the land of the Gamecocks
told us by the master
The Silver Grey Swamp Fox.

* COASTAL HEROES, that is.

15 BY FOOT, BY CAR, BY BOAT, BY PLANE

You are never too old
to learn a new way
to do something stupid!

BILLY WHISKERS (1990)

(12 August 1990, Sunday - On a rock off Mitkof Island, Southeast Alaska)

I had been lying on that rock for about two hours after having crawled up on it to escape the rising tide. About 50 or 60 meters from where I lay, the tail of the float plane I had been flying across this area in on a mapping mission earlier in the morning was sticking up out of the water, which was rising in response to the 5 meter tide. Under the water and still in the plane was my pilot, Todd Hatfield, who had died during the crash. Though nursing a broken left leg and a badly mangled right foot, I had luckily managed to get out of the plane and swim to the base of the rock island, on the top of which I now waited, covered with dead bunches of grass, hoping to be rescued. The location of the crash site is shown in Figure 21B.

During that long wait–it took two hours for anyone to find me–I had time to think about many things, including the necessity for using such conveyances as float planes to do these coastal research projects. For example, one of the incidents that came to mind took place sometime in the late 70s. At that time, the Chinese Cultural Revolution had eased up to the extent that some of their chosen scientists were permitted to make a tour of U.S. institutions in order to, hopefully, catch up to speed on the latest developments in their field. I guess this trip was supposed to compensate for the ten previous years they had spent working in the rice paddies. When the Chinese were making contact with the different programs around the country, they sent our coastal geology group at USC-East a letter in which they said they "worshiped our research," and would like to pay us a visit. How could we possibly have turned them down after that? We didn't.

When a group of six Chinese coastal geologists showed up in Columbia, my students and I endeavored to showcase our work on the South Carolina coast for them by means of an elaborate field trip to many of our study areas, following more-or-less the same itinerary as the one used for the field seminars discussed in the previous chapter.

There was also a lecture session or two in Columbia. Being very proud of our historic data on the South Carolina coast, I showed historic maps that dated all the way back to 1661 (see Figure 62). They smiled and clapped and apparently thought that was pretty good stuff. Then their coastal geologist, who did similar studies on the coast of China, gave a talk on the Yangtze River Delta, showing maps that went back to 200 years B.C. Boy, was I ever humiliated!

At the end of their visit, the guests bought out one of the local Chinese restaurants for the evening and supervised the cooking of a banquet dinner, which was fabulous. During the usual round of toasts at the end of the meal, their leader praised our efforts on the field trip by pronouncing in wonder how we had gone "by foot, by car, by boat, and by plane" to see the different localities.

Remembering this gave me pause to contemplate as follows on how I had used these four modes of transportation during my scientific career.

By Foot

I mentioned the long walks through the dunes to do my dissertation work on the Texas coast in an earlier discussion in Chapter 1 – THE HURRICANE, an example of doing coastal research "by foot." Also, during the months prior to the plane crash, RPI scientists, under the sponsorship of NOAA, had been surveying large stretches of the Alaskan coast in order to determine the extent along the shoreline of the oil spilled by the tanker *Exxon Valdez.* Our major transportation to the field sites was either by boat or by helicopter, but in the summer of 1990, areas within half a mile of the bald eagle nesting sites were off limits to the helicopters. Thus, we had to climb and claw our way over large boulders and rock outcrops and walk for long distances along the shoreline in order to check out all of the potential oil-accumulation sites. The physical conditioning resulting from that work may have been part of the reason I was able to survive the crash.

As I was writing parts of an early version of this book on an airplane ride between Paris and Geneva (27 March 2003), we had just finished a 5.5-month project we called "Walking to Kuwait." During that project, we "walked" over 800 kilometers of the Saudi Arabian coast while mapping the oil spilled during the 1991 Gulf War, 12 years earlier. This work is discussed in Chapter 21 – NIGHTMARE IN JUBAIL CITY.

By Car

Trashing out cars and trucks during field exercises was a fairly common occurrence by our field teams, for example the demise of my new Jeep as Steve crashed into the Power Wagon on Kiawah Island (see Chapter 5 – KIAWAH ISLAND). During our first Alaska field surveys in 1969-71, while I was still at UMass, we bought two pioneer versions of the vehicles that they now call ATVs (all terrain vehicles) to use in our studies of the outwash fans that drained the glaciers east of Cordova. All of the fans were roadless, so we thought these ATVs, which were called Coots, would be ideal for hauling equipment, sediment samples, and so on across the fans. During one of the early days of the field study, one of the students ventured too far out into the standing waves of one of the main channels, and–WHOOSH! away went the Coot rolling over and over sideways into the deepest part of the channel, where it immediately became buried in the gravel.

No doubt it is still there, getting buried deeper and deeper as the fan continues to grow as a result of the glacier melting under the influence of global warming. What happened to the student? Only minor cuts and bruises.

In the heyday of the early growth years of RPI, I decided we should have a fleet of three pick-up trucks to use during training courses and local field projects. They were all used vehicles when we bought them, but in good working condition. The logistics for the training courses on the South Carolina coast are complicated and, among other modes of transportation, we used a catamaran to transport the students, coring devices, and other equipment in the rivers and tidal channels (see Figure 85). At the end of one of the courses, the boat captain at that time, who was named Cookie, pulled the catamaran up to a boat ramp so the equipment could be unloaded into one of my prized pick-up trucks for transport to storage at the office in Columbia. One of my field assistants at the time, who shall remain nameless, backed the truck down the ramp, so that its bed was pressed firmly against the vessel, which we called the *Megalops*, the genus name of one of the gamest fish that swims–the noble tarpon. After Nameless and Cookie had loaded all of the gear into the back of the truck, and while Nameless was still standing in the truck bed arranging things, the impetuous Cookie pulled on the throttle and sped away from the dock. Oops! Nameless had left the truck in neutral and had neglected to set the emergency brake. Down the truck and Nameless went to the edge of the ramp and then on down 30 feet to the bottom of the channel–GURGLE, GURGLE, GURGLE!! A wrecker did come after a while, finally managing to retrieve the truck but, because of that saltwater bath, it was delivered straight to the junkyard. What happened to Nameless? Only minor cuts and bruises.

FIGURE 85. One of the catamarans we used to conduct the field seminars on the South Carolina coast. Photograph taken in May 1997.

By Boat

The adventures aboard the *Ciencia* and the *Lorene* on the Texas coast were only the beginning of a long and spotty history of ours with boats. During the span of 30 years teaching the AAPG training course on the coast, we had built three different catamarans for the job (see example in Figure 85). One sank during the course, fortunately while none of the course participants were on board, while it was being moved for a pick-up the next day. Once during the earlier years, when the catamaran did not have a cover, Tim Kana, the course logistics officer and co-trainer at that time, who was a skilled carpenter, decided to build a cover for the boat himself, so that the participants could be protected from the sun, heavy rains, and so on.

I was a stickler about the timing of the course, time and tide waits for no man, you know, so the boat captains made every effort to be waiting at the pick-up spot when the troops arrived. During the very first course after Tim had built his cover for the *Megalops*, Cookie was running a little late and, damn, the drawbridge over the Stono River was down. Furthermore, the operator refused to raise the bridge during rush hour traffic for an insignificant little 45-foot catamaran. Cookie waited a while, looking feverishly at his watch until, in desperation, he zoomed under the bridge, ripping Tim's cover right off the boat.

Just as we were driving up to the dock with the vans to meet the boat, Cookie pulled up as well, *sans* Tim's pride and joy, the new roof.

Cookie said, "Here I am boss, right on time. Sorry about your cover Tim, I had a little disagreement with the draw bridge operator back there on the Stono River."

However, one of the least enjoyable incidents with the boat during one of the courses occurred more recently, in 2003. We were heading back from a field stop on an eroding barrier island called Sandy Point on Cape Romain (see Figures 28 and 60), going down a major tidal channel on our way to the dock at a small fishing village called McClellanville (the area discussed in Story 2 in Chapter 10 - PRICE INLET). As we were moving along at a good clip, the boat captain at that time, Anton, and I noticed that the darkest and blackest storm cloud you could ever imagine was right down against the water across the channel where we were headed, and the clouds were moving straight toward us. We could hear thunder in the distance and see numerous lightning bolts playing across the surface of the marsh.

Anton, who looked about as petrified as I felt, said, "What do you think?" as he eased back on the throttle.

"Well, normally, I would say let's get everybody off the boat and have them scatter out over the marsh, but with the tide low like this, everybody would sink up to their ass in the mud flat before they could get there. So, I guess, its full speed ahead."

I have never heard louder thunder as we sped on through the gale and heavy downpour. I was counting, noting that the lightning flashes, which were only a few seconds apart, and the crashes of thunder were getting closer and closer together. Until, finally, they were simultaneous as the lightning hit the boat–WHAM! I could see the lightning bolt run around the top of the somewhat elaborate cover we had on this latest model.

The students on the boat, most of whom were from Europe or Canada, were sopping wet to the skin, regardless of the cover, because of the strong wind. Looking to the boat captain and me for encouragement, these poor drowned rats didn't know whether to shit or go blind, as they say, or to laugh or to cry. Neither did I. Anton and I both stared straight ahead without even slowing down, as if nothing were happening, but I tell you, I don't want to go through something like that ever again.

So, what happened to the students? Nothing. Just a good soaking and possibly the scare of their lives; not even a few cuts and bruises.

By Plane

Back to the rock. Needless to say, this was the worst thing that had ever happened to me in an airplane. There had been a few close calls, such as when Gayle Ranney landed our wheeled plane in the surf up by the Malaspina Glacier back in 1975. I had flown with the best of bush pilots–Pat James, "Mudhole Smith's" son, Jim Foodie, and the skillful Cowboy Pierce–with a few noteworthy incidents, but nothing like this. Cowboy flew Chris Ruby and me around the entire Alaskan coast north of Anchorage in the summer of 1976, all the way to the Canadian border. He even had me flying the plane upside down.

But now, lying on the top of that rock island, my partially naked body covered with dried grass in an effort to keep from freezing to death, I was feeling terrible about Todd, who was still in the plane. I thought that somehow maybe I was responsible for getting us into a situation, where we would be turned sideways by turbulence and plunge 500 feet straight down into the water.

Luckily, for me anyway, two hours or so after the crash, I was discovered by a fisherman, named Jamie Debore, and his wife, who were out on a Sunday afternoon fishing for salmon. About an hour after that, I was taken in a U.S. Coast Guard helicopter first to Wrangell, about 40 miles away, and then on over the high mountains to Ketchikan (Figure 21B). I spent ten days in the Ketchikan hospital, where Dr. Bendowski patched up my legs and foot during three operations. JM, who was mapping in a different plane at the time of the crash, stayed with me the whole time, even sleeping in a chair by the side of my bed every night.

Finally, on 22 August 1990, JM and I boarded a jet headed south and, after spending one night in Seattle, went on back to Columbia, where I stayed in a Residence Inn for a couple of days. Being rolled around in a wheel chair in the muggy heat in Columbia was troublesome for all concerned but, for me, it was considerably better than being in Alaska lying on that rock freezing to death.

16 DOOMNESS

The older we get,
the greater we were.

SOURCE UNKNOWN supplied by DAN DOMERACKI

(24 August 1990 - Columbia, South Carolina)

A day after our arrival in Columbia, on a Friday, JM took me for an appointment with an orthopedic surgeon named Dr. Bobby Peele, who had been recommended by my pal Walter J. Sexton's wife, Emily, a surgical nurse in one of the local hospitals. Dr. Peele, who doubles as the bone doctor for the USC-East Gamecock's football team, was a family relative of Walter J. and Emily's. Walter J., the first employee of our original science technology company, Research Planning Institute, Inc. (RPI), was also a former graduate student, who by then had his own geological consulting firm in town, Athena Technologies, Inc. He also lectured and provided logistical support for our field training courses on the South Carolina coast. Inasmuch as there were two courses still scheduled for the fall of 1990, his services would be needed more than ever this particular year.

JM rolled me down the hall past the pictures of all the football players the doc had patched up over the years to one of the patients' rooms, where I would wait for the arrival of Dr. Bobby. When he got there, he leaned up against the door sill, looked over at me and my legs as I sat in my wheelchair inside the room, shook his head, clucked several times and said, "Oh my, oh my!" Not exactly a confidence builder on my part.

Not much inspired by Dr. Bobby's bedside manner, I said to JM, "Let's get out of here!"

However, she talked me into at least staying for the examination, which they carried out with considerable care. He looked at the x-rays and commented on the excellent job Dr. Bendowski had done putting the bones back together in Ketchikan, Alaska. I was beginning to like Dr. Bobby.[32] After all, I was once a football player myself. Not a great one like the ones he was treating these days, just a rag-tag, 155-pound center/line backer for the Oakley/Fairview High Schools, North Carolina consolidated team. But, as the saying goes, "the older we get, the greater we were." He said he would schedule an operation to take the pins out of my right foot sometime in mid-October.

JM took me out to The Bluff for the weekend on 25/26 August. When we got to the house, she had to call Gene Crim, our neighbor who lived 5 minutes away, owns a plumbing contracting business, and can do anything mechanical, to come down and help her haul me into the house, which he did with alacrity and his usual competent skill. Our original schedule had called for us to leave this particular Saturday morning for Yellowstone Park for a few days of fly fishing. After that, we were to go on to the *Exxon Valdez* oil spill site in Prince William Sound, Alaska to re-survey 18 permanent stations that had been established as part of NOAA's monitoring program for that spill. The late-summer fishing trip to Yellowstone had become a tradition that was about to be broken, needless to say. However, JM would leave in a few days to do the survey in the sound without me. She had lined up some NOAA scientists to take my place on the field team.

We left The Bluff at 0730 Monday morning and went in to our business office, which was located in a renovated warehouse building in downtown Columbia, one block from the state capitol. They had fixed up my office for my arrival, even bringing in a hospital bed where I could rest periodically, if I so desired. I went in to work every day after that, working at about half speed. What else was I going to do?

The next day, JM took me to see a skin doctor, Dr. Campbell, to discuss a skin graft for a large hole in my shin that had been gouged out when I lunged against the back of Todd's seat. I was riding in the back of the plane as I did the mapping. If I had been in a front seat, I would have suffered the same fate as Todd. I looked around the waiting room, which was full of beautiful women, and asked JM, "What am I doing here?"

"He must be very good at what he does, or a real charmer one," she answered.

He turned out to be both. He prescribed some detailed treatments for the hole in my leg, which had to be administered twice a day.

"It should fill in and be ready for the graft about mid-October," he opined.

You would have to look a long time to find two people as different as the suave Dr. Campbell and the rough-hewn Dr. Bobby, both matching their primary patients perfectly–that is, beautiful ladies versus banged-up, bloody football players. It was pretty obvious to me, however, that my good luck with the medical profession was holding steady, after the good luck of Dr. Bendowski doing a little moonlighting in Ketchikan, and that they were probably two of the best anywhere at what they do. Let's hear it for the good old American medical profession!

On Thursday, 11 October, I went into the Baptist Hospital in Columbia to have Dr. Bobby take the pins out of my right foot and Dr. Campbell do his thing with the skin graft, both operations being performed while I was asleep this one time. If I had any dreams, I'm sure they would have been some bizarre tales about those two exact opposites working together. Jack Moore, RPI's Chief Financial Officer, later told me that Dr. Campbell was a little too laid back to suit Dr. Bobby, but he had said Campbell was the best, which is why he recommended him in the first place. These operations seemed to go okay. Jack was looking after me again, because JM was out of town, at another oil spill, I believe.

On Saturday, 13 October, around noon or so, a new nurse stuck her head in the door, a jovial-looking black lady about 45 or so. As she walked up to the bed and before she could even introduce herself, I blurted out, "I'm out of here at 2:30 this afternoon!"

She chuckled and said, "Well I'm Dorothea your new day nurse, and we can't let you go until I have thoroughly checked out your vital signs."

As she started into the process of listening to my heart beat and other things, she stopped abruptly, looked aside, and said, "Oh my God! I should call the doctor about this."

Then she stood up, went to the phone, decided not to use it, and sat back down and went to checking something else.

"Unh unh," she said, "When did you say you were leaving?"

"You mean I can't leave at 2:30 as promised?" I asked desperately.

"Leave at 2:30? You could be here for weeks in this condition," she answered as she continued to shake her head.

"What?"

"Yeah, with this heart and these lungs you appear to me to be doomed. Unh unh. Not just doomed. No, doomness, that's what you are, DOOMNESS!" And then she couldn't keep it up and burst out laughing.

Not too long after that, folks started coming in to see me. My first wife, Barbara Anne, and my daughter Mya had driven down from North Carolina. My sister, Norma, was there visiting from Atlanta. Jack and a couple of people from the office were also there.

As we were talking, Dorothea came in to inform me that, actually, my vital signs were okay and that, indeed, I could leave at 2:30. We both laughed and laughed, and everybody looked at us wondering what in the world was going on.

Promptly at 2:30, as Jack was pushing me down the hall toward the elevator, and the entourage followed closely behind, Dorothea breezed by and said, "See you later Doomness, you'll be back, oh yeah."

Everybody was shaking their heads as we crowded into the elevator.

"What did she mean, Doomness," my sister finally asked.

"That's my new name," I answered.

By the time of this last operation, I was walking on crutches pretty well, because I could put some weight on my left leg. After the skin graft, however, Dr. Campbell said that I should not put any weight on the left leg for another 6 weeks, which meant back to the wheelchair. The cast didn't come off my right foot until 22 October, two days after my 56th birthday.

All through the months of October and November I went into the office every day, starting to work on our first major report on the results of the surveys at the *Exxon Valdez* oil spill site for NOAA. I worked at about half speed. The nights were miserable, as I could never seem to get comfortable. At one point in my journal notes, I referred to myself as "turboman," because I was spinning around in the bed so many times at night. I caught some grief from Dr. Campbell during my 30

October visit because I was "doing too much," and the skin graft was not healing properly.

On 17 November, I experienced one of the most excruciating toothaches I had ever had in my entire life. It lasted for days. I finally visited JM's brother, Steve, who is a dentist in Charleston, South Carolina, and he did a root canal on that tooth, the lower left rear molar. He said that the x-ray showed that the tooth had been cracked right across. We both surmised that it had probably happened during the crash.

The quasi-celebrity status that I had achieved in southeastern Alaska as a result of the crash continued in South Carolina. It was strange observing all of the different reactions to me after that accident. Want to get someone's attention? Just tell them you've survived an airplane crash.

The first time I hobbled into the Sandy Run country store, which was located on our way home about 5 miles from The Bluff, Pete Tucker, the owner, said, "What in God's name happened to you?"

"Oh, I was just in a little airplane crash up in Alaska," was my casual reply.

The old boy leaning over on the counter by the wall said, "No airplane crash is a little crash!"

He no doubt spoke from experience. Probably in some war somewhere.

In mid-November, my friend Murray Felsher, my UT office mate, called. Strangely, he had only just heard about the crash at a national meeting of The Geological Society of America.

"The rumors ranged from bizarre to the ridiculous," he said.

I told him a short version of the story.

"It was just basically a superhuman effort, that's all," I joked.

Murray's from New York City, so you have to be thinking pretty quick to get one on him. You may recall the poem he wrote about me when I was getting ready to leave UT for UMass (see footnote number 7).

When we finished talking, Murray said he would have the great Robert L. Folk, the one and only, give me a call, which he did. When Folk called we joked some about my burned behind (from the aviation gas that soaked into my pants) one of the side effects of the crash, and he told me about some of his latest theories. Still working after all those years.

17 GRAVEL BEACHES AND FAN DELTAS

ON COUNTING PEBBLES

Looked across the river
to the glacier
big and fine.
Guess there
must be a
hundred million pebbles
in a straight line.

But just can't
get into them
pebbles anymore.
Though they
come in many
colors and
shapes galore.

Because, I measured
them dutifully
once upon a time,
but found
out the papers
printed weren't
really worth
a dime.

Though they got
me fame and
notoriety,
weren't
good enough
to bring
immortality!

So who cares
about them pebbles
anyway?
It's only just
another way
to pass
the day.

....

Guess I'd better
go fly
to another place,
and take some pictures
go through
the motions of time and space.

Maybe an insight
pure and simple
will come to me,
then I could
jump and run
and holler and
stomp with glee.

Because I'd found
another way to
pay my fee,
without them
God-forsaken
pebbles
worrying me.

The water in
the stream
is gray.
The glacier is white,
the mountain is black,
the sky is blue,
I am too.

BILLY WHISKERS – BY THE MALASPINA GLACIER,
ALASKA (SUMMER OF 1975)

The Science – GRAVEL BEACHES

In this discussion, the term gravel is defined according to the Wentworth (1922) scale, which includes four classes: granules (median diameter = 2-4 mm), pebbles (median diameter = 4-64 mm), cobbles (median diameter = 64-256 mm), and boulders (median diameter >256 mm). I received my Ph.D. without ever having seen a gravel beach, except for the beach composed of gravel-sized shells on central Padre Island, Texas. However, since then, we have mapped and surveyed gravel beaches in numerous localities, including much of the Pacific coast of Alaska (Hayes, 1980b; Hayes and Michel, 1982, 1989; Hayes et al., 1976a; NOAA/RPI, 1983, 2000, 1993-2001), New England (Hayes, 1969; NOAA/RPI, 1999), California (NOAA/ RPI, 1994, 1999, 2006a), and the 1,500 kilometers of the shoreline of Oman (Hayes and Baird, 1993). Be that as it may be, it was not until the problem of the lingering subsurface oil at the *Exxon Valdez* (1989) oil spill site evolved that it became clear to us that any coastal geologist involved in oil spill response should have a thorough understanding of the characteristics of gravel beaches (Hayes and Michel, 1999; Michel and Hayes, 1999).

Gravel beaches are the most difficult of all the beach types to clean and restore after an oil spill (Hayes, Michel, and Fichaut, 1990). Specific characteristics that make them so include: (1) High porosity and permeability that allow deep penetration of oil from the surface (>100 centimeters in some places); (2) Potential for deep and rapid burial by clean sediments; (3) Complex patterns of sediment reworking during storms (in contrast to sand beaches); and (4) Slow rates of natural replenishment (should the contaminated sediment have to be removed during cleanup). Therefore, of the different beach types, gravel beaches rank the highest (most likely to retain oil and most difficult to clean) on the Environmental Sensitivity Index (ESI) scale, a mapping system used worldwide for oil spill contingency planning purposes.

[NOTE: Porosity is defined as the ratio of the volume of openings (voids) to the total volume of material. In gravel deposits, the primary porosity represents the spaces between the individual clasts that comprise the deposit. Permeability is a measure of the ease with which fluids will flow through the gravel deposit (e Notes, 2011).]

[ANOTHER NOTE: The ESI of gravel beaches is 6 on a scale of 1-10; wave-cut rock cliffs = 1, and salt marshes = 10; Hayes, Gundlach, and Getter, 1980; also discussed in Chapter 20 – ENVIRONMENTAL SENSITIVITY INDEX (ESI).]

During the study of the world's shorelines and inner continental shelves, discussed in Chapter 1 – THE HURRICANE, that I did at the U.S. Navy Defense Research Lab in the early 1960s, I recognized that the common occurrence of gravel in continental shelf and shore zone sediment budgets (see Figure 86) in the higher latitudes (>40 degrees north and south) is primarily the result of glaciation at those latitudes (Hayes, 1967c). To extend that idea a bit further, the shoreline classification data from the ESI maps produced by our team in many parts of the world verify the fact that gravel beaches are very common along coastlines that have been subject to glaciation, either under present conditions or during the Ice Ages of the Pleistocene Epoch. Glaciers have the competence to carry huge volumes of coarse material to the coastline, where it is redistributed along the shore by wave action (Hayes et al., 1976a). Accordingly, the ESI shoreline data for the Southeast Alaska coast show that gravel beaches are common along that

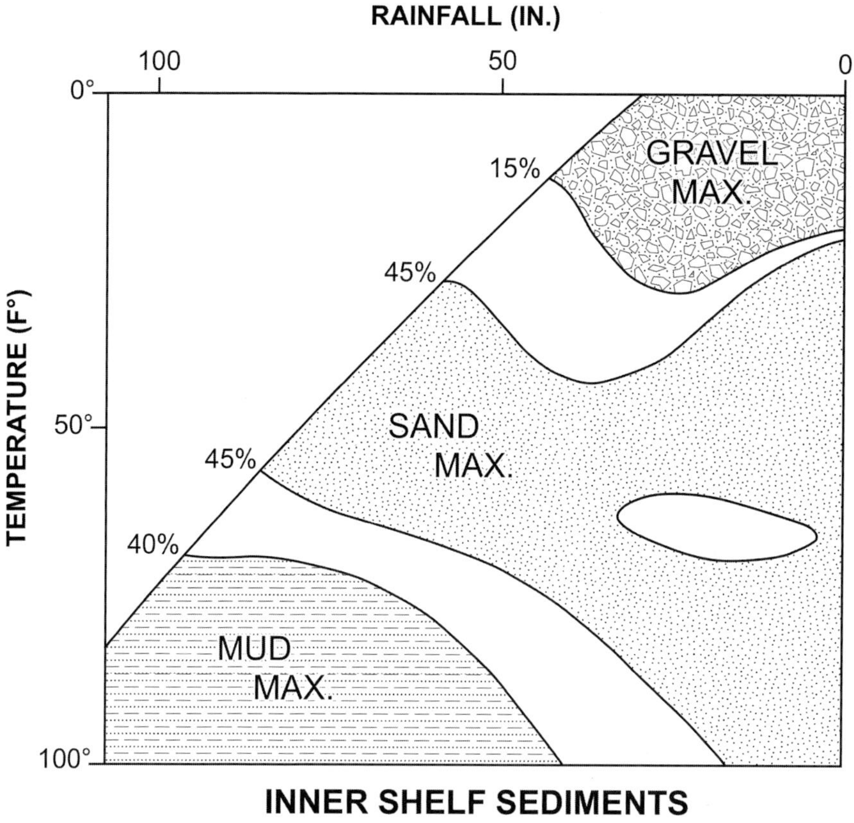

FIGURE 86. Dominance of sediment types in the sediments of the inner continental shelves of the world. Note the dominance of gravel in the areas with the colder climates (nearer the polar latitudes), and the dominance of mud in area with warm climates and abundant rainfall (in the more equatorial latitudes). From Hayes (1967c).

coastline, most notably off the large Malaspina and Bering piedmont glaciers (NOAA/RPI, 1993-2001; discussed in Chapter 4 – ALASKA – *SUZANNE,* EARTHQUAKES AND TSUNAMIS, and Chapter 7 – THE MEN THAT DON'T FIT IN). Glacial outwash streams and their associated fan deltas are also important contributors to the coarse fractions of beach sediments in southern Alaska (Hayes and Michel, 1989; Hayes and Ruby, 1994). Lower Cook Inlet, Alaska (Figure 87) contains numerous large fan deltas, most with glacial sources that are bordered by spectacular gravel beaches, some containing abundant boulders (Hayes and Michel, 1982; Hayes, 2005a). ESI shoreline data for New England (NOAA/RPI, 1999, 2004) and Puget Sound (NOAA/RPI, 2006b) also delineate gravel beaches where waves are eroding into Pleistocene glacial deposits (e.g., the outer coast near Plymouth Harbor, Massachusetts and Rye Beach, New Hampshire; Timson, 1969). Polar regions subject to ice gouge along the coast also tend to have gravel or mixed sand and gravel beaches (Rosen, 1979). The barrier islands of the North Slope of Alaska are an excellent example of this (Hayes, 2009; NOAA/RPI, 2005; Pilkey and Fraser, 2002).

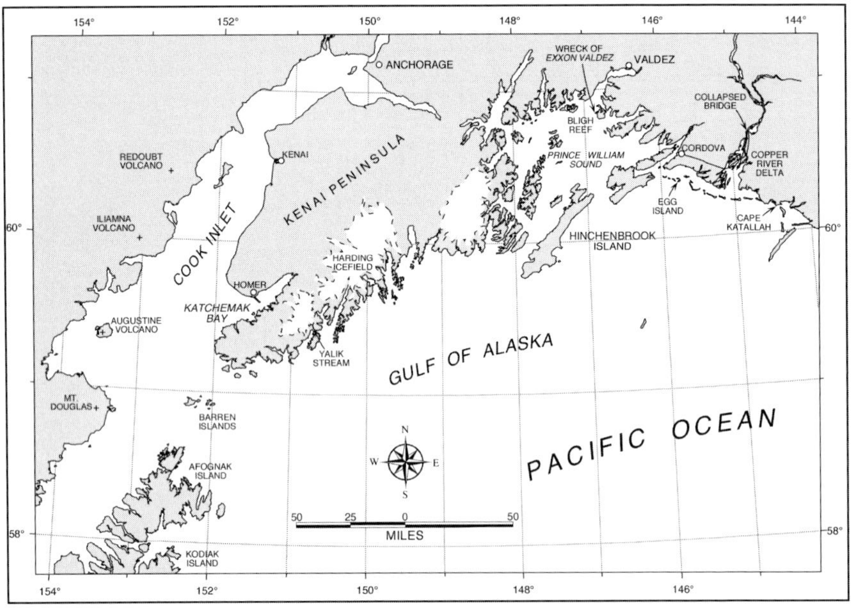

FIGURE 87. Southern Alaska coast showing the locations of Prince William Sound and Cook Inlet.

We recently published a paper on gravel beaches in the *Journal of Coastal Research* (Hayes, Michel, and Betenbaugh, 2010) that summarized our key findings on gravel beaches, with special emphasis on those occurring in Prince William Sound, Alaska (Figures 87 and 88). Some of these findings follow:

1) A general model for exposed (to wave action) coarse-grained gravel beaches, on the basis of studies in New England, California, and Baja California, consists of the following components from land to sea: A relatively permanent high storm berm composed of cobbles and pebbles; Small ephemeral high-tide berms along the lower offshore face of the storm berm; A zone of imbricated cobbles at the toe of the storm berm and; A low-tide terrace covered with an armor of cobbles and boulders. The cobbles and boulders in the imbricated and armored zones have undergone structural strengthening to the point that they can be entrained during only the most extreme storms. This general model is illustrated in Figure 89.

2) The landward portion of six intermittently exposed gravel beaches in Prince William Sound, Alaska (subject area of the paper; see Figure 88) consists of a storm berm with small high-tide berms perched near the high-tide line. The middle and lower portions of these beaches, termed the upper and lower platforms, are typically gently sloping (4-6 degrees) surfaces of bedrock, which are probably uplifted, wave- and possibly ice-carved shore platforms, covered by a veneer of gravel (Hayes and Michel, 1997). The grain size of the gravel on these beaches typically increases in an offshore direction, with pebbles dominating the high-tide berms

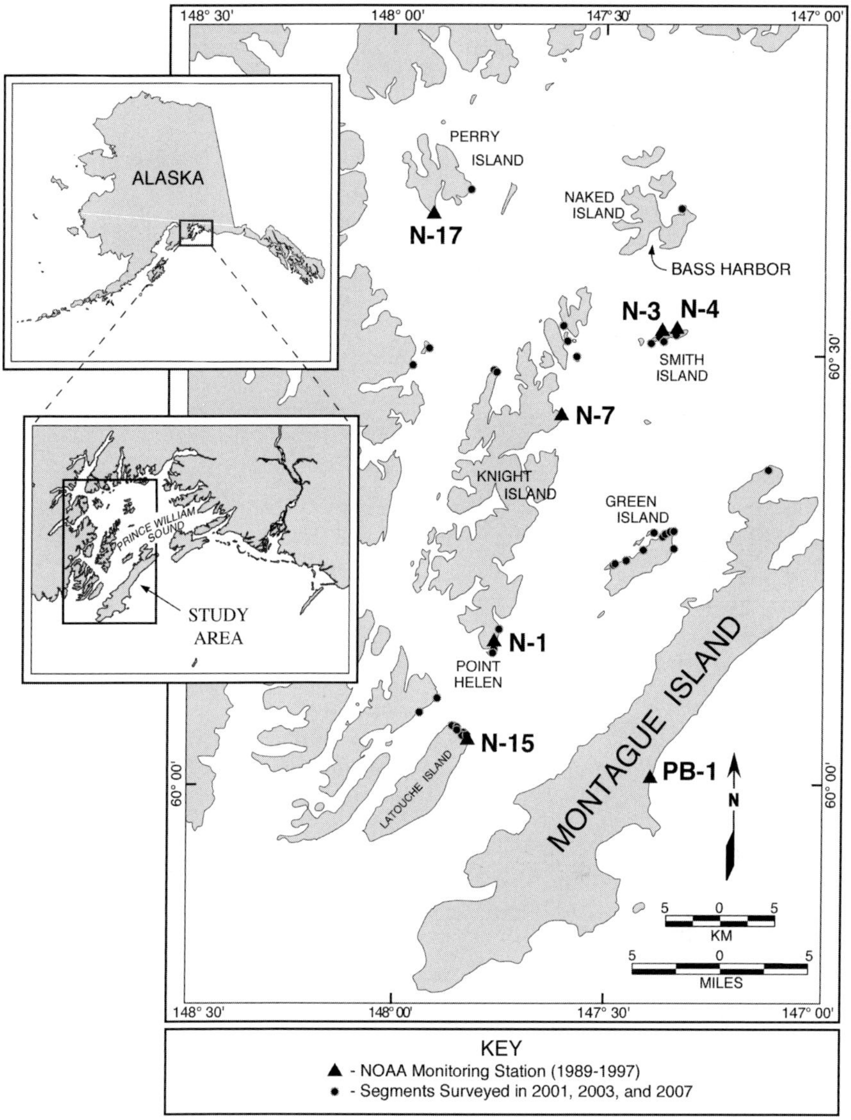

FIGURE 88. Location map for study sites in Prince William Sound, Alaska discussed in this chapter. Specifically shown (in dark letters) are six of the permanent NOAA coarse-grained gravel beach transect sites surveyed by our teams between 1989 and 1997. The black dots show the locations of the 34 coarse-grained gravel beaches of the type discussed in this chapter that were surveyed during the summers of 2001, 2003, and 2007. From Hayes, Michel, and Betenbaugh (2009).

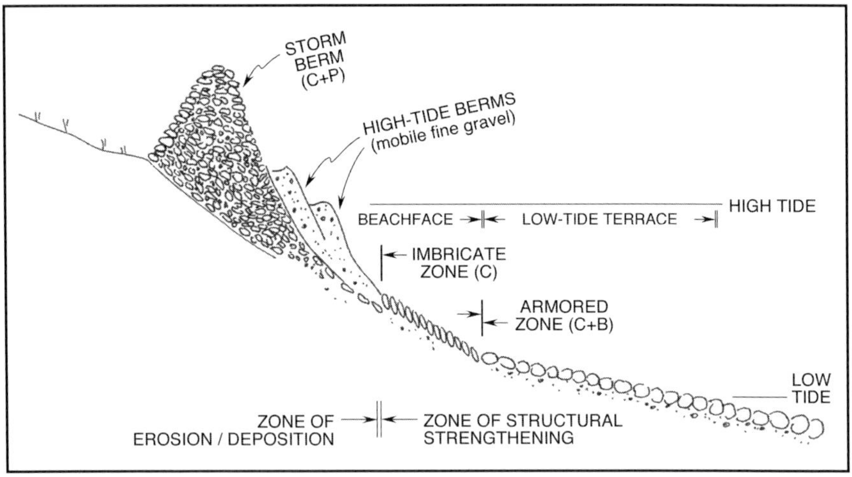

FIGURE 89. Typical beach profile and sediments of an open-ocean (exposed) coarse-grained gravel beach. The landward half of the profile is primarily depositional, with the storm berm being relatively stable and the minor, accretionary high-tide berms eroding during storms and reforming after they pass. The seaward half of the profile is an area where structural strengthening occurs by either clast imbrication or by the formation of a surface armor of coarse gravel. B = boulders; C = cobbles; P = pebbles. From Hayes, Michel, and Betenbaugh (2009).

(except for some cobbles on the storm berm), cobbles dominating the upper platform, and boulders being more abundant on the lower platform. A general morphological sketch of these beaches is given in Figure 90.

3) The relatively flat surfaces of the upper and lower platforms are armored on the surface with the coarsest gravel clasts available (typically cobbles and boulders). This armoring is the result of selective sorting of a range of particle sizes, with the size just below the threshold of entrainment (i.e., too large to be moved) by the swash-generated currents during maximum storm conditions forming the armor. The finest particles are either swept away by the strong currents generated during storms or left behind beneath the armor as a result of sheltering, infiltration, or other mechanisms. Intermediate-sized clasts may also make up part of the finer-grained substrate, but more commonly, once the spacing between the coarser clasts becomes narrow enough, they are rejected and overpass the armor selectively, because of the absence of secure resting places. This process is illustrated in Figure 91. Once established, this armor is quite stable and immobile, as evidenced by the fact that it provided a protective cover for oiled subsurface sediments resulting from the *Exxon Valdez* oil spill for at least 18 years. This process of the protection of the subsurface oil by the surface armor is illustrated in Figure 92. Photographs of both the surface armor and the subsurface oil are given in Figure 93.

4) Gravel sediments of the high-tide berms and portions of the lower face of the storm berm were relocated onto the upper platform at 29 sites impacted by the *Exxon Valdez* oil spill in the late summer of 1990 to

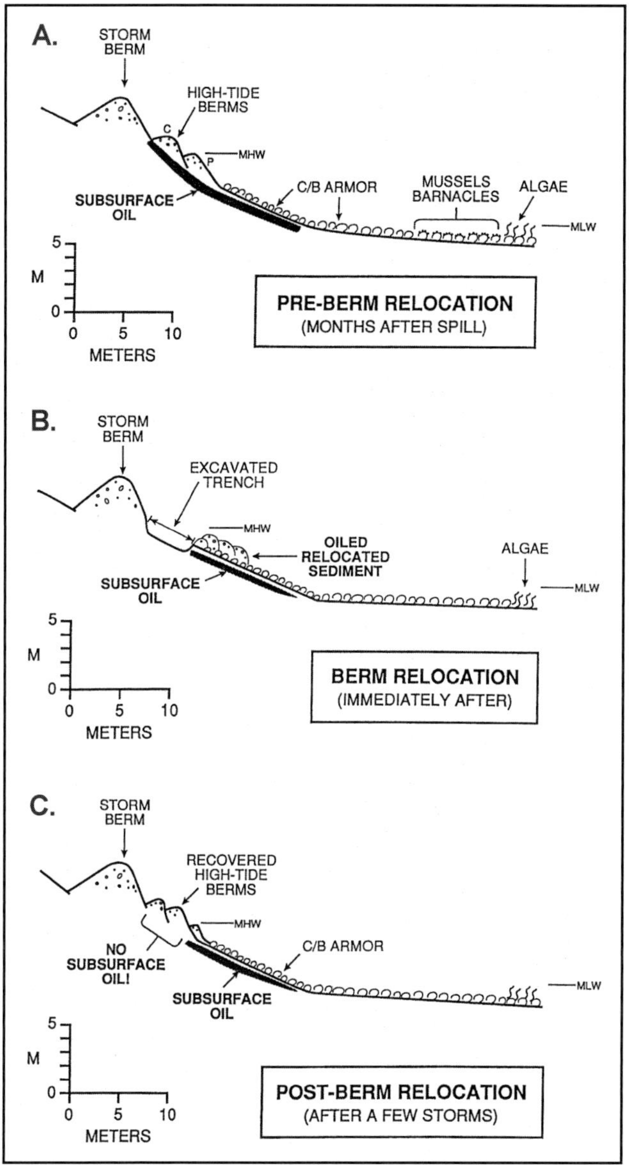

FIGURE 90. General model of the process of berm relocation that was carried out at 29 sites in Prince William Sound in the late summer of 1990. (A) Subsurface oil is located in the gravel sediments below the lower portion of the storm berm, the upper portions of the high-tide berms, and the armor on the upper platform. (B) The oiled subsurface sediments associated with the face of the storm berm and the high-tide berms were excavated and deposited lower down the beach. (C) After a few storms, the relocated sediment, which had been cleaned by wave action, was returned to the excavated trench by cross-shore sediment transport, and the sediments along the profile quickly returned to their pre-relocation configuration (both morphology and sediment distribution patterns). Subsurface oil remained under the armor along the upper platform, as is illustrated in Figure 92. From Hayes, Michel, and Betenbaugh (2009).

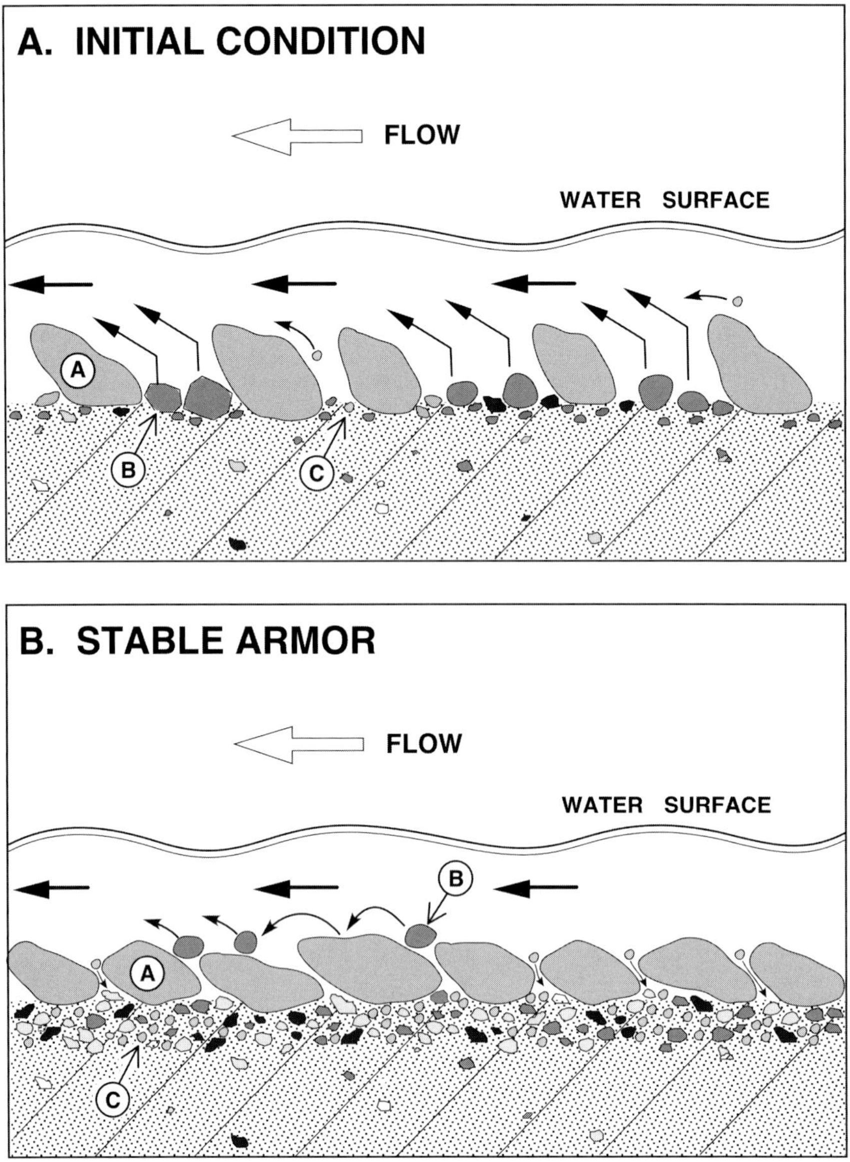

FIGURE 91. Schematic model of the processes involved in the development of an armored surface of coarse clasts on a gravel beach. (A) During storms, the largest clasts (size A) are too large to be removed by the prevailing swash-generated currents, the intermediate-sized ones (size B) are readily transportable, and some of the smallest ones (size C) are sheltered by the larger clasts and are not removed. (B) At a later stage after the surface is completely armored. With the coarser clasts so close together, the intermediate-sized clasts in transit are rejected and pass over the armor. In some cases, finer particles in transit may filter between the armored clasts to become part of the substrate. From Hayes, Michel, and Betenbaugh (2009).

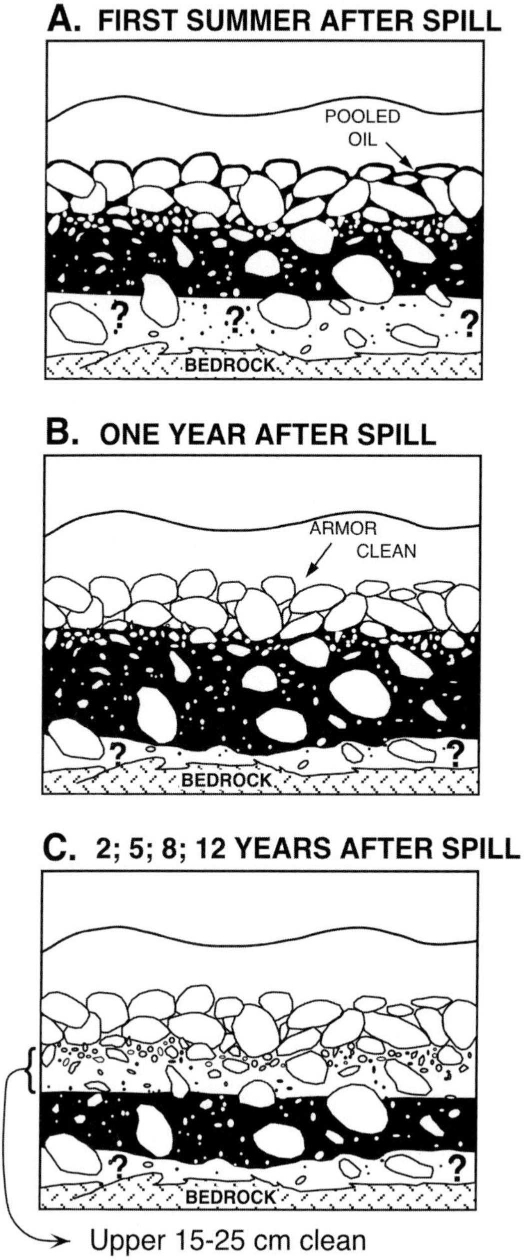

FIGURE 92. Schematic model of effect of well-developed armor on retention of subsurface oil in the upper platform at stations N-1 (Point Helen) and N-3 (Smith Island) in Prince William Sound (Figure 88). (A) Initial oiling in summer of 1989. (B) One year later, cleanup efforts and natural processes had cleaned the gravel in the armor. (C) After 2 years, natural processes had cleaned the oiled sediments under the armor to depths of 15-25 centimeters. The deeper oiled sediments remained relatively unchanged after 12 years. From Hayes, Michel, and Betenbaugh (2009).

FIGURE 93. Station N-3 on Smith Island (Figure 88) on 21 July 1997. (A) The armored surface of the upper platform. Arrows point to excavated pits. (B) Oil in sediments (heavy oil residue; HOR) and on the water table at bottom of pit A at a depth of 52 centimeters below the armor. The lingering subsurface oil was still there eight years after the *Exxon Valdez* (1989) oil spill, because of the protective cover of the heavy armor. From Hayes, Michel, and Betenbaugh (2009).

expose the oiled sediments to cleaning by wave action. The disrupted sediments were returned by cross-shore transport, and the beach profiles were restored to their former morphology and sediment distribution patterns by wave action after the passage of a few storms (usually over a period of less than 2 or 3 months). This process is illustrated in Figure 90.

5) A ranking of coarse-grained gravel beaches based on wave height and frequency is given in Figure 94.

The Science – FAN DELTAS

In April 2005, I submitted an 82-page report titled *Fan Deltas as Exploration Targets* to some clients who worked for Devon Energy Corporation in Oklahoma City. The report also included an Appendix containing summary discussions of 25 of the most critical papers published on fan deltas up to that time. Several of the Devon clients had attended our AAPG Modern Clastics field seminars on the South Carolina coast (discussed in Chapter14 – THE ROCK RECORD), during which I lectured about braided streams. I probably also mentioned the paper JM and I had written that discussed the fan deltas in Lower Cook Inlet, Alaska (Hayes and Michel, 1982). The completion of this study for Devon Energy took place while JM and I were living through our year and a half experiment of residing in Uptown, New Orleans.[33] I also spent two days of lecturing and discussion of fan deltas in the Devon offices in Oklahoma City. All of the remaining figures in this chapter are excerpted from the lecture I gave there.

In the classic sense, fan deltas are defined as alluvial fans that build into a standing body of water, although that definition was expanded in my report to include other types of braided streams as the source for the deltaic sediments (e.g., braidplain deltas). Holocene fan deltas occur in many parts of the world in tectonic settings such as forearc embayments and divergent plate margins, where steep, typically geologically young mountains form the shorelines of oceans and lakes. Because of the steep gradients of the source streams, the sediments of fan deltas typically contain abundant sand and gravel.

In our 1982 paper on Lower Cook Inlet (located on Figure 87), which was titled *Shoreline Sedimentation Within a Forearc Embayment,* we presented the diagram in Figure 95, which relates the depositional shorelines of the inlet to wave energy and sediment supply. In this diagram, we showed two types of fan deltas–arcuate-cuspate fan deltas on the shorelines exposed to large waves and lobate fan deltas in more sheltered areas.

Detailed stratigraphic and sedimentologic studies of Holocene fan deltas are rare, but many such studies have been carried out on Tertiary and Quaternary fan deltas in Spain, Italy, Spitsbergen, and southeastern Korea. The researchers conducting those studies have most commonly described two classes of fan deltas with regard to their stratigraphic makeup–Gilbert-type fan deltas and shoal-type fan deltas. Gilbert-type fan deltas, which build into deep offshore waters, have a tripartite configuration with respect to their general stratigraphy (see model in Figure 96): 1) A flat delta-plain surface, upon which flat-lying topset beds are deposited; 2) A steeply sloping (20-30 degrees) delta slope, where highly inclined

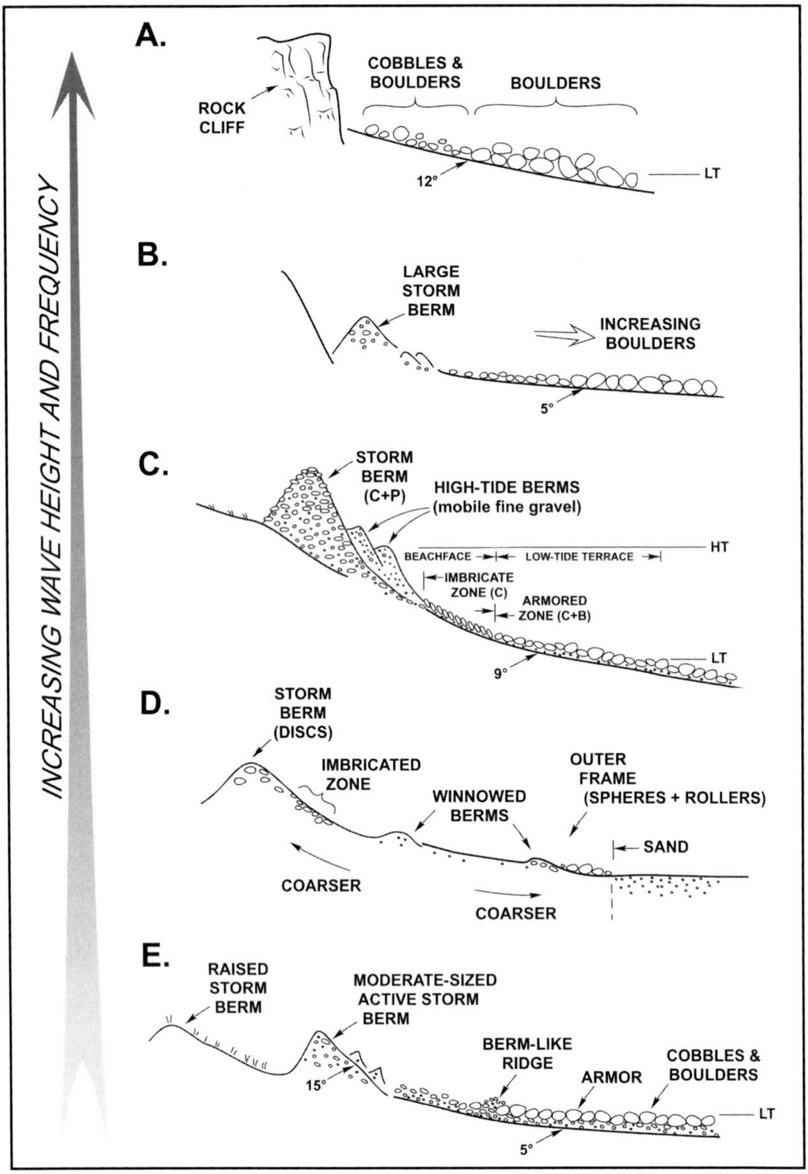

FIGURE 94. Ranking of coarse-grained gravel beach types with regard to increasing wave height and frequency, with the most energetic site at the top of the diagram and those with lesser levels of wave action at the bottom. (A) Large boulder beach in front of a wave-cut rock scarp on the southeast coast of Australia that was described by Oak (1984). (B) A gravel beach exposed to open-ocean waves on the Pacific coast of Montague Island, Alaska (station PB-1 in Figure 88). (C) The open-ocean gravel beaches similar to the ones on the Pacific coast of Baja California (see also Figure 89). (D) The gravel beaches of Ireland and Wales, modified slightly after Bluck (1967), Carter (1988), and Orford, Forbes, and Jennings (2002). (E) Intermittently exposed coarse-grained gravel beaches of Prince William Sound, Alaska (see also Figure 90). From Hayes, Michel, and Betenbaugh (2009).

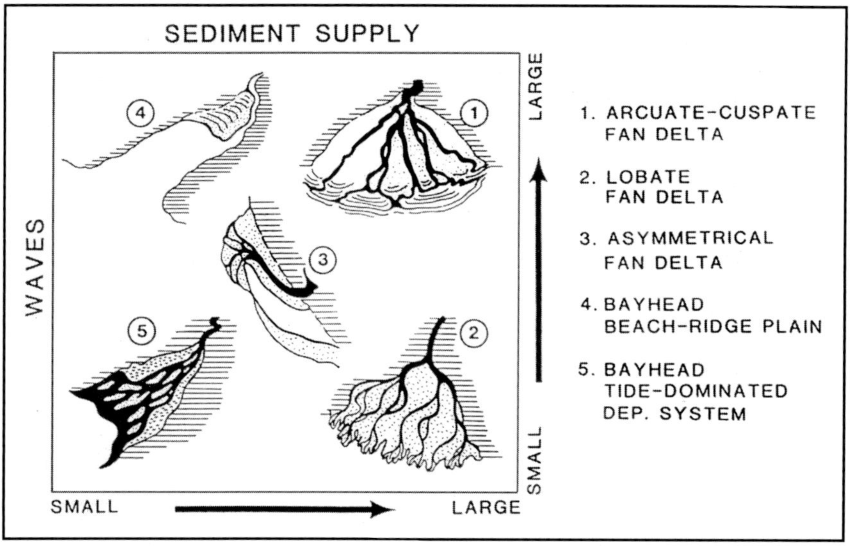

FIGURE 95. Relationship of wave energy and sediment supply to the morphology of depositional systems along the shoreline of Lower Cook Inlet, Alaska (located on Figure 87). Note the three types of fan deltas. From Hayes and Michel (1982).

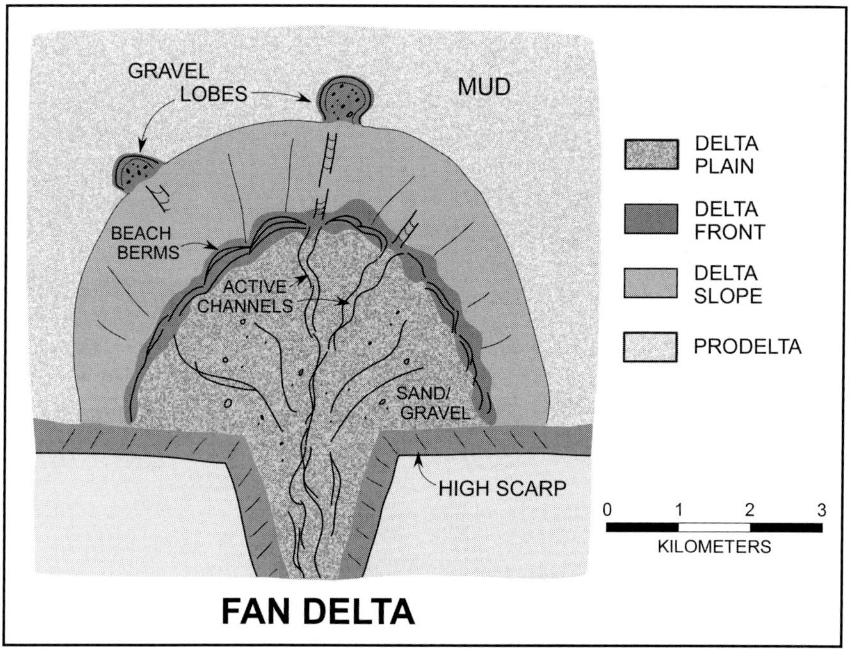

FIGURE 96. General model of Gilbert-type fan deltas. The four main morphological subdivisions of such fan deltas–delta plain, delta front, delta slope, and prodelta–are shown.

foreset beds are deposited; and 3) An offshore prodelta region, where flat-lying bottomset beds occur. Shoal-type fan deltas, which build into shallow water, do not have steep delta fronts. Gilbert-type fan deltas are by far the more common of the two classes.

In the Devon Energy report, general geomorphological models were given for Gilbert-type fan deltas that show the differences between those exposed to heavy wave action (wave dominated; see example in Figure 97), and those in areas sheltered from wave action (sheltered; see example in Figure 98). Wave-dominated fan deltas typically contain extensive buildups of beach berms composed of gravel or sand-and-gravel all the way around the delta front, whereas the more sheltered fan deltas have digitate margins with numerous stream outlets. In shoal-type fan deltas, sandy mouth bars are said to develop off the stream mouths similar to those on river-dominated deltas. Such mouth-bar deposits are not thought to be common on Gilbert-type fan deltas, however, because of the precarious nature of the steep delta fronts where the streams discharge their sediment load. Sediment deposited at that location is usually carried immediately down the slope by gravity-driven processes.

With respect to their general geomorphology, typical Gilbert-type fan deltas have four major geomorphological components, each of which has distinctive sedimentological characteristics:

1) Delta plain – This area slopes gently seaward and typically contains numerous channels of braided streams. Sediments are typically coarse grained and can show a wide range of textural properties, including poorly sorted mud flows in arid settings, well-sorted imbricated gravels and cross-bedded sand in humid areas, and debris flows with complex bedding in both climatic settings.

2) Delta front – This area, where the delta plain abuts the sea or lake, most commonly consists of extensive gravel beaches on wave-dominated fan deltas, which contain well-sorted gravel clasts (usually predominantly disc shaped) and steeply dipping beachfaces. Imbricated cobbles and armored cobble and boulder zones within the gravel beaches have high values of porosity and permeability. River outlets, which produce scour-fill sand-and-gravel deposits, dominate along the front of the more sheltered fan deltas.

3) Delta slope – This is an area of high sediment instability, with a wide range of gravity-driven, sediment-transport processes at work, including creep, slide, slump, liquefied flow/debris flow, turbidity currents, and debris fall. These processes combine to produce accumulations of massive coarse-grained deposits at the toe of the slope. The understanding of these processes and their products has been greatly augmented by study of side-scan sonar images and other criteria of Holocene fan deltas in the fjords of British Columbia (Prior, Wiseman, and Gilbert, 1981; Prior, Wiseman and, Bryant, 1981; Prior et al., 1982).

4) Prodelta – A zone where the sediments are mostly fine-grained silt and clay that have settled out of suspension from sediment plumes emanating from the delta front. Turbidity-current deposits may be present in some areas.

FIGURE 97. Wave-dominated fan delta at the outlet of Silver Salmon Creek on the western shoreline of Lower Cook Inlet, Alaska. Note the large intertidal bar on the lower left, which is composed of sand and gravel. Photograph taken during low tide in May 1976.

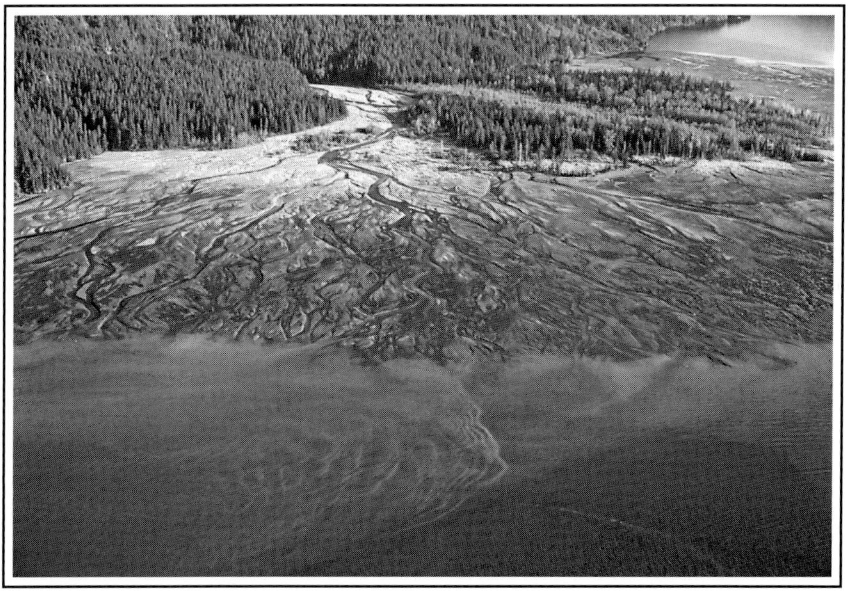

FIGURE 98. Halibut Creek fan delta in an area sheltered from heavy wave action in Lower Cook Inlet, Alaska. Note the absence of wave-formed, coarse-grained beach deposits. Photograph taken during low tide in May 1976.

In the report, general models were presented that ranked the different components of fan deltas on the basis of their potential as prospects for containing sufficient quantities of oil and gas to be targets for exploration (the zones with the greatest porosity and permeability). For example, as shown in Figures 99 and 100, on wave-dominated Gilbert-type fan deltas, the top-ranked prospect is the ring of beach berms that line the delta front. If the delta contains gravel, a common occurrence on fan deltas, the beach gravels would be ideal targets. Second priority ranking goes to the massive gravel lobe and coarse debris fall deposits that accumulate at the toe of the delta slope. The third priority was assigned to sediments in the main outlet channels, and fourth priority to the potential for permeable beds being deposited in some parts of the upper delta plains.

The Story – Gravel Beaches and Fan Deltas

(5 March 1991 – International Oil Spill Conference, San Diego)

The year of 1991 was extremely eventful for us in several ways, not the least of which was our beginning to present what we thought was the most detailed summary available on the topic of oil spills on gravel beaches. In April of 1990, I had participated in my first fieldwork at the *Exxon Valdez* oil spill site, where almost all of the beaches oiled were gravel beaches. This field survey was followed by my

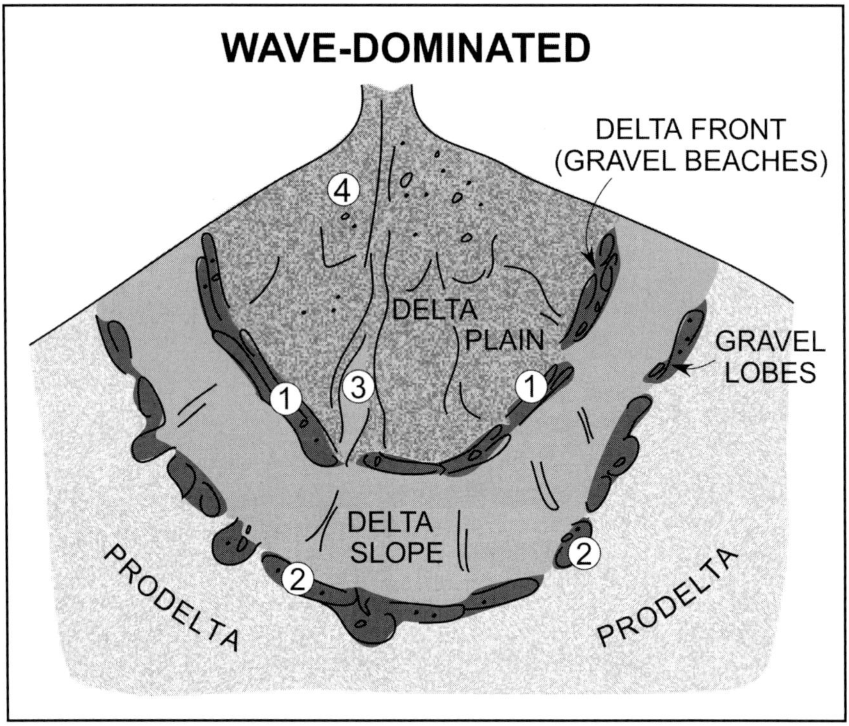

FIGURE 99. Plan-view sketch showing the parts of a prograding, wave-dominated Gilbert-type fan delta that highlights the sediments that have the greatest porosity and permeability (i.e., most likely to be productive of oil and gas if drilled into). 1 – best prospect; 2 – second best; 3 – third best; 4 – fourth best.

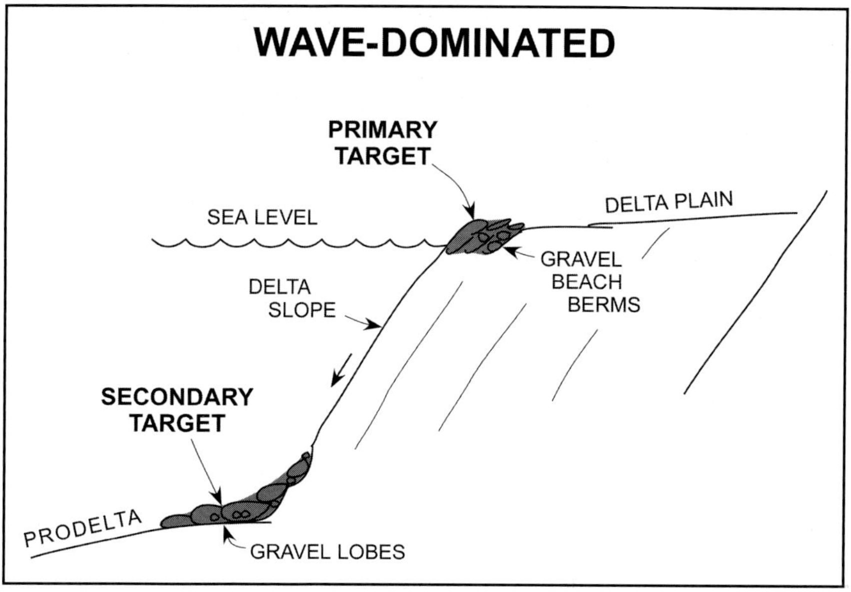

FIGURE 100. Cross-sectional views of prograding, wave-dominated, Gilbert-type fan deltas showing the two primary exploration targets: 1) Gravel beach on delta front; and 2) Gravel lobe at base of delta slope. Compare with Figure 99.

participation in the "Great Rock Washer Debate" as a representative of NOAA, at JM's request. She was the chairperson of that particular debate, which became a shouting match between representatives of Exxon and the State of Alaska, one of the major issues being how to deal with the heavily oiled cobble and boulder beaches in Prince William Sound.

Early on, in May 1990, we published our first paper specifically on oiled gravel beaches entitled *Oiled Gravel Beaches: A Special Problem*. (Hayes, Michel, and Fichaut, 1990). Then we did a survey of our permanent stations at the *Exxon Valdez* spill site again in May and June of that year, followed by the airplane crash on 12 August (discussed in Chapter 15 – BY FOOT, BY CAR, BY BOAT, BY PLANE, and in Chapter 16 – DOOMNESS).

My next field trip to Prince William Sound was in late January 1991 aboard a U.S. Coast Guard vessel the *U.S.S. Sweetbrier*. Despite the waist-deep snow in places and the cold weather, we were able to resurvey the 25 stations we had originally planned to do, most of which were gravel beaches, of course. A photograph of this field crew is given in Figure 101. I had some trouble getting my injured leg and foot in and out of the survival suits. Steve Lehmann, a NOAA Scientific Support Coordinator on the trip, helped me do this, commonly referring to himself as "Miles' Gentleman's Gentleman." Despite all this extra maneuvering, I enjoyed being there.

Later that year, in the first week in March, JM and I were in San Diego to attend the International Oil Spill Conference, where we both gave papers. The weekend before the conference started, we had a good visit with my daughter Joy

FIGURE 101. Field crew on the deck of the U.S. Coast Guard ship *U.S.S. Sweetbrier* in Prince William Sound in late January 1991. From left to right, the team members are Joe Goulin, David Sarena, JM, Debra Simecek-Beatty, Steve Lehmann, Ivan Nance, and me.

and her boyfriend Steve, both Ph.D. candidates at UC San Diego in La Jolla.

I observed in my journal that in attending this conference, "I accomplished exactly what I came for," by doing the following:

1) Gave a paper and showed my face so as to let people know that I was still in the oil spill business, which appeared to be booming at that time. JM also gave a paper on the natural trends of clean-up of the *Exxon Valdez* spill.

2) Learned as much as possible about the status of oil spill research up to that date by attending papers, reviewing all the exhibits, and so on.

3) We were also supposed to do a little marketing. JM said she had obtained lots of leads, so the marketing also seemed to be going pretty well.

There are some days when it seems like you just can't lose for winning. Tuesday, 5 March 1991, was one of those days. It was 2:00 p.m. that Tuesday, and I was sitting on the podium with the four other speakers in one of the afternoon sessions at the Conference. The title of my paper was "*Factors Controlling Initial Deposition and Long-term Fate of Spilled Oil on Gravel Beaches*," which was published in the Proceedings of the Conference (Hayes, Michel, and Noe, 1991).

I had spent that entire morning in our motel room with JM, who was violently ill with food poisoning. She finally got better, and we went to the day's luncheon at the Conference. I was not feeling all that well either, but I ate what was served me, including a big piece of roasted red meat.[34]

Ann Hayward Walker was the first speaker,[35] and as she spoke, my stomach ache got progressively worse, until I was virtually doubled over in pain. As she talked, I kept thinking, "My God! I'm going to have to tell the Chairman that I can't go through with this." The pain just kept getting worse and worse.

The second speaker, Blair Humphrey, one of Ed Owens' associates, talked about a spill exercise in the Arctic. About two thirds of the way through his talk, I began to feel much better.

I was up next, so when I stood up to start speaking, I felt pretty good. However, my fourth slide didn't drop and the stupid projectionist's solution to that problem was to take all of my slides out of that tray and put them into another one. I usually have at least 40 slides in my talks, so that was going to take a while. Normally, this kind of delay would have shaken me up, but the fact that I was no longer doubled over with pain influenced me to stay in a jovial mood.

So, as the projectionist fumbled with the slides, I said, "Well, this gives me a chance to tell you about my fishing trip to the San Juan River, New Mexico last week."

[NOTE: Before coming on out to the west coast, I had stopped in Boulder so Dan Domeracki and I could go trout fishing. I had not been fishing since the airplane crash, and the guys, including Dan, had given me a new, very fine fly rod as an award for not dying in the crash, I guess. Therefore, I was anxious to try out my new fly rod. Anyway, at that time of year, to be successful you need to fish a tailrace fishery like the San Juan, where the temperature of the water draining from the bottom of the lake stays warm enough for the fish to feed. When we were there, the big trout in the river were feeding on tiny midges, and I tied a fly, we called the MoDan Midge, which matched the hatch perfectly, so we caught several of the big ones.]

Back on the podium, I said, "I caught a bunch of big rainbows!"

To demonstrate my point, I held my hands wide apart with the microphone in between them. About half of the audience laughed. The other half looked confused.

I went on, "Yeah, caught them on a #24 midge pupae, that small!"

Then I held my right hand with the thumb and the first finger about 3 mm apart. More laughter this time.

Then, while I was trying to think up a good joke to tell, some guy stood up and asked if he could direct a question to Ann Hayward.

"Oh no! I don't want you eating into my time." And at this remark, everybody laughed. No wonder, there I was trying to think up a joke, and I didn't want to give up any of the 20 minutes allotted for my talk? Anyway, I did relent, and he started a discourse with Ann Hayward.

As they wound down their discussion, the projectionist finally flashed up my fourth slide on the screen. But, three or four slides later, he had put two slides into the same slot, killing my "don't use the word rock for gravel" joke.[36]

I guess I sort of recovered after that, but the damn chairman didn't reset the clock, so the light turned red about 10 minutes after I started. Then, I reluctantly decided not to do the last third of the talk. There were a variety of questions, but my 20 minutes were up soon, so I sat back down beside Ann Hayward, who said, "You're a natural-born teacher, Miles."

"I didn't think I had a prayer of making it through that one," I whispered back to her, explaining the upset stomach.

After the session was over, I starting feeling ill again; therefore, I decided to go back to the motel, and walked out of the entrance to the Convention Center. A lady named Stacy from the Chugach Native Corporation, with whom we had worked with some in Prince William Sound, joined me and we walked away together. I told her about being sick.

"I thought you were looking bad," she said. "I was mentally rehearsing my CPR."

We walked along silently for a while, and then she said abruptly–"Your talk was wonderful. You warmed everyone's heart."

18 THE SOUTHERN CROSS

No - The stars are close and dear and I have
joined the brotherhood of the worlds.
And everything's
holy - everything, even me.

JOHN STEINBECK in GRAPES OF WRATH

The Science

(17 April 1991 - Muscat, Oman)

Sitting in my palatial hotel room in Muscat, Oman, I was composing a fax to JM and Jack Moore while nursing my sore right foot. This was day two of a 21-day excursion to begin the description of the 1,600-kilometer coast of Oman as part of a project to establish legal guidelines for the future development of that essentially pristine shoreline along the Indian Ocean. My job was to map the geomorphic provinces of the coast, which would include detailed descriptions of 70 individual stations (eventually turned out to be 85), located at evenly spaced distances along the shore.

The next morning after sending the fax, I headed toward our first field site at Ras al Hadd, a major headland southeast of Muscat, the country's capital (see Figure 102). I traveled under the guidance of one Rod Salm the Turtle Man, whose main mission appeared to be to prove to me the foolishness of my plan to survey 1,600 kilometers of shoreline and establish 70 stations in 21 days during this trip, and a slightly shorter follow-up visit in the fall. Rod was a native white East African, from Mozambique I think. He was a professional, turtle-expert-for-hire, who worked all over the Indian Ocean. As we moved along, he drove like a madman over the gravel roads, getting lost here and there, turning around several times, and then detouring to look for dead turtles. I guess I was supposed to be impressed with how hard it would be to work in Oman or something like that. As you already know, if you have read the earlier part of this book, previous to this project I had worked in some pretty exotic and difficult places (e.g., Antarctica, Magnetic North Pole, Nigeria).

Anyway, I was glad to get back to Muscat, get rid of old Rod, and set off toward the south with my own crew, which consisted of Steve Sturm, a geologist employed by RPI/Colorado, which was still holding on by a skimpy thread, Doug Kneale, the logistics officer for the project, and Freddie, the Indian driver.

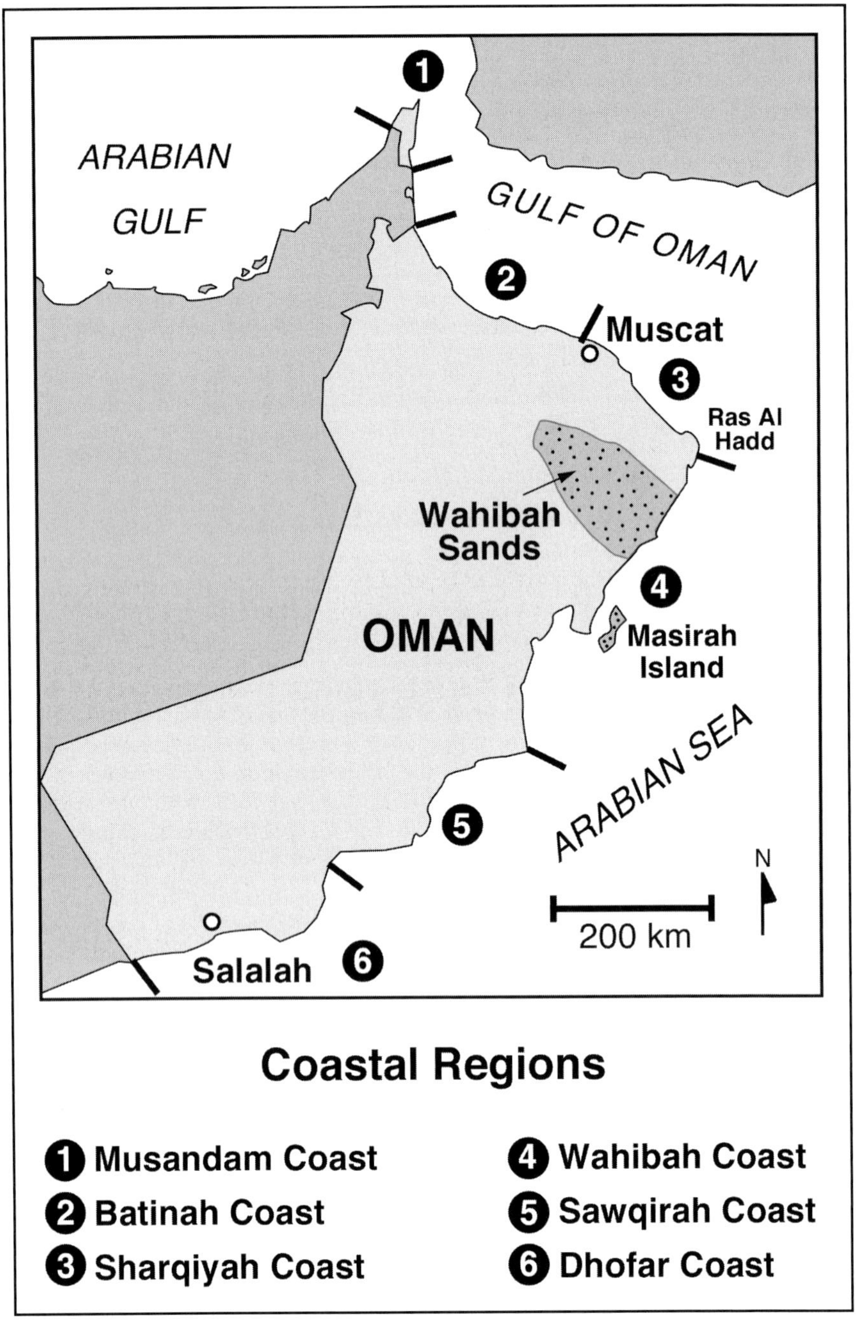

FIGURE 102. Map of the shoreline of Oman showing the location of the six geomorphic compartments.

I was back in October for an 18-day field expedition, this time with associates Doug Kneale, who had assisted with the earlier survey, and Bill Baird, a coastal engineer from Canada, founder and President of Baird and Associates, headquartered in Ottawa. Bill Baird, and a partner in his company, Rob Nairn, whom I introduced earlier in the discussion of Ghana, were a somewhat different brand of coastal engineer, in that they loved coastal geomorphology and were very good at it themselves. Bill and I did write a paper on the shoreline erosional/depositional patterns in Oman (Hayes and Baird, 1993), but I am remiss that we didn't do more publishing on this phenomenal shoreline.

With regard to the geomorphology of the Oman coast, in our preliminary draft report for the project (James Dobbin and Associates, 1992), we subdivided the shoreline into six major geomorphic compartments, which are located on the map in Figure 102. A brief description of these compartments, going from north to south, follows:

1) Musandam Compartment – Isolated to the north, this coast is bounded by two borders with U.A.E. and flanked by the Gulf of Oman on the east and the Arabian Gulf on the west. For the most part, this coastline is steep, rocky, and mountainous, and also spectacularly beautiful. The rocks of Musandam are a thick sequence of Permo-Mesozoic shelf carbonates that have been eroded into high stark cliffs along the promontories of the coast.

2) Batinah Compartment – This shoreline flanks the broad wadi plain lying to the northeast of the northern Oman mountain range. It is the most populated of the geomorphic compartments, and it is flanked by long, mostly sandy beaches. As shown in Figure 103, a subtle topography developed on the plain during the last lowstand of the sea, which peaked about 20,000 years ago. During the lowstand, the major wadi fans had a depositional, cone-like shape, with topographic lows occurring between the major fans. When sea level reached the present relative still-stand, the shoreline of the Batinah was quite irregular, with embayments occurring between the headlands created by the major fans. Since this relative still-stand, however, the shoreline has undergone a straightening process, whereby the headlands began to erode and the embayments were filled in with sediment. The result of this process has been the occurrence of interspersed erosional and depositional areas all along the Batinah shoreline. Another discovery we made was that the beaches there are backed by what we called the "old raised beach ridge," illustrated in Figure 104. Bill Baird and I concluded that the presence of this feature meant that probably sometime after sea level was near where it is now, around 5,000 years ago, there was an abrupt uplift of the land of about a meter of so, creating this raised beach ridge. We intended to do more research and write a paper on this discovery but, unfortunately, we never did.

3) Sharqiyah Compartment – The western boundary of this compartment is located just north of Muscat, and it extends in a southeasterly direction around the prominent headland on the Oman shoreline (Figure 102), Ras al Hadd, as well as a bit further to the south. This is mostly a mountainous coast with some embayments, fan deltas, and a major cuspate spit at Ras al Hadd. This is the most complex (in detail) of all the compartments,

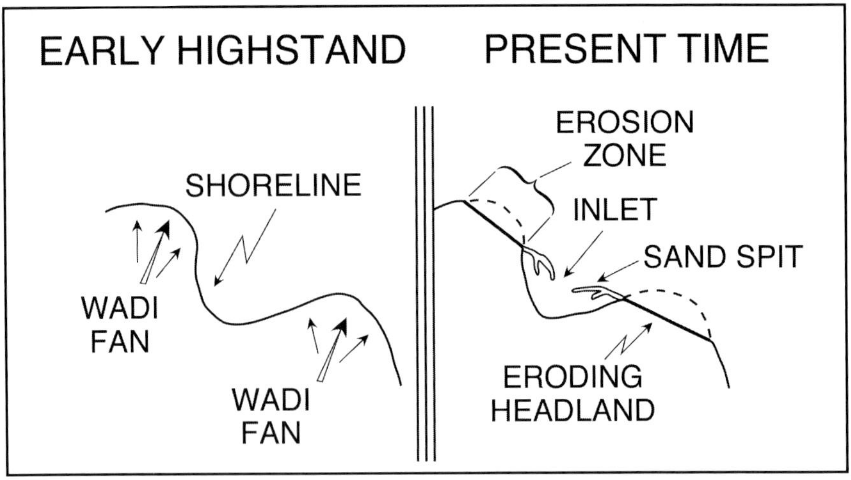

FIGURE 103. Evolution of the coast of the Batinah Compartment (Figure 102) from the time when sea level first came up to near its present level around 5,000 years ago and the present time. With the highstand, the outer margins of the alluvial fans built when sea level was out are being eroded, and the topographic lows between the fans are being filled in by longshore transported sediment.

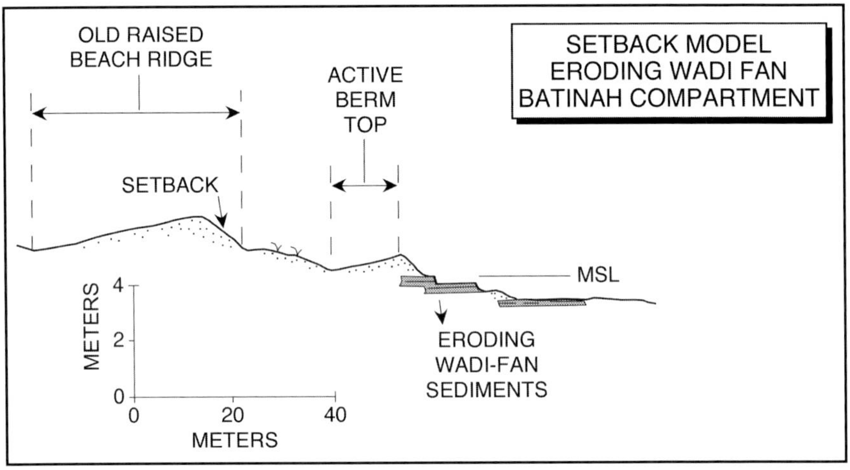

FIGURE 104. Configuration of the beach zone in the Batinah Compartment (Figure 102), showing the presence of an old raised beach ridge.

and it has a wide range of beach sediments from coarse gravel at the wadi mouths to pristine carbonate sand beaches in some minor indentions in the shoreline.

4) Wahibah Compartment – This coastline, the longest of the compartments that extends from north to south a straight-line distance of about 450 kilometers, includes the eastern border of the extensive Wahibah Dunes, Masirah Island, Barr al Hikman (these three areas are discussed in Story

Two), and three segments of shoreline dominated by large crenulate bays, the most spectacular examples of which I have seen anywhere. Accordingly, this compartment has a variety of geological components, as well an assortment of sediment sources and sinks.

5) Sawqirah Compartment – Perhaps the simplest of the compartments, this shoreline is essentially made up of three major wadi plains separated by broad headlands. It is characterized by a wide continental shelf, relatively small waves, and significant storm-surge flooding.

6) Dhofar Compartment – This, the southernmost of the compartments, extends all the way to the Yemen border, which we reached, but did not dare to cross. Most of the area contains a high mountain backdrop, except for the large wadi plain, upon which the beautiful town of Salalah is located. Of some note is the fact that our seasonal beach profiles showed a *bona fide* winter/summer beach at Salalah, the first well-documented one I had ever seen. In this case, however, the beach erodes in the summer, the monsoon season, and accretes in the winter. Our winter profile at Salalah showed accretion of many tens of meters in comparison with the much narrower beach in the summer. In a few localities, such as at Hasik, which is illustrated by the very high altitude photograph in Figure 105, fan deltas have been built at the mouths of the steep-gradient ephemeral streams in the V-shaped wadis, resulting in the development of extensive gravel beaches along the delta fronts. Another interesting fact we discovered about the white sand beaches in Salalah was that they are composed of >95% foraminifera tests.

In our preliminary report to the Ministry of Regional Municipalities and Environment (James Dobbin and Associates, 1992), we pointed out the key erosional areas, where shoreline development projects would have problems, the "evils" of mining sand from the beaches, a prevalent practice, and so forth. However, although Bill Baird continued to work on the Oman coast for years, I never went back after my second trip, so I do not know how well our recommendations were followed. However, if they did do what Bill recommended, they probably fared well.

Story One

(6 May 1991 – On a jet flying over the Atlantic Ocean)

Even though I had been away from the office during that first three-week survey of the Oman shoreline, our business (RPI) was still in deep financial trouble, but I had left it in the capable hands of JM for the environmental side and Jack Moore, our Chief Financial Officer, and Dan Domeracki for what was left of the energy wing. I still had a lot to worry about in that regard, thus it was not the easiest thing in the world to focus on the work we were doing in the field in Oman. Therefore, I wasn't in the mood to keep a journal. Somehow, it (such writing) just didn't seem to be all that significant.

I did, however, finally write the following discourse, part of my summary of the happenings during the first field excursion, while riding on a B-747 jet over the Atlantic Ocean on my way back to Atlanta.

In short, I was feeling pretty uneasy, as I wing-and-a-prayered my way back

FIGURE 105. High-altitude photograph of the shoreline near Hasik, Oman in the Dhofar Compartment (Figure 102), showing fan deltas being built at the outlets of steep-sided wadis. The beaches of the fan deltas are composed mostly of gravel carried to the shore by the streams in the wadis. Courtesy of the Omani Government.

home. While in the field, we spent a lot of time bouncing over 2,000 kilometers of gravel roads. We saw a big piece of the coast of Oman, my first detailed look at an exposed plateau-shield coast.

We slept under the stars five nights and there was something special about that. I experienced considerable discomfort and uneasiness most of the time (because of the injuries suffered in the airplane crash). But, I was still working in a fairly concentrated way.

On our last day there, we met with the Ministry of the Environment and some USAID/Embassy bureaucrats in two separate meetings. They said I was "the star of the show," which I guess means that they laughed at my jokes. We promised to deliver too much like a bunch of young kids who had never done this before. Oh well.

Some of the memories about that drive in the desert still stood out in my mind as I wrote in the journal. For example, one night when we were trying to find a place to camp, the wind was howling and the sand was blowing, and the only suitable place we could locate to place the tents was to snuggle them behind some bushes down in a huge gully.

Steve Sturm said, "Wait till I get back to the office and tell everybody that M.O. Hayes wanted me to set up camp in a deep gully in a flash-flood-prone desert!" We all slept like babies. It had been a long day.

In the late afternoon of day seven of the drawn-out drive, having spent all that time bouncing over ruts and coarse-grained gravel endlessly in our four-wheel-drive Toyota Land Cruiser, Freddie gave a shout for joy as he bounced up onto the paved road that we would follow all the way back to Muscat.

As we moved along, we all looked around at each other, saying almost simultaneously, "We're flying!" It felt as if the Land Cruiser was floating on a huge cushion of air, and that sensation didn't go away for fifteen or twenty minutes.

As we came within 30 kilometers or so of Muscat, we noticed that large numbers of people were lined up along the road holding flags and waving at us as we passed by.

Steve, a typical New Yorker, started making wisecracks right away, such as, "Hey, Miles, I thought you were pretty famous, but a welcoming committee fit for the President?" and "Just think, when you told me this would be an important project for the people of Oman, heck, I didn't believe you."

We waved and laughed our way on into town, as the crowds grew larger and larger along the road. Turns out the Sultan of Oman was entertaining the Emir of Kuwait in one of his palaces, and we were riding into town a few kilometers ahead of the Sultan's entourage. During our absence, hundreds of signs bearing pictures of the Emir had been erected along the median of the main thoroughfare.

The legend of the Sultan of Oman goes something like this. About twenty or so years earlier, the young man, who had just returned from college in England, bought for his father, who was apparently a backward and tyrannical leader, an airplane ticket, put him on a plane to elsewhere, and told him not to ever come back. Then he took over the reins of the country. In the following years, the new Sultan initiated many reforms in the country, building new roads and infrastructure, generally raising the standard of living for most of its citizens, and so on. I was most impressed with what he had accomplished in such a short time in a country that did not have near the wealth in oil-and-gas resources as was possessed by its neighbors to the north.

Later that summer, we were back in Prince William Sound, Alaska, continuing our follow-up studies of the *Exxon Valdez* oil spill. One night as the six of us were in our bunks in the crowded sleeping quarters of Captain Randy Becker's fishing vessel, *The Renown*, we started discussing what we would like to be in an alternate life, were such an option available. Pauline, the chemist, said she would like to be an Antarctic explorer (she was wintering over at the South Pole when I first wrote this). JM said she would like to have a great singing voice, and I said, "I would like to be the Sultan of Oman."

My comment caused a lot of merriment on the part of the others, especially Captain Randy. Next morning, he had printed the following in bold letters on the back of one of our blank data sheets and taped it on the door of the head - THE SULTAN'S THRONE.

Story Two

(1 October 1991 – In a jet flying over Iraq)

On the first of October that same year, I was in a jet flying over Iraq on my way back to Oman for an eighteen-day field excursion. I felt depressed again, this time about the wearing erosion of the oil-and-gas business and the unbearably slow pace we were making in affecting a turnaround. Being rather hard on myself, I thought, "I can sit in my office all day and not get a single phone call. That happened two or three times last week. Looks like nobody knows or cares where I am or what I am doing these days. I guess that's what six years of rot and decline of the business can do for you. The energy side now has three employees. Dan doesn't call much anymore."

At that time, I was into country music, because it seemed to fit the mood of the times better than any other kind. Songs like *I'm No Stranger to the Rain* sung by Keith Whitely were humorously rejuvenating.

As I looked down on the burning desert of Iraq, the words of a popular country song were running through my mind, something about the singer wishing his lover were still there so he could watch her walk away again.

Then I wrote in my journal, "Maybe it was better when the creditors were calling all the time. At least they cared what I was doing." And later I added, "Of course, JM cares what I am doing. If I don't get some more projects soon, she is going to collapse under the weight of carrying the company alone."

Next I wrote - "TURBULENCE, of course."

That was the bad news. The good news was that in a couple of days, I would be back in the field in Oman again, completing the survey we started in April. This time, my field associates were Doug Kneale, who had assisted with the earlier survey, and Bill Baird, a coastal engineer from Canada. One night in camp, we started telling each other our life stories. Doug, who led off the discussion, had lived a phenomenally nomadic life, though still only in his early thirties, managing one after the other failed aquaculture operation all around the world. And he was now the logistical officer of this rather complex coastal zone management effort. I was next, and I put emphasis on the science, downplaying to some extent our struggling business, but they were quite surprised to learn that at one time we had around 200 employees in six separate offices. Bill didn't talk much, except to say that his father was a famous and successful engineer, who traveled a lot and sent him off to boarding school at an early age. He grew up on the coast of England and was an accomplished sailor and swimmer. A very energetic and amicable person, he was the principal of his successful coastal engineering company. Unlike many engineers I have known, Bill has an unbounded curiosity about how the natural system works, including the geological aspects, which was one reason we had such an excellent rapport and working relationship. Both Doug and Bill were in fine physical condition, thus I had some keeping up to do while still hobbling a bit on my reconstructed leg and foot.

One morning during the middle part of the expedition, we were preparing to begin a trip to survey Masirah Island, which is located off the east-central shore (Figure 102). By 8:00 a.m. that morning, we were off to the national survey office to put the finishing touches on our study of the aerial photographs of the country's shoreline, slavishly finishing the most populated coastal area, the Batinah, which is located to the west-northwest of Muscat. By then, we had discovered the effect of inherited topography on the erosional patterns of that part of the coastline (see Figure 103). The villages that were losing the most houses were located on protruding headlands created by ancient alluvial fans. We felt pretty good about having "figured out the Batinah."

In the afternoon, we went to the airport and took a commercial flight to Masirah Island. The Twin Otter we boarded was somehow able to take off in the scorching air, and as we banked over one of the old alluvial fans on the Batinah called Seeb, I excitedly pointed out the outcropping of the eroded face of the primeval fan to Bill.

After turning back across Muscat and heading out over the water, I was treated to outstanding views of the shoreline for many miles as we traveled south. From a place called Sur, we flew up over the coastal wadis (dry river valleys) and mountains, doing some bouncing as we crossed the crests of the *jebels* (Arabic word for mountains). Then we passed over a magnificent dune field called the Wahibah Sands, which seemed to go on and on. It was the best view I had ever had of a desert dune field. The dunes became larger as we once more approached the shoreline, assuming a reddish hue that I didn't understand. We had taken a short cut behind the large headland located near the middle of the Omani coast (Ras al Hadd; Figure 102). Then we crossed the Masirah Strait, which separates the mainland from Masirah Island. The water in the strait contained linear current plumes running from northeast to southwest that were sending boils of suspended sediments to the surface. This was a very fierce and impressive current, which was generating large underwater waves of sand that were clearly visible from the air.

Our host, as it were, for this brief visit to Masirah Island was an anachronistic German biologist around 40 years old, who was conducting a long-term study of the biota of the island. I believe his funding support came from the European Community. In my notes, I called him Achtung Julian (AJ), because of the autocratic way he managed his field camp and ordered his Arab subordinates around. He was about six feet, five inches tall and slim, with rather long hair. He puffed on a cigarette hanging out of the side of his mouth as we were introduced. I thought, "This guy is a space wanderer, I have met him before." In a metaphorical sense, that is.

An Omani driver named Saud that Doug had chartered for our two-day stay on the island met us with his car. We piled into Saud's Toyota and followed AJ to his neatly furnished bungalow, which did not have a wooden or sawdust floor. No, the floor was covered with sand. Native sand, AJ called it. We went inside and sat around a large dining table.

Enter the lovely Sylvie, AJ's "field assistant." She was lean and tan and moved over the sand like a dream, pouring us cups of coffee and tea. As we lingered around the table sipping our drinks, it was blazing hot outside, Bill scooped a handful of sand off the "floor" and suggested that I examine it under my hand lens, so as to discover what riddles the sand might pose for us.

My eyes nearly jumped through the lens when I saw the large, coarse-grained, perfectly spherical and frosted quartz in the sand, which, judging by the shell material in it, was beach sand. Frosting is caused by tiny chinks in the surface of the quartz caused by impacts from other grains of sand. The formerly transparent quartz is thus rendered translucent, having a milky appearance.

"That quartz has to have come from a dune field," I told Bill, who appeared to be fascinated to learn what frosting was and its significance. Frosting cannot be developed under water, because of the cushioning effect of the water weakens the strength of the impact between the grains. Such texture can only be formed in the air, thus the sand must have obtained its frosting in a dune field like the one we had flown over at the Wahibah Sands area. This was another one of those facts about sediments that I had learned from the great Robert L. Folk at UT. Therefore, I concluded that, though the sand had been collected from a beach by AJ's staff, it originally inhabited a desert dune field.

Doug was anxious to get under way, urging us back out into the sun. As AJ's boys loaded some camping gear into Saud's Toyota, first Bill and then I wandered over to view Sylvie's lizards she was studying. She was telling Bill about how she traveled to the "large inland seas" in Canada when she was a child.

We finally exchanged pleasantries with AJ and Sylvie, making ready to leave.

Doug had plans to rush right out into the field to survey our first station. The sun was dropping and it was getting cooler. Saud had other plans. First, he pulled into a service station and filled his dead-empty gas tank.

"They all do this," Doug muttered.

Then we had to get a tire. Then we had to stop at Saud's home so we could exchange him for his son, an exchange that never took place, for some reason.

Then he wanted to get some fish for dinner.

"No, dammit," I said, "We won't make the station on time."

Finally, we were on our way to do the survey. The first stop was about one third of the way down the east side of the island, where we had spotted a clearly defined old raised beach ridge on the aerial photographs. Sure enough, it was the best example of the feature we saw anywhere in Oman (see Figure 104).

As mentioned earlier, Bill and I had come up with a theory that there had been an abrupt elevation of the Oman shoreline, probably caused by a major earthquake, sometime between 3-5,000 years ago. This had left a distinctive imprint on the geomorphology along many parts of the shoreline, including stranded, high-level beach ridges like this one.

It took us a while to complete the survey line, which was considerably longer than the average profile, because it included the raised ridge. And would you believe there was frosted quartz in the sand?

While we were running the profile, Saud said, "If I only had my net, I could catch us some fish." He was very concerned that we should have a fine evening meal, probably at AJ's urging.

It was 5:00 p.m. when we reached the next profiling station, and we completed it just at dark. We headed on south, to set up yet another field camp by feel.

Not to be thwarted in his efforts to have fish for dinner, however, Saud pulled into his mother's place, a forsaken shack of poles well off the highway, to get one from her. We also loaded some wood for a fire.

AJ had established a permanent campsite, namely a large tent on a rocky terrace, just at the southern tip of the island. We pulled up to the tent, flashing the headlights on it so we could see to unload our gear. The first thing Saud did was grab our large flashlight ("torch") and search the tent floor meticulously for scorpions. And, not finding any, before long he had a fire going and was happily cooking his fish, while Bill and I struggled with AJ's stubborn camp cots.

At first, I was reluctant to try the fish, but I did eventually, and it was quite good. After dinner, we sat around AJ's table for a long time and talked. Saud worked at the American air base on the island. He mentioned how hard the numerous blacks and women that came to the base during the Gulf War had worked, among other things.

In answer to my question concerning what he thought about the war, he said that too many people had died and that there was "too much crying." He also expressed a pretty low opinion of the Kuwaitis in general, an assessment that appeared to be universal in Oman.

I also asked him to tell us about the tropical cyclone of 1977, which we had read about in our research, and he animatedly related how the wind blew the tops off all the houses on the island. His grandmother had led a bunch of people to safety in a cave that apparently only she knew about.

"Nobody died in the storm," he said.

Then he added as an afterthought, "Only two women."

I decided to carry my bunk outside the tent and sleep under the stars. After all the lights were out, it looked as if you could reach up and touch the Milky Way. I watched for a long time, the feeling of awe sometimes overwhelming. We were camping out in the middle of the Milky Way. I woke up several times and watched the stars. About 4:00 a.m., I noticed that the Southern Cross was out, with the brightest star in all the heavens at the top of the cross.

We were on the beach again at 6:00 a.m., profiling a station in a beautiful fishhook-shaped bay by the camp. An example of one of the Masirah Island's beaches is shown in Figure 106. After a quick breakfast, we worked our way up the west side of the island, mostly taking notes on the man-made structures, one of Bill's specialties.

Saud pulled the Toyota into AJ's place shortly before noon. Doug had forewarned us that AJ would want a report on our studies, so I laid out the maps on the table. First, I showed him under the hand lens the frosted quartz that I took from the sand on his floor.

FIGURE 106. A beautiful wide, fine-grained sand beach on Masirah Island that was decorated with an abundance of preserved antidunes. Photograph was taken at low tide on 9 October 1991.

"Hmmmm."

"Take notes!" he barked at James, his Omani assistant, who wrote furiously as I quickly reviewed the major findings resulting from our brief tour of the island. I also expounded upon my theory about the Wahibian-type dune field that apparently occupied the Masirah Strait when sea level was lowered during the Wisconsin Glaciation. Achtung Julian appeared to be duly impressed, and said he wanted to learn what we found at our next stop, Barr al Hikman, the most remote and sparsely populated section of the entire Oman coast.

We said our good-byes to Sylvie and James. AJ wanted Bill to take a quick look at the jetties on the north end of the island, so we followed him in Saud's Toyota.

"Awesome, Miles," Bill said as we rode along.

"What?" I asked, and he answered the succinct and poignant way I had summarized our findings to Julian.

I had been surprised that Bill didn't have anything to add when I finished. I didn't respond to his comment, never having thought of myself as being especially articulate. Quite to the contrary, I usually tried to make up for a lack of that ability with unabashed enthusiasm. However, I had gained some experience summarizing the day on our field trips for the oil-and-gas industry on the South Carolina coast. Then I flashed back to UMass, when I was in my early 30s. I thought about how the newly appointed, 55-year-old Sea Grant Director had stood up in a meeting and summarized his new plans for the program in two brief minutes without a note or a pause or a flaw. I had felt tongue-tied at the time. There was no way I could ever have done that. Maybe that ability comes with age, probably one of those few things about a man that improves with age - so there!

After checking out the structures at the radio station on the north end of the island, we said good-bye to AJ.

"Ve vill see you again somevere in ze vorld," he said to me as we shook hands.

Doug had made plans for a driver from Muscat to meet us at the ferry landing on the other side of the strait, so we three boarded the small ferry carrying only our field packs and data. The ferry was a canoe-shaped affair about 40 feet long that held around 20 passengers, three white skins and the rest a mixed bag of Omani citizens.

"I bet there's some old dune sand down there on the bottom," I said to Bill while looking over the side as we were underway.

Near the middle of the channel, we asked the ferry driver to stop, and Bill dove overboard, headed for the bottom, which was about 30 feet down.

"Wow, look at that current!" I exclaimed to Doug, as I watched Bill being swept past the boat as he kicked to the bottom.

The driver turned the ferry around and went down current to catch up with Bill so he could clamber back on board. We looked at the sand he had collected with the hand lens and yelled excitedly about the rounded, frosted quartz and the chemically precipitated carbonate material in the sample.

"I bet these folks have some interesting stories to tell at the dinner table tonight," said Bill, referring to the gawking passengers on the ferry.

I felt good about the lowstand dune field theory as we finished the ride to the mainland, where we were met by Doug's driver to begin our survey of Barr al Hikman.

"When you write your memoirs, Miles, make sure that you record that I was the one that dove for the sample," Bill called to me from the back seat, as we headed north toward the hamlet of Nuqdak.

19 INLET PROTECTION STRATEGIES

If you send a bunch of clowns
out into the field to collect data
(and collect samples),
the laboratory work on the data (and samples)
they collect will produce
funny results.

BILLY WHISKERS (1996)

The Science

As is probably apparent from the discussions of oil spills in Chapter 8 – OIL SPILLS, and of tidal inlets in Chapter 10 – PRICE INLET, our group had a pretty good record of experience on these two topics fairly early on. Question was–how to get the two together? Of course, we thought we had been successful meshing our tidal inlet expertise with the U.S. Coast Guard Strike Team's experience with oil spills when our two groups successfully restricted the passage of oil into the lagoons and bays of Texas during the *Ixtoc I* oil spill in 1979. However, we didn't try that again for many years.

I believe it was in early 1992 that JM suggested we submit an unsolicited proposal to the Marine Spill Response Corporation's (MSRC) office in California to provide a synopsis of the relevant characteristics of the coastal inlets in the state of California, of which there are 170. In that proposal, we broadened the definition of tidal inlets a bit to include any channel-like opening that connects the sea with the mainland. Of the 170 tidal inlets (thus defined) that are present on the California coast, the most common inlets are the ones where a relatively small stream mouth, and its associated valley, meets the sea. However, there are a number of classic tidal inlets in California like the ones already described for the coasts of New England and the Georgia Bight.

The MSRC managers were interested in our proposal, and the Office of Oil Spill Prevention and Response (OSPR) of the California Department of Fish and Game also became involved, so we were awarded the project. However, early in the project these clients suggested that we also try our hand at doing some protection strategies (oil boom configurations, sand dikes, etc.) for each of the inlets as well, while we were at it. Their reasoning was that these 170 tidal inlets are the most important sites for designing strategies to protect the vital resources of the state's coastal water bodies from oil spills (e.g., salt marshes, key bird feeding areas, etc.), because it is through such conduits that oil spilled on open ocean waters could reach these resources.

This project was completed in 1993, with Volumes I and II of the final reports (Hayes and Montello, 1993a, b) covering all of the inlets on the Central California coast, providing some useful information for a recent book of ours, *A Coast to Explore: Coastal Geology and Ecology of Central California* (Hayes and Michel, 2010).

After that initial project in California, we were off and running with this activity, doing most of the tidal inlets in Florida (with MSRC support again), New Jersey, Rhode Island, and a few other areas. The field team that did this work usually consisted of the following individuals:

1) Two coastal geomorphologists or otherwise experienced coastal scientists from RPI. The team always included me. Todd Montello assisted in California, Florida, and New Jersey, and Linos Cotsapas in Rhode Island.

2) One or two responders from oil spill response co-ops or companies, who would probably conduct the actual strategy should that become necessary.

3) A representative or two from the state's environmental and/or natural resources agency.

Before going into the field, we would research all the scientific literature on the inlets in question, of course, including data from NOAA or otherwise on tidal current velocities in the inlet. For the most part, this tidal current information was pretty unreliable, so I just mostly used the geomorphic evidence available. You could almost always count on maximum tidal current velocities of 3-4 knots in the inlet throats of most inlets. Wave data were even more unreliable, except for a very few exceptions. However, we knew enough about the waves to predict whether the booms would work under those conditions or not. The coastal geomorphology was a good clue for that as well.

Once while doing protection strategies for the inlets on the east coast of Florida, a Florida state biologist with us kept nagging me about asking all the marina operators to tell us what the maximum tidal currents occurring in their areas were. Finally, I gave in and said, "Okay, go ask this one."

She did, and came back all enthusiastic and full of smiles.

"What did he say?" I asked.

"Thirty-five knots!" was her delighted answer.

The maximum tidal current I had ever heard of anyone measuring anywhere was nine knots. This and many other similar experiences persuaded me to postulate another one of MOH's laws about preparing for oil spills–A Little Local Knowledge Is A Dangerous Thing!

We also obtained the most recent vertical, low-tide photographs available for each inlet. These were usually pretty valuable, except for those inlets that are migrating rapidly. Also, once in the field, we would hire a fixed-wing, high-wing airplane and circle each inlet at different altitudes (during low tide) taking a multitude of color oblique aerial photos. Then we took the film to a photo shop to have all the rolls developed that night (this was before the era of digital cameras). Therefore, in the field, we had both vertical and oblique aerials of each inlet.

Then we got the gang together and visited every single inlet in the study area during low tide. When we finished the inspection of each inlet, I would make a sketch of it, upon which the proposed protection strategy was superimposed. We never left the inlet until everybody on the team agreed upon what we considered to be our best option, and that it had a good chance of working if executed properly. It was very critical that the response people on the team, whose group might be responsible for executing this strategy, also agreed with it.

In making a decision on which protection system or systems to use, the following hierarchy of controls dictated the final strategy:

Physical processes of the inlet
↘ Protection priorities
↘ Effectiveness of response

For example: a) If the waves are too big or the tidal currents are too strong at the entrance to the inlet, you should move further back from the entrance to where the conditions are more amenable to success; b) If the estuary you are protecting is large and complex, it may be necessary to select specific areas to deal with first when the spill response is under way. Such a ranking of specific areas would depend upon their ecological sensitivity and importance to the entire ecosystem, or possibly human-use priorities as determined by the governmental unit you are working with; and c) Choose the response technique that will work under the conditions present. In other words, don't just throw out a bunch of boom helter skelter and watch the waves wash it up on top of the marsh, which has happened at oil spills way too often.

[NOTE: With regard to protection priorities, when dealing with tidal inlets of a manageable size, clearly the first effort is to stop the oil from passing through the inlet throat. However, under adverse physical conditions (e.g., during a storm or onslaught of exceptionally high swell), this may not work. We always constructed second and sometimes third "lines of defense" further back inside the estuary in the eventuality that these unmanageable conditions might occur at the entrance to the inlet. In that case, protection priorities become important in choosing which strategies to emphasize.]

Both the state of Florida and New Jersey did actual field tests of our proposed strategies on a couple of their most relevant inlets. They worked.

The New Jersey project was summarized in the Conference Proceedings of the International Oil Spill Conference in 1999 (Hayes et al., 1999). My co-authors on that paper were Todd Montello of RPI, Rob Schrader of the New Jersey Department of Environmental Protection, and Ed Levine, the NOAA Scientific Support Coordinator whose area included the New Jersey shore.

In our final reports, we classified the inlets studied on the basis of the degree of difficulty for containment and recovery of spilled oil once it reaches the inlet. The ranking we used in California, and pretty much stayed with it in the other states, is summarized in Table 1. The scale ranges from A to D, with the inlets classed as A's being the most difficult, and consequently, the most expensive ones to deal with.

Table 1. Ranking scale for California coastal inlets, based on estimated degree of difficulty for containment and recovery of spilled oil.

Type	Description
A	Extremely difficult because of large size and extreme physical conditions. Large expense because of magnitude of resources to protect.
B	Difficult because it is permanently open; subject to strong currents and/or large waves.
C	Inlets that can be closed with a sediment dike under mild conditions, but that may be reopened under adverse conditions (i.e., normal storms, 10-year floods, exceptionally large spring tides).
D	Close with sediment dike under normal adverse conditions.

Some of the general guidelines and assumptions for the construction of the protection strategies follow:

1) The most important consideration is keeping the oil on the seaward side of the inlet, because of the abundant natural resources (e.g., salt marshes and sheltered tidal flats) and human developments (e.g., marinas) usually found landward of the inlet.

2) I should point out these strategies we did only worked if the oil floated on the surface of the water, which was the case with most of the big spills in the early days. Later, oils heavier than water were manufactured, and RPI has responded to a couple of spills for this type of oil. JM became an expert on this subject, on which she has published several papers. Another exception would be similar to what happened during the *Deepwater Horizon* spill in 2010, when the oil release occurred 5,000 feet below the water surface, and was dispersed extensively at those depths. Of course, some of that oil did reach the surface, where it floated and was collected with skimmers, as well as burned and chemically dispersed.

3) When the oil is floating on the water surface, the first priority is containment and the second is recovery.

4) When dealing with inlet types type C or D, if at all possible (e.g., a ready sediment source is available), use a sediment dike to prevent oil from passing to the landward side of the inlet. The first time we used this technique was during the *Ixtoc I* spill in Texas in 1979.

5) In the case of inlet types A and B, we followed guidelines established by the U.S. Coast Guard Strike Team by recommending the use of deflection

booms (i.e., those set at an oblique angle to the current) as the best means of controlling oil in the vicinity of tidal inlets, which usually have strong tidal currents. The threshold velocity for entrainment of oil past a boom set at 90 degrees to the current is 0.7 knots. One of the most common mistakes in utilizing booms that we have seen over the years is for inexperienced responders to lay out a line of boom straight across a channel, perpendicular to the current. They always fail.

6) Even if oblique deflection booms are used, they may fail if the current velocity exceeds about 3.5 knots. This is the number we normally recommended. However, we later learned that deflection booms were used successfully in the Mississippi River where the currents exceeded much higher velocities than this; however, those booms were set at a very small angle and extended for an exceptionally long distance.

7) Collection of oil from the water surface with skimmers is a desirable method of recovery but, in many situations, it is better to deflect the oil to a usable collection point along the shore. The preferred collection point would be a fine- to medium-grained sand beach, which is easily cleaned because penetration of oil into the sediment is minimal. Trapping oil against vertical rock scarps or concrete seawalls is also acceptable. Coarse-grained gravel beaches, riprap, tidal flats, and marshes should not be used as collection points except as a last resort. In the field while doing our protection strategies, we visited all the accessible proposed collection points for each inlet and described them in detail. Obviously, the most amenable collection site would be one you could drive to.

8) The protection strategies depicted related only to spills located seaward of the inlet and the strategy recommended applied only to flood conditions. If ebb currents carried the oil out of the inlet, that would be a good thing.

9) All collection sites, predicted current patterns, approximate angles to set deflection booms, and other details were presented on the sketch of the protection strategy. Figure 107 gives the strategy suggested for a Type A inlet in Rhode Island.

Story One

(Monday; 2 November 1992, Election Day; Tijuana River mouth, California/ Mexico)

On Election Day in November 1992, Todd Montello and I started our study of the 170 tidal inlets on the California coast, beginning at the southernmost tidal inlet in the state at the mouth of the Tijuana River, the Mexican/U.S. border. It was just the two of us on this first visit; the rest of the team would join us later. Todd and I got along great, but we didn't talk politics much because he was a hard-core, conservative Republican born and raised in Connecticut, and I had already sent in my absentee ballot, voting for Clinton.

The low tide was early, so we walked down the beach to the tidal inlet in the dark, waiting for the sun to rise. At the first rays of daylight, we began to take notes. Suddenly, out of the fog across the channel on the Mexican side of the inlet, three vague shadows emerged. These forms turned out to be three men, who walked up

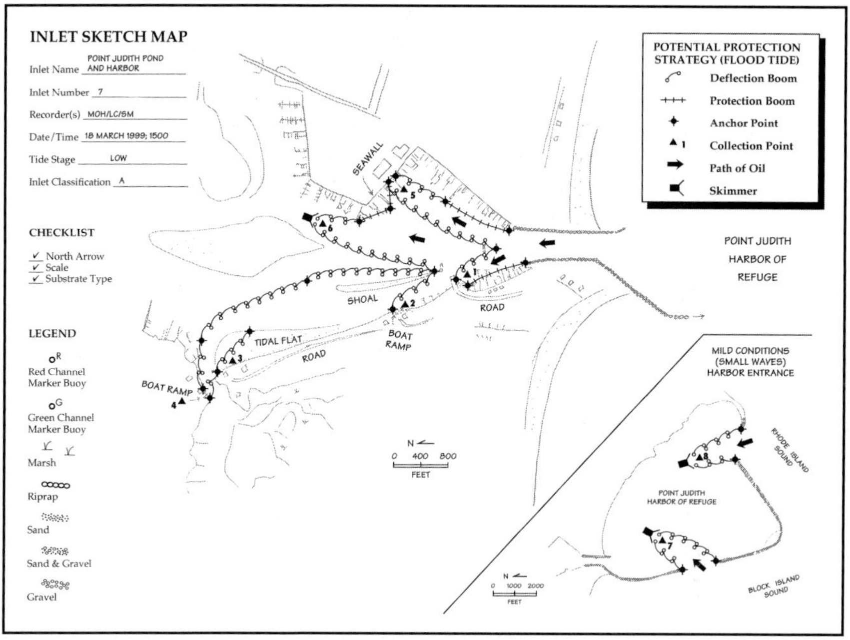

FIGURE 107. Protection strategy proposed for the entrance to Point Judith Pond and Harbor, a Type A tidal inlet in Rhode Island.

to the edge of the channel and started undressing, taking off all their clothes and holding them in a wad high over their heads.

Paying us little heed, they walked upstream a ways and then waded across the channel onto the sacred ground of the United States of America. In the deepest part of the channel, the water was up to their armpits.

Soon, we watched their three bare rear ends disappear into a fog bank as they headed on toward San Diego. No doubt, they soon put their clothes back on, just in case you were wondering.

In response to this scene, Todd said, "If Clinton wins the election today, I am going to take my trousers off, wade across this channel, and you can stand here and watch my bare ass go the other way."

Story Two

(Probably a couple of years later at a meeting in the San Francisco Bay area)

We completed the outer shoreline survey of the California tidal inlets by 1993, and relevant to this story is the fact that, if the three volumes of our final report were stacked one on top of each other, they would be about seven inches thick. It takes a lot of paper to cover the photographs, data sheets, the individual strategy sketches, and detailed descriptions of 170 tidal inlets.

Don Lollock, one of the chief administrators of California's Office of Oil Spill Prevention and Response (OSPR), asked if we would be interested in also doing protections strategies for a number of critical areas along the shoreline of San Francisco Bay and, naturally, we said we would. Shortly thereafter, Don called a meeting of groups interested in doing this, which I attended. One of the groups at the meeting was a private co-op funded by the oil and shipping industries. In this case, their function was oil spill response and cleanup for San Francisco Bay. While doing our outer coast inlet protection strategies, individuals representing a couple of these co-ops had been members of our field teams, but at the time of the meeting, I was not familiar with this particular one (I won't give its' name).

Don started the discussion by explaining his desire to have protection strategies produced for the entire shoreline of the bay. After this short briefing, the head man of the co-op at the meeting held up his hand.

Don called on him, and he said "We have already done that!"

In answer, Don exclaimed excitedly, "Wow. No kidding. Could OSPR have a copy of your strategies? This would save us a lot of time and money. Your tactics would become a key part of our response plans. How can we get copies?"

The answer to Don's request, "Nah, we can't do that, because as sure as you write that stuff down, it won't work!"

I heard this with my own ears. Would you believe that we were awarded the contract to do the strategies?

Story Three

(A few more years later)

To introduce this story, I will point out that many of the inlet protection strategies we proposed were somewhat complex, a few calling for using thousands of feet of boom. And, as I mentioned earlier, we commonly also gave backup strategies, in case the primary scenario for the inlet entrance should fail, because the waves were too big that day or some other problem. Once in a meeting in Florida, run by Brad Benggio, the NOAA Scientific Support Coordinator for that region, I reviewed our strategy for protecting one of the key tidal inlets on that part of the Florida coast. After I had finished my presentation, some irate person in the audience piped up, "Nobody could afford that much boom! Where would it all come from?"

To which I replied, "I was not responsible for creating this inlet, the forces of nature did that." My comment probably would have had more impact if I had blamed God for the creation of the inlet, but I don't do that anymore.

I continued, "I only presented the strategy you would have to use if you want to keep the oil from going through the inlet and oiling the marshes and the four marinas in there."

"But it would just cost too much," the disgruntled one replied.

Costs too much to save the estuary? Hmmm.

Now let us revisit the protection strategies for San Francisco Bay. Jim Hardwick, another key oil spill responder for OSPR, with whom we have worked for years, was having a conversation with JM during the *Cosco Busan* oil spill in 2007. It turns out that his team was particularly interested in protecting some key area in San Francisco Bay. Therefore, they called on the local private co-op again, asking for their detailed response strategy for this particular area.

Their reaction to his request was less than satisfactory, so Jim said, "Oh well, that's okay, we will just use the Hayes/Montello strategy then."

Jim's response built a fire under their tails, so to speak, and the co-op boys got busy trying to do a strategy that would cost less money.

Costs too much to save San Francisco Bay? Hmmm.

20 THE ENVIRONMENTAL SENSITIVITY INDEX (ESI)

But you are the American.

SALEH – THE QUIET ONE [37]

The Science

During a phone call from Pat Wennekins of the Alaska Department of Fish and Game in early 1976, I was able to land a $14K contract to do a general study of the shoreline of Lower Cook Inlet, Alaska, in order to make an assessment of potential impacts of oil spills along that shoreline. I believe he had heard a talk I gave on our studies in the Strait of Magellan, which clearly showed the similarity between many aspects of the strait and Cook Inlet. This revelation, plus the fact that I had been doing coastal geomorphological studies of the Alaska shoreline for many years, was probably why Pat called me. Therefore, in the summer of 1976, JM, Jeff Brown, and I did the usual geomorphological survey of the Lower Cook Inlet shoreline, including Katchemak Bay (Figure 87). However, in the final report, I added a concept that we called the Vulnerability Index of shoreline types based on potential impacts from oil spills. This index listed coastal habitats on a scale of 1-10, with rocky shores exposed to large waves ranked 1, and sheltered salt marshes ranked 10, obviously based primarily on what we had learned at the two big oil spills we had worked on earlier that year (*Metula* in Chile and *Urquiola* in Spain). The results of this project were presented in two publications (Hayes, Michel, and Brown, 1977; Michel, Hayes, and Brown, 1978).

The only major change we made to that primary scale from then on was to move exposed sandy tidal flats from 5 to 7 on the scale, based on the tremendous kill of burrowing clams and urchins on an exposed sandy tidal flat in Brest, France as a result of the *Amoco Cadiz* oil spill in March 1978. By the time that spill occurred, we had obtained a contract with NOAA to become a part of their team of oil spill responders. As noted earlier, we have maintained that relationship with NOAA to this day, with JM in charge of this work at RPI. John Robinson, to whom this book is dedicated, was our first contract monitor at NOAA.

We soon changed the name to the Environmental Sensitivity Index (ESI), concluding that sensitivity is a more appropriate designation than vulnerability (Hayes, Gundlach, and Getter, 1980). As we now use the term vulnerability, it refers to whether or not a habitat is susceptible to its being impacted by oil moving on the surface of the water because of its location. On the other hand, the way we use the term sensitivity has nothing to do with the habitat's position in space. For example, a salt marsh located far away from a potential source or pathway for the oil to follow (e.g., located in a remote part of a system) would be less vulnerable to impact from oil entering an estuary than a marsh located near the estuary's entrance, but the marshes in both locations could be equally sensitive to oil spill impacts should oil reach them.

Shortly after forming RPI in 1977, we hired some biologists, beginning with Chuck Getter, who developed biological criteria to add to the maps. We also placed on the maps human-use features, such as marinas, boat ramps, special protected areas, and so on. By the time of this writing in the winter of 2011, we have conducted ESI mapping projects of the entire U.S. shoreline, including all of Alaska and the Great Lakes. Usually, these mapping projects were carried out on a cooperative basis between NOAA and the individual states, with NOAA being our primary client in most situations.

Since 1988, all of these maps have been produced in a digital format by our GIS group at RPI. Therefore, most of the ESI maps and databases are available in digital format from NOAA at http://oceanservice.noaa.gov/dataexplorer/; technical information about the ESIs and how they are produced is available at http://response.restoration.noaa.gov/. We have also carried out ESI mapping projects in numerous foreign countries, including Nicaragua, Honduras, Guatemala, Panama, El Salvador, Nigeria, Kuwait, Israel, Jordan, and the Gaza Strip. Our early ESI map of the Nigerian Coast is given in Figure 72.

One of our more recent ESI maps, for Moss Landing on the Central California Coast, is given in Plate 11. The different shoreline types (or habitats) of the Central California coast were mapped in detail by our group and published in 2006 (NOAA/RPI, 2006a). The shoreline types encoded on these Central California maps were created by inspection of available imagery, visual inspection flying over the area at 500 feet altitudes, oblique aerial photography, and ground inspection of critical areas. As usual, the different habitats were ranked on a scale of 1-10 based upon their sensitivity to oil spills and other considerations, with those ranked 10 being the most sensitive. The 1-10 scale we used for Central California is shown in Table 2.

This type of field mapping is usually done by a coastal geomorphologist, with JM and I probably having done more of this than anyone else in our group. A considerable amount of this mapping has also been done by Erich Gundlach, Walter J. Sexton, Todd Montello, Colin Plank, Chris Ruby, Dan Domeracki, Zach Nixon, Tom Freeman, and Thomas Minter. The biological and human-use information is collected meticulously from a variety of local sources by the RPI biologists. Before final publication, early drafts of these maps are scrutinized carefully by the biologists working the local area, who are usually associated with nearby universities and federal and state natural resource agencies.

As already noted, ESI mapping continues to be a major activity of our company. As I write this, our group is working on updating the ESIs for North Carolina and Florida, with prospects of soon beginning on Delaware Bay, Texas, and Louisiana. The first story that follows is based partly on a project to create an ESI map of the Gaza Strip in 1999 (Figure 108).

Table 2. Shoreline types mapped for Central California coast (NOAA/RPI, 2006a).

ESI Scale	Description
1A	Exposed rocky shores
1B	Exposed solid man-made structures
2A	Exposed wave-cut rock platforms
3A	Fine- to medium-grained sand beaches
3B	Scarps and steep slopes in sand
4	Coarse-grained sand beaches
5	Mixed sand and gravel beaches
6A	Gravel Beaches
6B	Riprap
6D	Boulder rubble
7	Exposed tidal flats
8A	Sheltered rocky shores
8B	Sheltered solid man-made structures
8C	Sheltered riprap
9A	Sheltered tidal flats
9B	Vegetated low banks
10A	Salt- and brackish-water marshes

Story One

(25 July 1999, Sunday - Herzlia Airport, Israel)

"Do you have a gun?" the teenage Israeli security officer at the smallish Herzlia airport asked, grinning sheepishly as he conducted his cursory security check before I was allowed to board a waiting dilapidated Cessna 172 for an overflight of the shoreline of the Gaza Strip.

After the frisking, he said "Okay" and I made one more trip to the men's room before Ellik Adler, an Israeli oceanographer with whom I had worked before and was joining me on this project, and I walked over to the plane, observing that the right door was missing.

"Great, great!" Ellik exclaimed. He thought the door being off would allow for better photography, which was true.

"You sit in the front, I insist," he said.

"But Ellik, you'll be blown away in that back seat."

He climbed into the back anyway, and I took the front seat. Then the pilot walked around to my side of the plane and tied with a granny knot a single scrap of a broken seat belt across the open space where the door used to be, to keep me

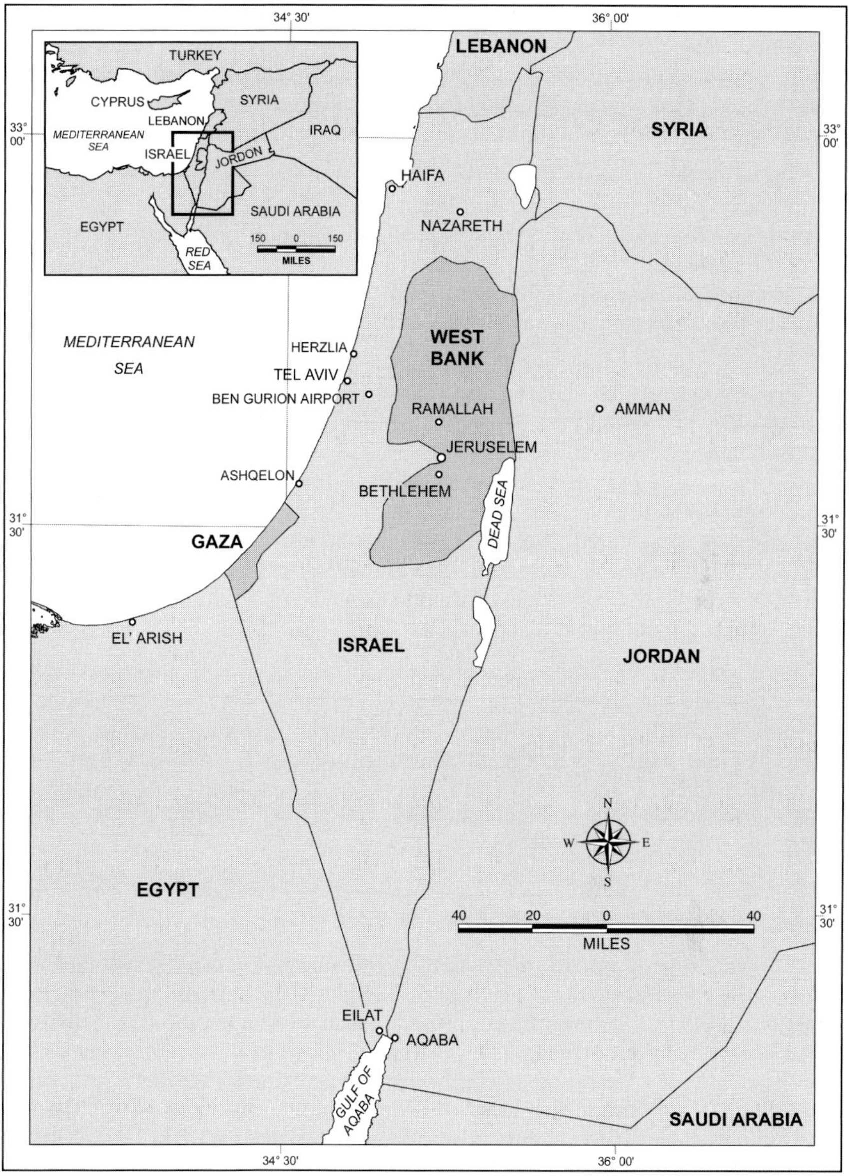

FIGURE 108. The Gaza Strip and vicinity.

from falling out, I suppose.

With my "safety strap" tied and the pilot back in his seat, we were soon up and over the shoreline of Tel Aviv, flying at 1,000 feet, headed toward the Egyptian border. We bounced our way south, as the pilot maneuvered around the skyscrapers and smoke stacks of Tel Aviv and finally across some open fields in order to avoid crossing the runway of the busy Ben Gurion International Airport. Once inside the Gaza Strip proper, we continued south, keeping 200 feet or so landward of the shoreline. Consequently, we flew over some of the refugee camps that house 70% of the occupants of that place the Palestinians call 'Azza. This spelling, which is printed on numerous road signs in Gaza, is an English bastardization of the Arabic word. There are also several other versions of that spelling on the road signs.

The contrast between the refugee camps and the middle class, suburban communities in Tel Aviv was astounding. The camps were all composed of small, gray box-like houses surrounded by huge piles of garbage. Likewise, the pollution along the shore was overwhelming, especially the garbage littering the backbeach and the numerous sewage outfalls turning the otherwise clear, blue nearshore waters of the Mediterranean Sea into a foul brown color.

After we turned around, giving the Egyptian border a wide berth, and moved offshore, we flew at elevations of 500-1,000 feet back to the north, looking away from the sun so I could take photographs. One of the sewage outfalls near the southern border was a fountain spewing vertically into the air at a height rivaling the Old Faithful Geyser in Yellowstone National Park. I had the pilot circle back around it at 400 feet so I could take some more detailed pictures of this amazing spectacle. Then we continued on up the coast, taking more pictures of the nearshore bar formations, the breakwater at the Port of Gaza and, coincidentally, of Yassar Arafat's beach house, which was located just south of the Port, more sewage outfalls, fishing boat haul-out areas, and so on.

Once past the fortified northern border of the Strip, with its many barbed wire fences and border guards, we continued up the Israeli coast, taking photos of the man-made harbors at Ashqelon and Ashdod. Finally, we had to head back inland as per instructions by the air controller at the Ben Gurion Airport. Over the mainland, the air was rough again, and the sun was shining on the bright tan buildings of Tel Aviv. But the gray, ghost-like aura of 'Azza stayed with me all the way to the rough landing at the Herzlia airport.

"Cheated death again," I muttered to myself, as I struggled with the granny knot in the scrap of seat belt tied across the space for the missing door.

I dined alone that night on my standard vegetarian pizza while sitting on an open outdoor terrace on the waterfront in Tel Aviv. I was at peace there, watching the sunset over the Mediterranean, but it was hard to forget about the sights in 'Azza. This was my fifth trip to Israel within the past three years, working on the Red Sea mapping jobs and on an engineering project on the Mediterranean coast with Bill Baird. In the process, I had developed an excellent working relationship with Ellik, the head of the country's marine pollution arm, which included an oil spill response unit stationed in Eilat on the Gulf of Aqaba (Figure 108). Ellik is a marine geologist, having spent some time working at the Scripps Oceanographic Institute in California. An excellent swimmer in top physical condition, he had just recently retired from his part-time duty as a SEAL in the Israeli military.

For some reason, I have always felt "at home" in Israel. Maybe it is because of the Bible stories my Grampaw Worley used to read to me during winter evenings as we popped corn over his open fireplace–stories about David and Goliath, Samson and Delilah, and so on.

The goal of this project was to develop an ESI map of the Gaza shoreline, as part of a continuing effort by the State Department and other international funding agencies to encourage cooperative environmental activities among the countries in the area, namely Israel, Jordan, and Egypt. Employed as a NOAA contractor, RPI had already made maps of the Red Sea coast of Israel and Jordan as part of the Peace Process funding by the State Department. I did the fieldwork for those projects, with help from local experts, such as Ellik. Once these maps were published, the Palestinians showed interest in having similar maps made for the shoreline of the Gaza Strip, and the State Department was anxious to comply with their request. In order to accomplish this task, Ellik and I purchased an excellent set of color photographs of the Gaza coast from an Israeli private firm, and the overflight we did earlier in the day was carried out to clarify details of the coastal habitats and sediment types not readily discerned on the photographs. After completing the overflight, we were scheduled to finish the mapping by means of a ground survey.

One time when Bill Baird and I were in Israel together, we rented a car and drove to Jerusalem. Once we had finished our tour of that marvelous historic city, I said, "You know, Bill, I would like to go to Bethlehem, so I can tell my mother we went to the place where Jesus was born." And even though that holy town is now in Palestinian territory, we did go there, but it was a little scary with all the young boys crawling over the car and asking for handouts while we drove along the crowded, narrow streets. The next morning after that visit to Bethlehem, a Sunday (their first day of the week), a bomb went off in a bus traveling along one of the streets we had walked along in Jerusalem, killing several of the occupants.

The day after the flight over the Gaza coast, Ellik and I left Tel Aviv early enough, we thought, to meet our Palestinian counterparts at the Gaza Strip at the appointed time, 8:30 a.m. At the border, we stood in the wrong line for about 20 minutes. As we were waiting, Ellik told me he had not been in Gaza for twenty years. During that last visit, he and his men raided what they thought to be a terrorist's house and killed four Palestinians. He told me the story in a matter-of-fact way, but he didn't seem to be very proud of his part in it.

As we stood in line, various young Israeli soldiers in their teens and carrying M-16s checked us out. I thought, "Thirty-two holes in you in the blink of an eye."

Finally, one of the youngsters recommended that we drive around to the VIP entrance, where we stood in several more lines before we left the Israeli side for the Gaza territory, 1.5 hours after we arrived. No problem on the Palestinian side, where our associates, Na'im and Husam, were patiently waiting for us.

We started working our way down the shoreline right away. The Palestinians provided us with two four-wheel drive vehicles and six police and military personnel as security escorts armed with M-16s and other weapons. We first drove on the beach some, but after one of the vehicles became stuck in the sand, I recommended that we stay on the road, which parallels the beach on the rim of a bluff the entire 40 kilometers of this part of the coastline we were mapping. We made many stops along the way, eventually establishing over 50 study sites, where I took photographs; we also dug trenches in the sand, made measurements of the sediment grain size, and took some sediment samples. We plotted the changing habitat types on the aerial photographs and maps as we moved along. Also, with the assistance of our Palestinian associates, Ellik located on the maps important "human use" features, such as archaeological sites, fishing boat haul-out areas, bathing beaches, sewage outfalls, and so forth.

Being the oldest guy there (I was 64 at that time) and an American to boot, the Palestinians were especially friendly and helpful, saying "Okay, Dr. Miles, what next?" and things like that. Any time I pointed down to the beach from the cliff, indicating that we should take another sample, two or three of them would scramble down to the bottom and come running back to us with another bag of sand. A young policeman named Waled took special interest in me, being sure he was close by to help if I did something stupid like go down the cliff myself to collect a sample. He was from Egypt, having only been in the Gaza Strip for a few months, and was always asking me questions, such as "How much would a beach house like that cost in Miami Beach?"

By the end of that field day, we had completed the mapping of all of the shoreline of the Gaza Strip that was then under Palestinian control. The work went well, with a very cordial, cooperative, and enthusiastic spirit being exhibited by all participants, even "Dr. Miles."

As we drove back toward the border (Ellik was not allowed to stay overnight in the Strip), he and Na'im kept pointing out for me the cars that had been stolen in Israel and transported into the Strip. They could tell by the license plates, which the thieves didn't bother to change once the cars were in Palestinian territory. This was a major smuggling operation, and I wondered how it persisted, considering how difficult it was for us to cross the border. I suspect bribes work anywhere, if the price is high enough.

Ellik and I spent that night in a cheap hotel in Ashqelon, on the Israeli side of the border. Much to my chagrin, Ellik was fond of cheap hotels, but the fish dinner was great!

The next day was devoted mostly to meetings at the Ministry of Environment in Gaza. These meetings were pretty tedious, but we worked out some of the details for a training course on oil spill response that Brad Benggio of NOAA and I would teach for them in September. After that meeting, Ellik and I mapped the short section of the Israeli-controlled territory at the northernmost part of the Gaza Strip. Then we spent a second evening in Ashqelon, which included watching the sun set over the Mediterranean Sea, as we enjoyed another fine fish dinner.

On 28 July, a Wednesday, we mapped the shoreline of the Israeli-held Gush Qatib area, which contained 13 kilometers of shoreline at the southern end of the Gaza Strip. Ellik and I met two of his pollution inspectors, Itzik and Ronen, and drove around the western boundary of the Strip so as to have no delays at the border. I rode with Itzik, who was listening to western songs such as *One More Night* on his radio, as we sped through the lush orange groves in southern Israel. At the beach, we met the same Palestinian contingency that worked in the field with us on Monday. Then the caravan of three vehicles and 9-10 people worked its way south along the coast, establishing twelve more stations. At one of those sites, several of us were walking in a group separated from the others, and a strange group it was, led by a hillbilly from North Carolina and a SEAL-fighter oceanographer from Haifa, Israel. Also in the group were two Israelis raised in a kibbutz (Itzik and Ronen), a Palestinian raised in a refugee camp in Gaza, who through hard work and perseverance had become a computer mapping expert (Na'im), and a young Palestinian policeman recently imported from Cairo (Waled). I thought, "Maybe this is what the Peace Process is all about."

Another stop was made at the spewing fountain of the sewage outfall that I had photographed from the air. I was standing by Waled, who was grumbling that the "settlers" were responsible for this. Then he looked at me and said, "Our beaches will never be clean."

I touched him on the shoulder and said, "No, no, Waled, you guys can do it, it will just take some time."

He shook his head and turned back to the vehicles, which we boarded and continued south to complete the fieldwork for the sensitivity-mapping project. At the border, while observing the Egyptian flag flying over the next border crossing, we had a little completion celebration and took a group picture with a white Gaza donkey pulling a cart loaded with Palestinian children in the background.

After the photo session, Ellik and I said goodbye to everybody and rushed back to Tel Aviv for a meeting at the domestic airport with Dr. Reuven Yosef, Director of the National Center for Bird Watching in Eilat, who had assisted us with the Red Sea work. He told us that, indeed, the beaches of the Gaza Strip are important nesting and feeding stop-off points for wading birds and shorebirds on their migrations between Europe and Africa. Dr. Yosef noted that, in the fall, a large flight of quail-like birds, I believe they are called "manna" birds, come crashing to the ground in Gaza after their long, exhausting flight across the Mediterranean Sea. Manna from heaven! He also told us that he would be glad to contribute bird data for the Gaza Strip sensitivity maps.

My last day in Israel was spent in Jerusalem with Ellik as we enlisted the services of a number of additional Israeli biologists who would supply various biological data for the Gaza maps. On the drive back, Ellik gave me a geography lesson about how the Israeli "settlers" had inhabited the tops of many of the highest hills in the West Bank, so they "could look down on" the less affluent Arab population.

After one last meeting with a marine mammals expert at the railway station in Tel Aviv, Ellik headed to Haifa for a reunion with his family, and I caught a cab to the Tel Aviv jewelry district to buy JM a present for her birthday, which was coming up in two days.

Next morning, I caught the jet to Zurich-Atlanta-Columbia, completing a trip in which I had encircled the globe and spent time in four different time zones. It started twenty days earlier, when JM and I devoted 7 days to mapping the Southeast Alaska coast. By the way, during that excursion, I had a chance to finally complete the map I was working on when my plane crashed in 1990 (discussed briefly in Chapter 15 – BY FOOT, BY CAR, BY BOAT, BY PLANE, and Chapter 16 - DOOMNESS). From there, I went to Tokyo for a three-day oil spill training course at the U.S. Navy base, and then across Asia to Tel Aviv. Needless to say, I was ready for a break when I reached South Carolina, so JM and I headed to our new place in the mountains of North Carolina for a long weekend.

Story Two

(7 September 1999 – RPI office in Columbia, South Carolina)

At the end of August, JM and I took off more time than just a long weekend, something unusual for us, and spent it at the mountain house, doing a little fishing, fence building, and hiking. The high-altitude flowers are in full bloom that time of the year, and the meadow by the house was also full of butterflies and fledgling songbirds.

Tuesday, 7 September was our first day back at work for almost two weeks. Brad Benggio, a Scientific Support Coordinator for NOAA, with whom JM and I had worked often at oil spills, on mapping projects, and in training courses, was in Columbia preparing for the training course he and I would give in Gaza in September. It took me a while to get back into that. In other words, I had forgotten almost all of the details I had worked out before heading to the mountains. Anyway, I sent out some e-mails and packed some stuff to take to Gaza.

Next morning, I had a dental appointment to get a permanent crown for one of my rear teeth, another one I had apparently fractured during the airplane crash in 1990.

Young Dr. Sawyer asked, "How did the temporary crown work out and, by the way, where is it?"

"Doc, I believe I could write a short novel about the problems I've had with that tooth, and I suppose I swallowed the temporary while I was sleeping on the mountain, at least I couldn't find it the next morning."

When he stuck the permanent crown on the stub that was left of the original tooth, the pain from the glue was the worst of the whole ordeal. A small bit of advice, try not to be involved in any airplane crashes–you never know how long the side effects will last. The crash was nine years earlier.

At our weekly staff meeting the following morning, Jack Moore, RPI's Chief Financial Officer, gave a report on the company's financial standing, which was looking pretty good. In fact, a year or so earlier, we purchased a 17,000 square foot, two-story warehouse-type building located a block from the state Capitol, renovated it, and moved the office there. I was quite proud of our efforts at interior design, with brilliant red air ducts hanging from the high ceilings.

Thursday morning, at a brief Board Meeting, I informed the other three members of the Board that I wanted to cut back to 50% billable time next year, so I could concentrate on more fishing and on writing books. I also recommended that on my upcoming birthday, October 20, when I would turn 65, JM should take over the reins as President of the company, and that I remain as Chairman of the Board, noting that she was already responsible for bringing in the lion share of our sales and managed a good deal of the on-going projects. This suggestion received unanimous approval.

After that meeting, as we sat in my office gobbling down our sandwiches sometime around noon, JM said, "I'm really going to miss you. We've had such a good time." I think she was referring to our vacation on the mountain. She almost became teary eyed. Very touching.

Brad and I left the office for the trip to Israel at 12:30 p.m. that day, and we had the usual rather miserable flight through the late summer skies over to Atlanta. But that was just the beginning of the fun. Some of the thoughts that I had, and recorded, regarding the upcoming flight from Atlanta to Zurich follow:

Anybody that travels a helluva lot like I do knows that once in a while you're going to have a trip like this, sitting in our B-747 on the ground for four hours in Atlanta waiting for the thunderstorms to clear out of the D.C. and New York areas so we can fly the great circle route to Zurich. And going back to the terminal to drop off the guy who had a heart attack, which now leaves two seats empty in the whole damn plane. And to replace all that fuel we burned up sitting on the runway for four hours. The new route we are on puts us out over the ocean off of Cape Hatteras right over a hurricane or something, and this big old plane rattles, bounces, and shakes. Turbulence like this always scares the living shit out of me. Don't laugh. You would be scared too if turbulence had caused your float plane to crash off the Alaska coast–killing the pilot, breaking both of your legs, and nearly freezing you to death.

I was real glad to see clear skies over Zurich, even if we were four hours late, arriving around noon. As expected, we missed our connecting flight to Tel Aviv. Because the plane had been so packed, there was a long line at the transfer desk, and it took us quite a while to learn that we could catch a 9:00 p.m. flight to Tel Aviv.

Staying around Zurich for nine hours waiting for that flight wasn't that bad. I was just happy to be out of that wretched airplane for a while. Brad went into Zurich, falling in love with the place, while I stayed in the business class lounge finishing reading *Catcher in the Rye*, by J.D. Salinger. I truly enjoyed his stream-of-consciousness style of writing.

Of course, arriving in Tel Aviv at 1:30 a.m. of the Jewish New Year's night wasn't the best of planning. We walked out of the terminal with our luggage to catch a cab and none were there. Usually, there were dozens available. We waited for a long time and finally our Russian cab driver headed his Mercedes down the freeway at 160 km/hr toward our hotel, which was located across the street from the beach. I kept pleading for him to slow down to around 120 km/hr, but he paid me no head. Finally, I pointed to 120 on his speedometer, and he looked at me and said, "Okay, boss!" which, as far as I could tell, was the only English he knew. Then he slowed down. When we reached the city proper, we spent more happy time with our Russian driver working our way through the throngs of New Year's revelers that packed the streets between us and the hotel. They were probably too drunk to realize their eminent danger as our mad Russian driver cavorted down the open lanes the wrong way, scoring several near misses.

"So, this is where all the taxis are," I observed to Brad, who was cringing in the back seat.

But, yet, there was even more fun in store for us. When we at last reached the hotel, sometime before 3:00 a.m., I was handed a fax from one Bill Crane of the Environment, Science and Technology Office of the U.S. Embassy, which said, "Oops, sorry, the Israelis have closed the Gaza border and you can't cross until Monday. Why don't you extend your stay?" The title of the fax was UNFORESEEN GLITCH - CHECKPOINT CLOSED, at which I had to laugh, so far this whole trip had been nothing but an "unforeseen glitch."

Next morning, I called Na'im and told him we would have to start the four-day course one day late. He said, "That's okay, we can meet on Thursday." Thursday is the first day of their two-day weekend, and I didn't get the impression that he was overjoyed with the change, but what else could we do?

The logistical arrangements for this course were more difficult to organize than they had been for the field-mapping project. Neither the State Department, nor the U.S. Embassy, for that matter, was particularly concerned about the arrangements made for a low-life, consultant contractor like myself. However, Brad was a *bona fide* employee of the U.S. government and, of course, he could not spend the night in the infamous Gaza Strip, as I had arranged for us to do during the earlier trip. He would have to get a hotel room in Israel and go back and forth across the border each day. Furthermore, they would have none of this using Palestinian vehicles for our transportation, we had to be chauffeured around in an armored car supplied by the Embassy. Therefore, we had made arrangements for a driver from the Embassy to pick us up at 10:00 a.m., the day after we arrived, so we could move to our living quarters near the Gaza Strip border.

Joseph, our driver, was right on time. He was an Israeli citizen of Arabic descent, who was a Catholic and spoke four or five languages. His job was special tasks driver for the Embassy. He was very pleasant, making us feel at ease right away. When we reached the Embassy, which was only a few blocks away, he opened the hood of the armored car for inspection by the Israeli police guarding the entrance. Then they walked around the car with a mirror on a stick, looking for bombs underneath it. It was about 10:30 in the morning on a Saturday, the Jewish Sabbath, thus the Embassy was empty except for a number of tough-acting, highly conditioned young U.S. Marines and Israeli soldiers, who looked as if they could snap their fingers and break your arm in two. Once inside the gate, Joseph pushed a buzzer by the main entrance to the Embassy, and Sean, a security officer, came out to meet us. He ushered us inside and past the security booth with a wave of his arm.

"It's on the third floor," he said, motioning for us to follow him. I had expected to take an elevator, because Sean, who was about 35, partly balding, and with a bit of a potbelly, didn't look as if he could snap his fingers and break your arm in two. The truth be known, he probably could have, he just didn't look like he could. Anyway, I bounded up the stairs behind him, hoping I wouldn't be breathing too hard when we reached his office.

He ushered us in, and the four of us gathered around his small office table. Sean sat slumping over the table, rubbing his eyes. I apologized and thanked him for coming out on Saturday for our briefing and he said, "No problem." Then he seemed apologetic about us not getting across the border in time to start the course as scheduled.

"I'm going across tonight," he assured us, "but I guess you will have to do what Bill Crane says."

He then made a phone connection with Bill Crane himself, whom I had never met before, and who had never responded to any of my e-mails, faxes, and so on. That had been handled by one of his underlings. I asked Bill if we could go across the border as scheduled, because it would be an inconvenience for the Palestinians to attend the course on their weekend.

In response to my request, he said, "Start on Monday, when they open the gates."

After the call, Sean shrugged his shoulders, and went abruptly into a bizarre security briefing on how we should behave in the Gaza Strip.

His primary message was, "If anything happens, stay in the car. And head for a safe haven, such as the mayor's house. Joseph knows where they all are. In Gaza, it only takes five minutes for a riot to start."

And then, something as an afterthought, he said "Seventy percent of the population of Gaza is kids under fifteen."

I was thinking, "Seven seems to be a magic number for those folks who do statistics on Gaza - 700,000 people live in the Strip, 70% of those live in refugee camps, and so on."

Sean continued, "So be careful of the kids. If Joseph happens to hit one, get the hell away from there quick. They think a lot of their kids."

Brad leaned over to me and whispered, "If they think so much of their kids, why don't they tell 'em to stay out of the street." American values?

Sean also gave us some papers to read that spelled out the standard operating procedures for travel in the Strip.

Sean is one of those guys you just can't help but like, being so laid back and all. You could tell that Joseph was real proud to call Sean his "supervisor."

After the briefing, the three of us boarded our 10,000-pound armored vehicle, which looked like a suburban, and headed for the Holiday Inn in Ashqelon, where we would wait for the border to open. Oops, armored vehicle keeps trying to turn left up onto the curb. Fortunately, it wasn't turning to the right; else we would have run over the slope down into the ocean. Joseph was pretty upset about this, so he called Sean to tell him we needed a different vehicle.

Sean said come on back and get one. Why not? They had ten to fifteen more armored suburbans exactly like ours in the parking lot. No wonder Joseph didn't know if the one we were in was the same car that Secretary of State Madeline Albright had ridden in the week before.

Once back in the parking lot, after undergoing another elaborate security check, I helped Brad and Joseph move our 500 pounds of luggage and other stuff from the original armored conveyance into the new one. "Damn," I thought, "Somebody needs to teach Brad how to pack!"

While I was engaged in that activity, a rear door of the original vehicle, which weighs a significant part of that 10,000 pounds–no kidding!–whaps me across the ear and almost knocks me out! Jeez, wounded before we even left the Embassy.

As we drove in our new wheels back down the road toward Ashqelon, which for the most part was adjacent to the beach, Joseph would emit a grateful "unh-huuhh!" each time we passed one of the beautiful young Israeli women adorned in their holiday bikinis. He had on some *avant-garde*, oval-shaped sunglasses to go with his shaved head, Yul Brenner style, or should I say Michael Jordon style? And, oh yes, he could also probably snap his fingers and break your arm in two, but I didn't know that for sure.

"Hope I don't have to find out whether he can or not any time on this trip," I thought.

Presently, Joseph received a call from Paul, the head mechanic at the Embassy, who asked, "Why are you talking to Sean about a car?"

After essentially telling Paul to buzz off, I think, I wasn't sure whether he was speaking Arabic or Hebrew, Joseph said to us, "I don't work for Paul, I work for Sean."

After that little interlude, Brad and I both told Joseph we were well acquainted with how bureaucracies worked and that he shouldn't worry about it.

I spent most of the next day hanging around in our new quarters in Ashqelon, going over my lectures for the upcoming seminar and trying to change my travel arrangements for the trip back home. Unlike the hotel Ellik and I stayed in during the previous trip, this Holiday Inn was not a cheap one. It was quite fancy, being spherical shaped like a big basketball and positioned right on the beach.

After a few phone calls about the airplane tickets, I was thinking, "Wouldn't you know it, all this traveling and when you finally schedule one or two things in the whole year that you really want to do, somebody like a Bill Crane, who is having a nice weekend at home, asks you to extend your stay. Not that he lifted his hand to get us across the damn border on schedule."

The "one thing" that I wanted to make was the "reception and dedication" of a mural that the artist, Eric Lake, had painted on the side of our new office building in Columbia. I had been dreaming about having a mural like that on one of our office buildings for years, ever since JM and I painted our first one, the old motor cycle shop, Pepto-Bismol pink back in 1978. This new office of ours had a large two-story wall with no windows on the downstairs level that was clearly visible to traffic moving along Park Street, one of the major thoroughfares in the downtown area. It was perfect for the mural Eric painted, which depicted, in five panels, the history of the city.

The stark, primary colors Eric used knocked your eyes out from four blocks down the street. And that center panel with General Sherman on his big horse framed against the flames of the burning city in the background, what a masterpiece!

On 13 September, a Monday, we left the Holiday Inn a little after 5:30 a.m. and headed for the border. We were somewhat early, the border was open, and we passed right through with minimum hassle and met up with Na'im on the Palestinian side.

The course was held in a new, multi-storied building that housed the fledgling Ministry of the Environment. The recently appointed Director, whom I had met briefly on the previous trip, was there to make a few opening remarks for the seminar. I didn't like him much, all he seemed to be interested in was how I was going to arrange for them to receive funds to pay for "the logistics" of the course. Not being overly pleased that this whole thing had been organized by some other Palestinian officials in Ramallah, on the West Bank, he seemed to be more than a little unenthused to see us. He left immediately after his brief speech, and I was glad to see him go.

Also, rather than Na'im, whom I greatly admired and respected, the new boss of the seminar was unexpectedly a youngish, chubby scientist named Basem. He was a trifle uppity, noting that he had just recently received his advanced degrees from a University in Germany, kind of giving the impression that he thought American science was inferior to the European variety, which pleased me a great deal, as you might imagine. Chermans, bah, my nemesis in the Middle East, but that's another story.

Well, there was no carousel projector, which I had requested by fax about six times, the room wasn't even slightly darkened so they could view our slides, something else I had requested about six times, and the 24 participants in the course were crammed like sardines into the little conference room. Perfect conditions for conducting a seminar.

Nonetheless, they seemed to be an excellent class. Of course, they had a tough time understanding my accent, and many times I simply couldn't figure out what they were saying when they asked questions.

I don't know. I guess it went okay. I never feel real good about these things.

After the lectures on Monday, the first day of the course, we checked out the field stops we planned to make on Wednesday afternoon. We had no problems as we drove around Gaza.

Joseph said, "If there is trouble, I always drive in the middle of the road."

"And what is going to stop this 10,000-pound tank," I thought.

Ellik participated in the seminar the next day, giving over three hours of lectures. The class seemed to like him; however, the Director attended his lectures for a while and put Ellik through the third degree a couple of times, being exceptionally rude to him, I thought.

Later, I told Ellik, "I would have told that gentleman to kiss my ass!"

"Remember, my friend," he replied, "this is a part of the Peace Process."

On the ride back to the border, Ellik told me that he had recently returned from a vacation, swimming among the coral reefs along the Egyptian coast south of Eilat, where he had experienced a "love affair" with a female dolphin. This dolphin hung around with him for days, rushing to his side each time he entered the water. That encounter seemed to be freaking him out.

As the car was being searched that night at the border crossing, a rather attractive American woman in her early thirties, who was dressed in black, came over and quizzed Joseph as to why an American-tagged car was being searched.

"Because they don't have diplomatic passports," was Joseph's quick reply.

"This should be reported," she said, "I've never seen this before."

Then we boarded our 10,000-pound express to go on our way back to Ashqelon, and she, presumably, headed for Ramallah, where she was stationed.

Wednesday was the day we had scheduled the field trip, usually the most important part of such seminars, and I was worried about how it would go. I felt that much of the class had been a disaster. I was glad that Ellik had managed to come on the second day. It always seems worthwhile to have a local expert talk–the comfort zone of home. I talked that morning on beaches in preparation for the field trip.

We were off for the field about 1:30 p.m. Joseph, Brad, Na'im, yours truly, and some "groupie" clown, who had not attended the lectures, took the lead in the armored car, which was followed by a bus full of students, and a loose cannon named Sami in his own car.

We went to the fishery harbor first and discussed cleanup techniques for the sand and gravel fill on the beach, as well as the riprap jetties. This seemed to go well. Basem was already looking at his watch, but I decided to ignore him.

Next, we stopped at one of the rather spectacular sand beaches, where we did a cleanup discussion, dug a trench in the sand and discussed the beach's origin, demonstrated beach profiling, and taking current measurements, using "neutrally buoyant floats," otherwise known as oranges, and having COASTAL HERO Sami collect a sample of the sediment from the toe of the beach as the waves broke over his head. Amazingly, the whole beach system had changed dramatically since we had checked it out two days earlier. None of the massive rip currents, which had been so prevalent then, were present, and with the wind blowing from the northwest, the typically north-running longshore current had switched directions and was running south, as determined by throwing oranges out into the current and watching them blast off to the south, much to the delight of the students in the class, as well as to numerous other onlookers. The nervous one, Basem, disappeared after this stop, which I thought went very well.

Then, we stopped at a spewing sewage outfall on a shoreline where the beach topography was being completely rearranged by the waves approaching from the northwest. My main point at this, the last stop on the field trip, was that the dynamic process on a beach can change rapidly, and that the predominant conditions don't necessarily apply on any given day. Therefore, during an oil spill, one should always be cognizant of the local conditions at the time of the spill, and not rely entirely on background knowledge of the usual conditions.

One of the guys on the course, an older engineer who was really into it, said that one time two people had drowned simultaneously near here, probably caught in one of the big rip currents like the ones we had observed two days earlier, with one dead body going south along the coast and the other one north. So much for the science of longshore sediment transport, but that was a poignant illustration of what I was trying to tell them.

At the end of the field trip, we met at a beach club of sorts, which was located fairly near Yassar Arafat's beach house, to have a small, informal reception. While I played my usual "What did you learn today?" quiz of the participants, Na'im and Sami left to buy some soft drinks and a couple of cakes for refreshments. I was impressed, and very happy, with the group's responses to my questions. I thought the answers posed by the smart engineers in that crew were better than those given in similar classes in the U.S. One of the students said the course had been somewhat meaningless for him until he saw our major concepts demonstrated in the field, which pleased me no end, having been a proponent of such an approach to teaching this subject since day one.

It was very pleasant under the open shelter on a blue-sky day with a cooling sea breeze. For the first time on this trip since the long delay on the B-747 in Atlanta, I felt as if I were doing something worthwhile.

As Joseph drove Brad and me back to Ashqelon, he went into a minor rage about us paying the Ministry $800 for their logistical support for the course. When I told him that I thought their receipt showed $200 for the "reception" we had just enjoyed on the beach, he said he thought they had spent about $12.50 on it.

Next, I made a little speech about how I could have saved the government a bunch of money if they had let me handle these things on my expense account, rather than billing the expenses through work orders to the funding agency. And then I listed a number of reasons why I thought the Director of the Ministry was a flaming asshole and that Basem was aspiring to be the same.

Amazingly, almost the entire student body showed up for the class on their Saturday. First, I talked about how to protect important habitats during a spill, and then Brad went through a table-top exercise he had invented, which I thought went very well. The class was over at 12:15 p.m.

While we were waiting for preparations to be made for the "graduation," I walked out on the balcony. We were on the 7th floor, so I could look across the tops of the uniformly gray buildings and see the long, sharp horizon of the Mediterranean Sea. 'Azza looked so peaceful in the mid-day sun. I breathed a sigh of relief. We had done it and the course had gone fairly well, considering the logistical difficulties. Therefore, I was just cruising in neutral and enjoying the unique view.

They gave us a round of applause, and we gave them certificates.

The students came round and shook our hands, thanking us. The Director, who "was not in the Ministry," did not show up for the "graduation." Big surprise.

About 4:30 that afternoon, Brad and I were back in Tel Aviv at my favorite open-air restaurant by the shore. I ordered my vegetarian pizza. The breeze blew across our table as I explained to Brad how the offshore breakwaters that line the beaches at Tel Aviv are designed to slow beach erosion. Just cruising in neutral, and enjoying the fine weather.

On Friday morning, the wake-up call came at 1:30 a.m. I took a quick shower, dressed, left the Basel Hotel, and caught a cab for Ben Gurion Airport. Brad was going to stay behind to do some sightseeing with Joseph. The Swissair Airbus took off from Tel Aviv around 6:00 a.m., bound for Zurich. By 2:00 p.m. of that same day, I was flying in a Delta jet over Lake Michigan. We had flown a more northerly route than usual to avoid Hurricane *Floyd*. That jet landed in Cincinnati, and I caught a commuter plane on over the mountains to Columbia, where I took a cab from the airport to the RPI office. Normally, JM, or somebody else from RPI, would have met me, but I had insisted that everybody stay at the office for the mural dedication, which was already underway when my plane landed in Columbia.

The cab pulled up behind our Explorer in the parking lot across the street from the RPI office at 1121 Park Street. As I got out of the cab, I noticed that some people were milling around in front of the mural. By then, the reception had been underway for about an hour and fifteen minutes. It had been twenty-three hours since I boarded that jet in Tel Aviv.

Mark White, the head of our computer mapping group, who was pointing out the mural to some of his friends, saw me and called out, "Miles, you're back!"

Then he and a couple of his associates hurried across the street and carried my luggage into the building, so they could store it in my office.

I walked over and stood in front of the mural. Eric had put the finishing touches on it while I was away, and boy, did it look great!

While I was marveling at Eric's masterpiece, some people I didn't know came up, shook my hand, and congratulated me on having the company commission it.

"The mural is perfect," they said.

"Thanks."

I went around the corner and walked into the front door of the building. A crowd of people almost filled up the outer foyer and the conference room. The place looked good. Some folks were even admiring my photographs, which were mounted on the wall.

A large number of Eric's friends and disciples were on hand, but precious few of the friends and acquaintances I had invited.

But it was okay. What did I care? I was proud of the mural and of the building. And JM looked beautiful in her blue-striped blouse, and when she saw me walk in, she rushed over and gave me a big hug and a kiss.

"You look great!" she said.

What more could any man ask?

21 NIGHTMARE IN JUBAIL CITY

Strong hearts to the front,
weak hearts to the rear.

CRAZY HORSE

The Science

The following discussion in this chapter relates to our response to the largest oil spill in history (11 million barrels), the one initiated by the visiting Iraqi Army in Kuwait early in 1991, which was 40 times as big as the *Exxon Valdez* spill in 1989. The *Exxon Valdez* spill in Prince William Sound, Alaska was the largest oil spill in the U.S. until the more recent *Deepwater Horizon* spill, which is still far short of the Gulf War spill in total volume (at 4.9 million barrels).

JM was on scene during the U.S. Coast Guard response during the Gulf War spill in May 1991, at which time she took the photograph of the heavily oiled wide intertidal zone shown in Plate 12. That early excursion provided us with some basic information on the spill as the oil came on shore. Thanks to joint funding by NOAA and the Marine Spill Response Corporation (MSRC), we began a long-term study of the effects of the spill in January of 1992 that went on for several years, resulting in a number of scientific publications (e.g., Hayes et al., 1993a, 1993b, 1995; Michel et al., 1993a). The 2002-2003 project discussed here also resulted in a number of publications (e.g., Michel et al., 2005), and Tom Freeman, Joe Holmes, and I made a movie about the project called *Walking to Kuwait.* Figure 50 shows the area impacted by the spill, the shoreline between Kuwait City and Abu Ali, Saudi Arabia.

Of all the field projects I managed over the years, this six-month project was the most satisfying of all, perhaps partly because it was the last one of this scale that I plan to do now that I am semi-retired.

The story that follows took place while we were in the fifth month of the project, which involved surveying the 803 kilometers of shoreline oiled by the spill (part of which is illustrated in Figure 109). As a result of the spill, the Saudi government submitted claims to the United Nations Compensation Commission for environmental damages, including restoration of the oiled shoreline. Our company won the contract to assess how much oil remained on the shoreline

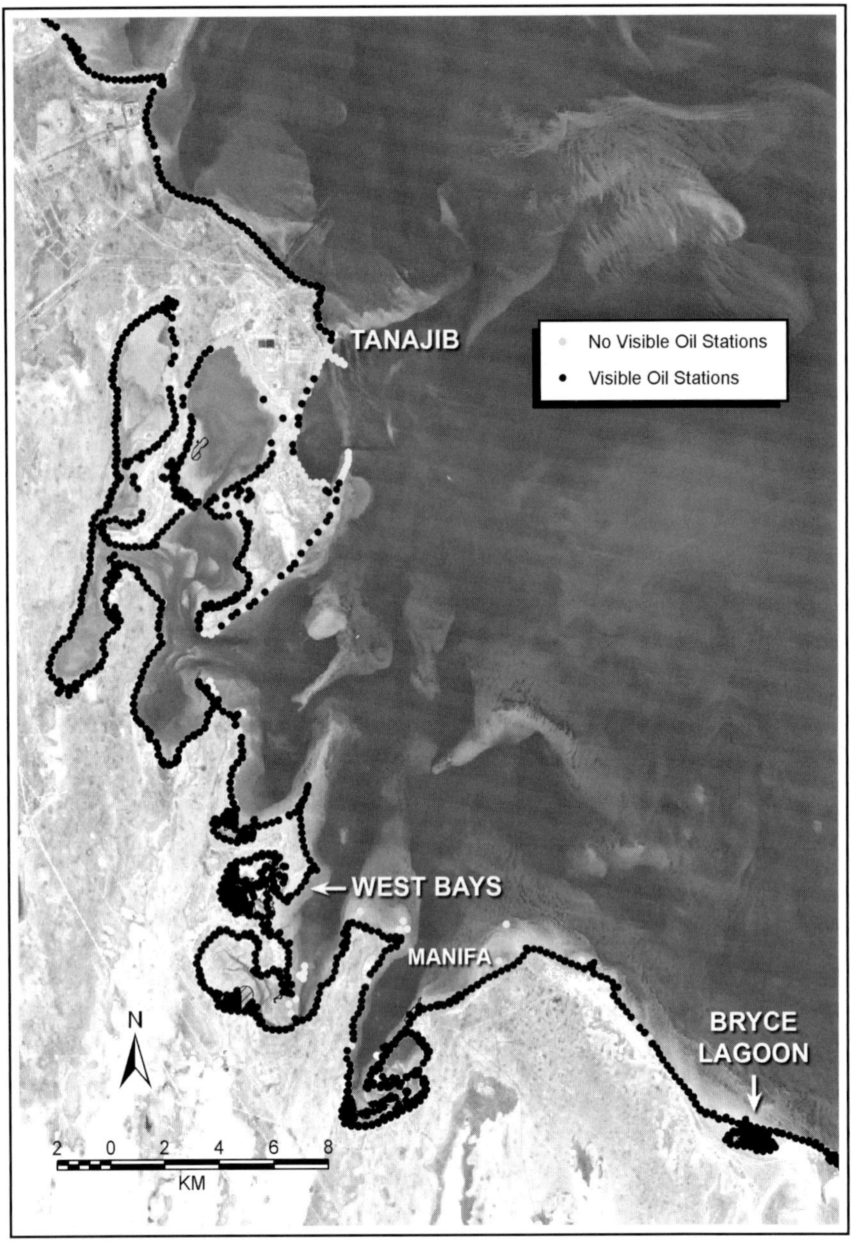

FIGURE 109. Map showing the oil extent in the Tanajib and Manifa areas on the Gulf Coast of Saudi Arabia in 2003, twelve years after the Gulf War oil spill. This area is located on Figure 50. The presence of oil is noted by the black dots, which represent our individual survey sites. This area includes the "butterfly bays" of Tanajib discussed in the text. Note that most stations contain oil, even along the outer shoreline. From Research Planning, Inc. (2003).

and to determine the long-term biological impacts. We assembled a team of 30 scientists and technicians, who eventually surveyed 3,107 transects across the intertidal zone, located at a spacing of 250 meters. Although the maximum tidal range in the area studied was only 1.5 meters, the intertidal zone is generally quite wide, exceeding 2 kilometers in places (see aerial view in Plate 12, and sketch map of a wide intertidal transect shown in Figure 110).

As already noted, I was the manager of this field project, during which we collected 26,000 sediment samples for chemical analysis. We also determined that eight million cubic meters of oiled sediment remained on the shoreline, mostly in oiled crab burrows in the sheltered marshes and tidal flats. Incredibly, we were able to collect liquid oil from some of the burrows twelve years after the spill. Biological recovery of the heavily oiled habitats was proceeding very slowly. Two of the more interesting findings, in addition to the incredible volume of oiled sediment still present twelve years later, follow:

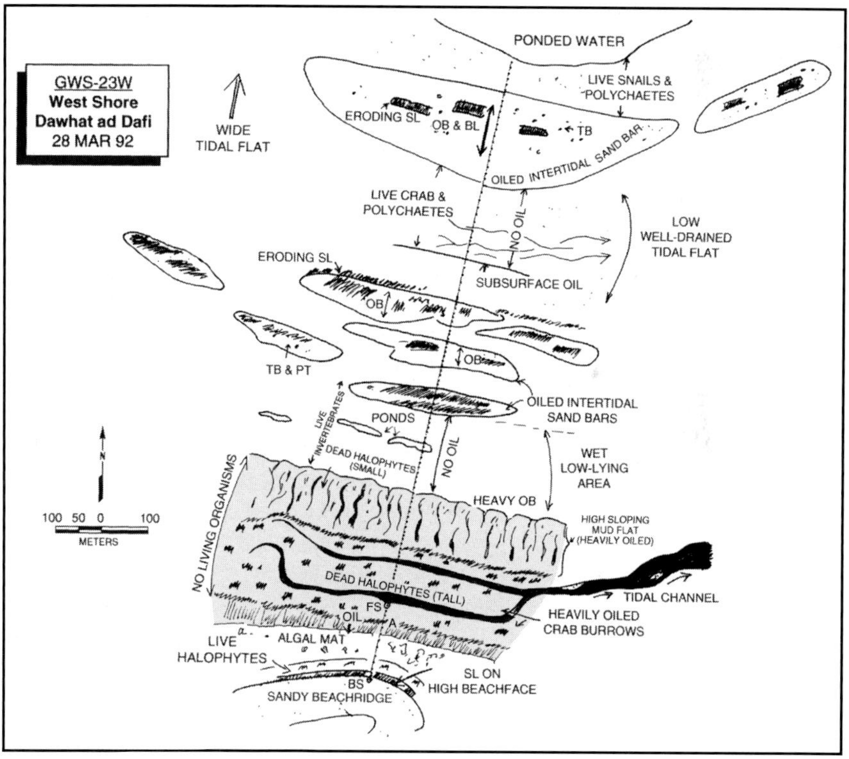

FIGURE 110. Plan-view field sketch, to scale along the intertidal transect, located on the west shore of Dawhat ad Dafi (Figure 50) and surveyed on 28 March 1992. The transect starts at a low-amplitude beach ridge at the lower center of the sketch and extends for 1,400 meters. The salt marsh in the upper part of transect was completely dead with no sign of living organisms at the time of the survey, a condition that continued for many years. BS = back stake; FS = front stake; SL = surface layer of oiled sediment; BL = buried layer of oiled sediment; OB = oiled burrows; TB = tar balls; PT = tar patties (larger than tar balls; diameter >10 centimeters). From Research Planning, Inc. (2003).

1) The fact that the oil spill impacted large areas of salt marsh, including the elimination of all forms of life within some parts of the marsh (see field sketch in Figure 110). Many of these areas have still not completely recovered as of 2011, twenty years later.

2) Our earlier studies allowed us to use a "rule of thumb" that oil does not penetrate very deeply (usually less than a few centimeters) in beaches and tidal flats composed of medium- to fine-grained sand. However, as shown in Figure 111, during this spill oil penetrated 30 centimeters (one foot) or more in several places. In fact, we found that 442 of the 1,336 stations we studied on these types of environments (sandy tidal flats and upper beachfaces) contained oil-residue thicknesses greater than 15 centimeters (~ 6 inches; in some places more than twice that deep). This deeper penetration took place in what is called bubble sand, which has the texture of a sponge (i.e., it is full of open, bubble-like pore spaces). Figure 111 shows a 1992 photograph of oiled bubble sand and an interpretive sketch of the photograph. The occurrence of these somewhat sheltered (from wave action) sandy intertidal environments along the Gulf shoreline provides the proper conditions for the formation of bubble sand, which allowed the oil to penetrate much more deeply than we had found in sand beaches at numerous other spills (where bubble sand was not present). When we reached this conclusion and published it, for the first time (Hayes et al., 1993b), there were within the scientific community some skeptics regarding this conclusion. However, those skeptics had not dug over 200 trenches in bubble sand in the sand beaches and sandy tidal flats of the shorelines of South Carolina and Georgia as we had.

The origin of bubble sand was investigated by Chris Reel, one of my graduate students at USC-East (Reel, 1982). Based on a number of studies of ground water flow in the beach at Price Inlet, South Carolina (an area discussed in detail in Chapter10 – PRICE INLET), plus a study of a large number of trenches, cores, and so on in the sediment, Reel concluded that bubble sand is best developed on upper intertidal sand flats and beachfaces where air is trapped between the water table in the sand and ocean water penetrating from the surface as a result of the rising tide. The water table in the beach and tidal flat sediments is elevated more slowly than the ocean surface is during the rising tide, which creates the trapping mechanism (air between two distinct layers of water-saturated sand). The resulting bubble sand has a sponge-like appearance, with porosities well over 50 percent. Also, with a little compaction by overlying sediments, some of the bubbles collapse and, in some cases, the escaping air distorts the original sand layers into convoluted, swirling bands (escape structures), such as the ones illustrated in Figure 110.

Using our basic data from the 2002-2003 survey, the Saudi environmental agency (Presidency for Meteorology and Environment - PME) submitted a claim to the U.N. Compensation Commission for several billion dollars to clean up and restore the shoreline. They were eventually awarded $465M for the shoreline part of the claim. At the present time (March 2011), we have a team stationed in Dammam, Saudi Arabia (Figure 50) to oversee from a management and scientific perspective the restoration projects that are being implemented.

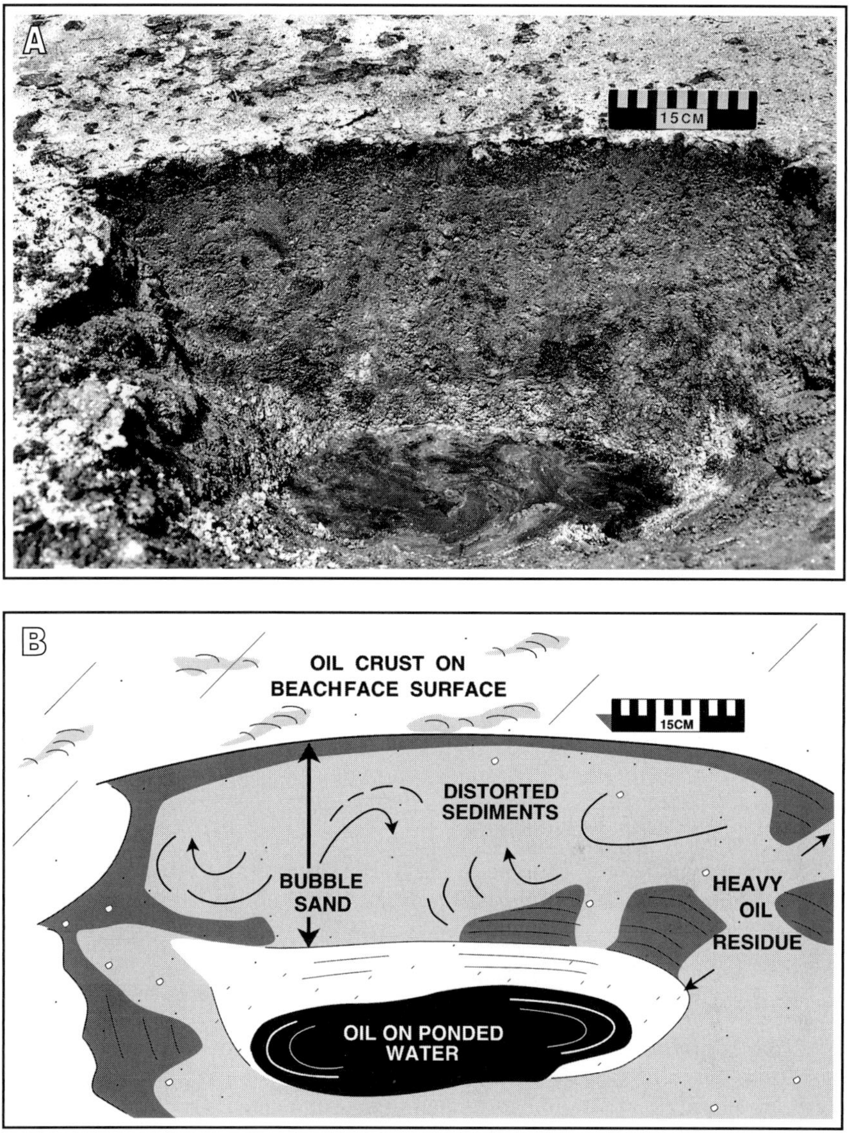

FIGURE 111. Example of bubble sand in sediments at a sheltered beach located near the boundary between Dawhats al Musallamiyah and ad Dafi (Figure 50) surveyed on 12 April 1992. (A) Trench in bubble sand showing penetration depths of over 30 centimeters (about 12 inches). (B) Interpretation of the trench in A, showing the distorted sediments typical of bubble sand. From Research Planning, Inc. (2003).

The Story

(22 February 2003, Saturday - Tanajib, Saudi Arabia)

I was one of the early arrivals in the dining area and got in line behind Freeman in order to prepare my usual breakfast of cereal and orange juice, eventually following him to a table where we joined Sami and Fathi.

"You going to miss this place?" I asked Freeman.

"Yes, I like it a lot better here than in al Jubail."

"Me too, the food's better."

"And the drives on the Highway of Death are a lot shorter," he added.

Fathi grinned and gave Freeman a gentle shove saying, "Why you worry? I'm the one who drives."

"That's what worries me."

Sami changed the subject and asked me, "Well Chief, think we can finish and close the gap today?"

"Yeah, sure, there's only 38 transects left, and we have 5 full teams today. Andrew will G-Tech for the 5th team. That's only 7 or 8 transects per team. There are some long ones in that big south bay, but quite a few of them should go pretty fast." This area that we were going to survey, the "butterfly bays" of the Tanajib area, is illustrated in Figure 109.

Freeman laughed, "Remember when we thought getting two done was a great day?"

"When we were in the mud," Fathi added.

Freeman continued, "Hell, man, I liked working in the mud. No long drives on the Highway of Death, among other things."

"Yes, I'll miss working in Tanajib too," I resumed. "It's a little tiring though, trying to keep up with Jason and Sami. But it's peaceful up here. Keep's me out of touch with Fahad and Aziz in Dammam."

The fieldwork on the project started in late September 2002, and after a Christmas break of a couple of weeks, we returned to our camp in al Jubail to continue the work. We started at the south end of the oiled area, near Jubail, and worked our way north, heading for and eventually reaching the Kuwait border, a total shoreline distance of 803 kilometers (see Figure 50).

We moved our field camp north to Tanajib near the end of January 2003. About the time of the move, rumors that the U.S. would soon begin to attack Iraq were rampant. In fact, we received a number of messages from the State Department urging all U.S. citizens to seriously consider leaving Saudi Arabia. After one of those warnings, I asked the Americans on the field crew if any wanted to leave, and ten members of the team were soon on their way home. They left Tanajib on 3 February 2003. This exodus left the team pretty shorthanded, and I was concerned that we may not be able to finish the project before the war started. But, by the date I am discussing here (22 February), we had worked things out and the "survivors," as we called ourselves, were just about to finish the last of the transects, which was the lion's share of the work.

I looked around the dining area. All of the "survivors" were now present, some gobbling down breakfast and some making their lunch of peanut butter and pita bread, and so forth. Before going any further, let me introduce the team:

American "survivors"

Freeman - One of my Master's students at USC-East, who had gone on to other things, including playing music professionally and teaching grammar school. The darling of the night life in Bahrain. Now the field team leader. Early forties.

Chuck - Ph.D. biologist. First biologist hired by RPI in the early years. Later obtained an MBA and formed his own company. Fairly recently back into the oil spill business. A fan of the mangroves, the subject of his dissertation at the University of Miami. Around 50 years old.

Andrew - Geologist jack-of-all trades who came on in November as a GPS technician. He was now chief field logistics officer, as well as geologist and GPS guru for the field teams as needed. Thoroughly devoted to the project. Handsome, blonde-haired fellow with a medium build. Late thirties.

Jason - GPS technician who came aboard after Christmas. Trained as a marine scientist, but more interested in world travel, especially climbing mountains. Works on fishing boats in Alaska during the summer. About six feet tall. Long blonde hair. Zero ounces of body fat. Mid twenties.

Travis - Geologist who formerly worked for "big oil," including a stint describing cores in Saudi Arabia. Prefers to run boats on the South Carolina coast (contracted by us to do field seminars). Medium build. Good field man. Early thirties.

Mikey - Geologist who took on the GPS task. Also came aboard after Christmas. A little over 6 feet tall. Late twenties. Likes to sing while in route to the field (sings off key, according to Freeman). Long goatee. Also a hard worker. First of the Americans to say he had no intention of leaving early. Then he walked out of the meeting.

Don - Chuck's step-son. Computer whiz. We traded books to read. Tall, about 6 feet, 4 inches. A nice guy. Very friendly with the Saudis. Late twenties.

Me - About 6 feet tall. Very gray hair. Turned 68 on October 20. Referred to by Freeman and Fathi as "the old goat herder." Likes things to run on time.

Saudi "survivors"

Dr. Sami - Ph.D. in biology from Texas A&M University. An employee of a university in Jeddah. The spokesman for the Saudis. Very personable. Handsome. A little under 6 feet tall. Early forties.

Dr. Faisal - Ph.D. in biology from Auburn University. Also an employee of a university in Jeddah. A real diplomat who could get us through any locked gate. Said he had "lost a lot of weight in the middle of nowhere" during our "Walk to Kuwait." Early forties.

Abu Khalid - This name means "father of Khalid." This is what the other Saudis called him. Like almost all of the Saudis on the team, one of his names was Mohammed. Therefore, I used that name for only one of the Mohammeds, an excellent field man who had to leave the team in mid-February in order to go to the hospital. Tough loss. Abu Khalid is very thin (tightly wound) and wiry. Outstanding worker. Tireless. A PME employee in his late thirties.

Fathi - Young biologist doing his first major field project. Very athletic. Winner of major bike races in Saudi. Versatile field man, who started out as a sampler, and graduated to GPS technician and field biologist. A natural comic. Mid twenties.

Abha - Another young guy doing his first field project. A PME employee in his mid twenties. Named Abha after his home town, a beautiful village located high in the mountains above the Red Sea. Shy and quiet.

Khalid - Educated in the states with a daughter still there. Spoke excellent English. One of the last Saudi hires. Sampler. Middle height and weight. Mid to late forties. When asked how he enjoyed the "Walk to Kuwait," he answered, "It's a tough job though."

Indian "survivors"

There were a number of Indian technicians and laborers involved in the miscellaneous jobs around the camps and in the delivery of the samples every night to the chemistry laboratory in Dammam. Fairly early on during the fieldwork in the fall, it was obvious that we could not get enough devoted Saudi workers to do the job of sampler, the hardest of the four positions on the field teams and one considered to be the "lowest on the totem pole." Therefore, I asked our logistics manager, Faleh, if he could find us some Indian technicians, and he came up with two excellent workers, who were with us the rest of the time.

Krishnan - Rather short and "strong as an ox," as they say. He would carry a cooler with its 60+ pounds of sediment samples on his head for several kilometers back to the truck at the end of the field day. Amazing. Tell him to dig a trench and a meter-deep hole would appear almost instantly. Great personality. Mid thirties.

Kumar - About 6 feet tall and also in his mid thirties. Another good worker with a pleasant personality. Quiet and efficient.

Other notables

JM - The brains behind the project. Did the bulk of the writing for the proposal and some of the reports. She was in the field for the first three weeks and helped train the field crews. Came back to write a report in late November. Did not return after Christmas, but kept in close contact with the project via e-mail.

Linos - A native of Cyprus, who has been a biologist with our company for over 20 years. He played a major role in making arrangements for and planning this project. He now serves as the Chief Operations Officer for our International Projects at RPI. He arrived in Saudi early in September to establish logistical operations and left in October. Although he did not participate much in the fieldwork, as soon as the ten Americans left the field, I put in an urgent call for him to join us and help with the completion of the project. A modest-sized ex-jock (played soccer for a major Canadian university) full of energy. Mid forties.

Aziz - Head man of the Eastern Province office of PME, who played an active roll in PME's response to the oil spill in 1991. He also studied oil spills for a month or so at RPI a few years after the spill. During that time, he came out to The Bluff and used our chain saw to cut down a tree we were going to use for firewood. Not many trees to cut down in the Eastern Province of Saudi Arabia. Very clever. Highly supportive of our efforts. Early forties.

Dr. Saad and the Prince - The ultimate PME authorities, who had offices in PME headquarters in Jeddah. Dr. Saad was the chief scientific advisor for the project. Prince Turki, an affable older gentleman, held several offices of importance in the Saudi government before being appointed to administer the funds and activities related to the Saudi claims against Iraq that involved environmental and health issues.

Fahad - Our Saudi partner for the project. Actually, we sub-contracted to his company. A business man who invented a machine for cleaning beaches. He studied in America, but is a traditional Saudi. Likes to disappear into the desert from time to time. Late forties.

Hamoud - Fahad's younger brother. Also studied in America. My driver and field scout throughout much of the project. Early forties. Much more about him later.

Faleh - Bedouin camel raiser who formerly worked for Aramco, being stationed in Houston for some time. Small and very energetic. Mid forties. In charge of the logistics for the project. I talked with him almost every morning and evening. Knew the field area extremely well. He and I scouted all the field areas before the teams started their work there. Good at getting us out of the mud when we got the vehicles stuck.

Elie - Lebanese former hotel manager in the U.S. Fahad's primary business assistant. With us the whole time in order to provide financial and logistical support. Very personable. Late forties.

Bill - American computer whiz and programmer. Long-time RPI employee. Responsible for management of the enormous digital database we created during this project. Down loaded all data at the end of each field day and submitted it to PME the next day. Worked at night. Very long hours. A little on the thin side. His nickname at RPI is "Stick." Mid thirties.

Truth be known, the ones left behind, who had to shoulder the leadership after the other guys left, namely me, Freeman, Chuck, and Andrew, still grumbled on occasion about their decision to abandon the project. However, as it turned out, and as noted earlier, the survey we carried out during those last three-plus weeks in February 2003 was the peak field experience in my career, even exceeding the fabulous tours in Alaska and southern Chile. At 68, it was a little tiring, but I loved it! Why?

1) Finishing a near-impossible number of transects and oiled polygon surveys before the war started.
2) The total focus required and exerted by everybody on the field team.
3) High quality of purpose–studying the world's largest oil spill and collection of such an incredible volume of data.

4) The coming together of the Saudi, American, and Indian "brotherhood" with a single-minded goal and belief that we could finish the job.

A little later that same morning at the Tanajib camp, while we were all standing in the parking area getting ready for the trip out, with Dr. Faisal handing out his usual round of tea, it occurred to me that Hamoud's car was missing.

I thought, "Oh crap! I forgot to call Hamoud last night to tell him that we were moving back to al Jubail today."

We had been moving along so well, we were a couple of days ahead of our latest projected schedule. So I did my usual slink behind our vehicle to pour Faisal's tea out on the ground, and then I went looking for Linos, who already had some of the Indian guys he had brought up from Jubail packing our gear into a truck.

"Linos, I forgot to tell Hamoud we were moving. Please apologize for me, and tell him that I look forward to seeing him back in Jubail tonight."

Hamoud had been my driver and associate during the earlier days of the project when I would commonly visit a number of field teams at their study sites each day. Now that we were short handed, I became the geologist/sampler on a team that included Jason, the GPS technician, and Sami, the biologist. We always traveled together to the field site in one vehicle (always in a caravan, however), so Hamoud and his vehicle were not needed nearly as much as in the early days.

"But Hamoud isn't here," Linos pointed out.

"I know. He'll probably be here soon. He sleeps in his family's camp over in the hills on the other side of the main highway."

"Okay, I will tell him. Have a good day out on the tidal flats."

"See you."

With Sami driving, we took the lead of the five-car caravan in our 4WD Nissan Patrol as we left the camp of scattered mobile huts and headed for the main highway that connects the Eastern Province of Saudi Arabia with Kuwait. This is the same road the U.S. troops drove over in February 1991 in preparation for the battle at Khafji, the first major battle in the first Gulf War. As we turned onto the highway, we were approximately 80 kilometers from Khafji, headed south.

[NOTE: This is the highway that Freeman referred to as the "Highway of Death" in our earlier conversation. Many drivers on this road zoomed along at >100 mph, usually driving without significant concern about the other travelers along the road. It was fairly common to see wrecked cars all along the highway, especially those that encountered a camel crossing it. A real scary place to ride along, especially on a Wednesday night, when many Kuwaitis headed south for a weekend in Bahrain were hogging the highway.]

After a five-minute drive on the "Highway of Death," we took a sandy track across the vegetated dunes and headed east toward the shoreline of the southernmost of the Tanajib "butterfly bays" (see Figure 109). Once we reached the shore, the different vehicles peeled off and drove toward the scattered transects that had been skipped on the earlier passage north. Our team headed for one of the most distant transects, and as we moved along the beach, we could hear members of the other teams talking back and forth on their radios.

"Andrew, this is Freeman, we'll start on the south end and meet you in the middle, at transect 0-75."

"Roger that."

"Abu Khalid, I've finished 0-65. Can't see you guys."

"Okay Chuck, I'm walking out to the end of the oil. Soon we finish 0-64."

Fathi called Sami, speaking in Arabic, with a question about some of the macro-invertebrates he was counting. And so on.

Once we reached our assigned section of the shore, the work went smoothly, and we were moving quickly. As soon as we finished a transect, I would load the 10-15 sediment samples we had collected into the back of the Patrol and drive the 250 meters along the dunes to the next transect. Sami did a "random walk" throughout the area between transects to make visual observations and counts of the biological components. Meanwhile, Jason, loaded down with gear, had run between transects and was already locating the start point for the next transect with the GPS he had strapped to his back. This equipment could locate a point in the middle of nowhere to within one meter accuracy. Amazing.

A couple of hours into the morning, I was inspecting one of our trenches while Jason was digging another one 30-40 meters away down the flat. Jason always carried a radio so we could keep in touch with the other teams. Our quiet efforts were interrupted abruptly by a loud, deep voice on the radio that shouted out, "Motherfucker! Motherfucker!"

"What was that all about, Jason?"

He turned away and said, "Oh, it was nothing."

I usually tuned out the more-or-less constant chatter over the radio, unless Jason told me one of the team members wanted to talk with me. In this case, I assumed we were picking up someone not associated with us, perhaps on a fishing boat offshore, and went back to collecting samples. Other than that little interruption, all was peace and quiet in our area.

However, things were taking place back at the Tanajib camp I didn't know about that would set the stage for the upcoming Nightmare in Jubail City. Well, when Hamoud finally arrived at the camp, apparently Linos and the Indians had already departed for Jubail. Hamoud eventually learned from some of the camp personnel, presumably, that we were moving back to Jubail. This information sent him into a fury, and he drove south looking for us. When he reached the bays, we were so scattered that he couldn't find our team. Separated from the teams and alone, he decided to engage in a little target practice with his "machine gun" (automatic rifle). Although I never heard the shots, some members of the teams did.

Somehow, I guess from radio contact or possibly by visual contact, Dr. Faisal, a biologist who always carried a radio because of his typical separation from the transect survey crew, radioed to Hamoud, "Hey Hamoud, you shooting rabbits over there?"

Anyway, I never thought about Hamoud, the mysterious voice on the radio, or any other distractions as we made good time, finishing up our assigned transects and moving south to finish one of the longer transects that remained to be done.

After finishing that last transect, and as Sami and Jason were loading the last of our gear and samples into the car, I picked up my cell phone, which I had left in the Patrol and forgot to check it all day. There were two missed calls from Hamoud.

"Damn, I wonder if Linos never saw him?" I thought, "I'll call him as soon as we're back on the main highway and the reception is better."

Then we headed for the final rendezvous point, where the teams were gathering. Everybody was there except Freeman's team, which I could see far out on the edge of the tidal flat.

After a few minutes, they came walking up, and we had quite a little celebration. We had done it! The eight American "survivors," thanks to the magnificent rally by the seven Saudis and two Indians on our crew, had done it, finished the remaining transects in the north (of the eventual total of 3,107) three weeks ahead of the original schedule, and before the second Gulf War started.

We were dancing about and taking pictures and congratulating ourselves. I was on a terrific emotional high and, judging by the smiles and jubilation all around, so was everybody else.

Finally, we loaded the vehicles and prepared for the drive back to Jubail. Since I knew a short cut to the main road, having scouted the area a couple of times with Hamoud and Faleh, we took the lead in our Patrol, soon crossing a major pipeline and within a few minutes turning south on the pavement of the main highway.

I picked up my phone and dialed Hamoud.

"Hamoud, what's up?"

"You left me! You left me!"

"Sorry, didn't Linos tell you? I forgot to call you last night, we were busy making preparations for the move."

"But you left me and I love you. You are my father."

I thought he had been drinking some moonshine or something. Freeman had told me earlier about Hamoud's mood shifts when under the influence, based on an experience he had with Hamoud one night in Bahrain.

"Where are you now, Hamoud?"

"In my room in Jubail camp."

"Okay, we're on the main highway now, nearing the turn to Ras Azor. We should be there in about one hour. Wait in your room, okay? We'll talk when I get there. Again, sorry about the mix-up."

"Yes, I will. I love you."

"See you soon."

All during the call, I was trying to keep the conversation private. In the back of the Patrol, Chuck, who had joined us for the ride home, was laughing it up with

Jason, and Sami, who was driving, joined in the frivolity from time to time.

I laid my head back against the seat, closed my eyes, and thought, "I only had 10 minutes to enjoy the celebration. Now this. How did I manage to forget Hamoud?"

"Want to hear some music?" Jason asked as he leaned into the front seat to insert an Alison Kraus CD into his portable CD player.

While still leaning over the seat, he looked at me and asked, "Hamoud?"

"Yes."

"Ummm."

Then I realized it had been Hamoud swearing over the radio and Jason had kept that bit of news to himself.

I turned to Chuck and asked, "Did you guys run into Hamoud today?"

"No, but we heard the gunshots and Faisal talking with him on the radio, didn't you?"

"No. I don't pay attention to the radio much, unless, of course, it's you calling, Chuck."

Then the subject was changed and we chatted about the day and listened to Alison Kraus all the way back. But, I was worried and feeling bad about my callous disregard for my friend and confidant Hamoud.

Like many moderately wealthy Saudi males of his generation, Hamoud was educated in America. He went to Colorado Springs while still in high school. After graduating from high school, he moved around, finally settling down in eastern Washington. In Spokane, he rented a space next to a church and established a bar that ran all night, just getting started good when the folks showed up for church service on Sunday morning. For that and other reasons, he was run out of the state of Washington, buying a ranch and establishing another bar across the border in Idaho. He told me that and many other stories about sinking boats, his blonde bombshell girlfriend, and finally being told by the Saudi Ambassador to head back home. Once back, he was married unsuccessfully a couple of times, and was talking about going to Syria to get number three while we made our numerous rides across the sabkhas and tracks along the shoreline.

He tried his hand at sampling for JM's team one day, and at the end of the day, proclaimed that he would not do that anymore. Thus, he became my personal driver. He was glad to follow my rules for not driving over 120 km/hr, not smoking in the car, etc., which was nice. I did notice that he liked to terrorize the diminutive Pakistanis and Indians if they threatened to cross the road in front of him. Otherwise, the work went fine with him up until the guys left, and I had to go to the field without him.

When we pulled into the parking area of the Jubail camp, Linos came quickly out to me, waving me aside. He was white as a sheet, despite his Mediterranean ancestry.

"We need to talk!"

"Why, did you see Hamoud when he arrived?"

"Yes, I spent almost an hour with him, and he threatened to kill me several times!"

He said, "I love you. You are my brother," followed by a headlock and saying, "I am going to kill you."

Would you believe that this statement got my attention? Then he told me that Hamoud, who probably outweighs him by about 50 pounds, had choked and pushed him around, taking time out now and then to empty a glass of what appeared to be vodka, Hamoud's favorite drink. Of course, it could have been moonshine of some unknown composition. He also showed Linos a couple of chunks of some good old Lebanese hashish.

"He had a handful of rifle bullets, and I guess he has a gun. I didn't see it. You need to talk to him."

"He still in his room?"

We had rented part of a new facility in the old town of Jubail. It was two separate buildings with three floors each. The parking area was located between the two buildings. Chuck said he had taken a picture of the camp to e-mail to his wife but, with the austere square exterior and barred windows, it looked too much like a prison to do so. Hamoud's pad was a corner room on the second floor, directly across from Linos. I was on the third floor in the other building, which also housed the dining area and our meeting room.

I was pretty muddy from the fieldwork, so I slipped off my muddy field shoes and walked up the stairs in my muddy socks to Hamoud's room and knocked on his door. The door was slightly ajar, so I walked inside and sat on his cot beside him.

"You left me. How could you?"

He was sobbing, red-eyed and staring dejectedly at the floor.

"My sister told me I have to take care of you. I love you like you are my father. I told her I would watch after you."

When she left the field in October, JM had taken Hamoud aside and asked him to look after me. I guess because he was my driver and she was a little worried about what would happen with the war coming and all. It was JM he was referring to as "his sister."

"That's why I always have my machine gun with me."

I had never seen this rifle, but he definitely had it with him in the field on that particular day.

"I am so sorry about not calling you. My fault."

After another long lament about my failing him, he abruptly switched into a rave about how horrible his two older brothers were, one the semi-wealthy local businessman, Fahad, his present employer and our prime Saudi partner, and the other a wealthy doctor in Germany.

"They want me to go to Egypt to manage a hotel there. They always against me. I hate both of them. I don't want to go! I will not go! I want to stay here with my friends and brothers the Americans."

"When my father was alive, I was favorite. He tell me I'm the smartest one. He expects me to do great things, better than both of them."

I listened for about another thirty minutes, not worrying about anything much, and then he changed, railing again about being left in Tanajib. As I watched his anger grow, I was gripped by a deep sense of unease and, for the first time, I wondered if he would lose it completely, and I would be trapped in his room with him.

But, he eventually began to calm down, and I moved off the cot to sit in a nearby chair.

"Why don't you relax? Lie down and take a nap."

He stretched out on the cot and closed his eyes as I left the room.

As far as I knew, all was quiet through dinner. However, Hamoud had awakened and decided to leave the camp. As Linos was coming across the parking area to go up to the dining area in my building, Hamoud was in his car and waved him over. When Linos walked up to the window, Hamoud sprayed him in the face with some aerosol deodorant spray. Just a little parting shot, I guess.

Dinner was still a rather pleasant affair. We had worked eleven straight days, and the mid-day high tides that were coming up allowed us to take a break for the next two days. Linos didn't mention Hamoud's good-bye salutation. As it turned out, almost everybody, except Bill, had to go to Bahrain the next day to drive across the bridge and get our Saudi visas renewed. After dinner, in addition to making preparations for that excursion, most of the team crowded into our office on the third floor of the Hamoud/Linos building to use our computers to read e-mails and send out greetings to the home folks. We did not have e-mail access in Tanajib.

After dinner, I read a while and turned in around 10 p.m. My phone rang sometime around midnight. It was Elie. Hamoud had gone on the rampage again, and Elie gave me the following report.

Hamoud had returned from his trip to God knows where around a quarter to eleven and found that his parking spot was taken. He immediately went to his room, retrieved his "machine gun," and went back downstairs with the intention of blasting the offending vehicle (or so one of the building managers later told Elie). The car in his space belonged to some PME scientists, which may have been the reason the car escaped being shot full of holes.

Anyway, after deciding not to blast the car, Hamoud returned the rifle to his room, and then went up to the third floor to Elie's room, and pounded on the door, shouting loudly, "Elie, whose car is parked in my spot. Move it now!!" Plus a few choice expletives.

Chuck, whose room was next to Elie's, opened his door and exclaimed, "Hey, hold down the noise. Be quiet. I am trying to sleep."

Hamoud immediately pushed Chuck back, forced his way into Chuck's room, and punched him in the face.

While this struggle was going on, Andrew and Jason, who were working on the computers in our office down at the end of the hall, came out and ran down to Chuck's room. Jason immediately grabbed Hamoud and pulled him back out into the hall.

Hamoud pulled free from Jason and paced around the hall screaming loudly, "Jason, you're a dead man! Chuck, you're a dead man! I will cut you with my knife." He ran his forefinger across his throat as he yelled this out, plus a ramble of obscenities.

Elie came out of his room and shouted at Hamoud, "Calm down, calm down."

Which seemed to have some effect, because after a while, Hamoud turned and ran down the stairs to his room, where he began to slam his door back and forth so loudly it could be heard throughout the building.

He also paced the hallway yelling "You're a dead man Jason," who apparently wasn't too worried, as he was back in the office sending e-mails.

Then Hamoud got his "machine gun" again and walked down to the parking lot, threatening to shoot the offending car once more but, again, decided not to.

Freeman and Linos, who were on the same floor with Hamoud, each took a different approach from Jason. As the ruckus started, Freeman was lying on his cot, which was exposed to the hallway by a glass window. When Hamoud came back to his room after threatening the trespassing car again, he started slamming the bolt of his rifle back and forth so loudly that Freeman could hear it in his room down the hall. Upon hearing this, Freeman jumped off the cot, out of the line of fire as it were, and retrieved a crude spear he had built for an upcoming fishing expedition. He built it from material collected from the garbage heap behind the camp–5 feet of PVC pipe with a piece of rebar and a broken hack saw blade attached. With this weapon in hand, he stood behind his door ready for the eventuality of Hamoud breaking into his room which, fortunately, never happened.

Linos, on the other hand, retreated to his bathroom, which was at the back of the complex, and also out of the line of fire from the door, thinking maybe he could jump out of his window from the second floor. But he couldn't get out, because he couldn't detach the heavy metal bars across the window. He was trapped. Hamoud was directly across the hall with his door open.

After several more minutes, all was quiet.

Linos went back to the bathroom window and looked down to the parking lot, where he could see the top of Elie's bald head.

"Elie, should you call Fahad?"

"No, he will not take the call."

"Then, you should call Dr. Miles."

"Let's give Hamoud time to fall asleep."

Which is apparently what happened before Elie called me.

After hearing the story, I said, "Should I come over there?" As noted earlier, I was in the other building.

"No, not yet. He appears to be quiet now."

I went back to bed and had a decent night's sleep. I was tired. Linos told me that several of the guys came out of their rooms later, and they sat outside and talked until 4 a.m. The first group left for Bahrain at 6:15 a.m. Linos and I, with Elie driving, were the last to leave, around 9:00 a.m. Hamoud was apparently still sleeping it off.

As we pulled out of the complex and headed south for our trip to Bahrain, Elie muttered under his breath, "How did you enjoy your Nightmare in Jubail City?"

A couple of hours later, we called Bill, the only American left in the camp.

Responding to our call, he said, "All quiet here. Hamoud is gone. And thanks a lot for leaving me all alone here with him. Have a beer for me."

The group assembled in a room in the hotel we always stayed in there in Bahrain. I initiated this strategy session with a word-for-word description of my conversations with Hamoud. This was met with a mix of reactions, as well as by some amateur psychological analysis. I told them Linos, Elie, and I would go back across the border and meet with Fahad to get assurances that Hamoud would be banned from the work for the ten days or so that were left. Travis became very disgusted with the talk, and left the room well before a final conclusion was reached.

As we were getting ready to leave, I took Freeman aside and told him to tell Travis I was pissed at him for not participating in the decision making.

As we crossed over the bridge headed back to Saudi, we tried and tried to reach Fahad, but he was not taking our calls.

Elie said, "Guess we will have to go on back to Jubail."

"Elie, pull over right here. We are not going back to Jubail until we have solved this problem!" At which he pulled out of the heavy traffic at a wide space on the shoulder of the road.

We debated what to do. Fahad had been in this habit of dodging tough issues ever since Ramadan in November. Most of my communication had been with Faleh, who had been a tremendous help and always responded to my calls. For some reason, I decided to call Faleh.

"Yes, doctor?"

"Faleh, do you know about the problem we are having with Hamoud?

"Yes, I have heard something."

"That jerk Fahad is not taking my calls. I need you to get a message to him."

"Okay."

"You tell him that if he has not solved the Hamoud problem within the next

20 minutes, I will call Dr. Saad and the Prince and tell them that his brother is threatening to kill my team members, and something has to be done about it right away if they expect us to finish this project!"

"Okay doctor, where are you?"

"With Elie and Linos sitting on the side of the road in Dammam, not far from Fahad's house."

I placed the phone on the dash and leaned back in my seat. Not more than 15 seconds later, the phone rang.

"Dr. Miles, this is Fahad. You don't have to worry, Hamoud has been restricted to house only, and I have the keys to his car. Please finish the job. I will be in Jubail day after tomorrow to meet with you."

"What about the machine gun?"

"Faleh will take care of that."

"Good. We are going back to Jubail now. We have finished all the transects to the north. We have a few more control transects to do to the south of the oiled area, and a bunch of oiled polygons to survey. Should be completely finished in 10 days or less."

The gang returned intact from Bahrain. Just before the science meeting that night, during which we made plans to finish the work, Faleh called and asked me to meet him in the parking lot. I walked down and he motioned me over to his car.

"Here is Hamoud's machine gun," he said, as he pointed to the rifle on the floor in the back of his Patrol.

After the meeting, Travis came up to me and apologized for splitting from the session in Bahrain.

"Just don't let anything like that happen again."

"I won't. This has been a great experience for me."

"Glad to have you here. You've done an excellent job."

I was around for a couple more days, meeting with Aziz and associates in Dammam, plotting future work with Fahad, reviewing the data, and so on. A photograph of the whole field team at the end of the project, holding up signs touting their accomplishments, is given in Figure 112. They finished all the surveys and headed for home on 10 March. The U.S. invaded Iraq on 16 March 2003.

Nigeria is another garden spot where we have spent some time, one of the toughest and most dangerous places in the world to work (see Chapter 13 – WEST AFRICA). In fact, as discussed earlier, we once had an office in Port Harcourt, Nigeria for five years. I had long since sworn that I would never go there again. However, JM told me about a very interesting potential project to solve a major problem with shoreline erosion threatening some of the rich oil fields leased by a major oil company. She had arranged for me to join her for a 4-day inspection of the erosion zone as soon as I could get away from Saudi.

FIGURE 112. The field team in Jubail camp at the end of the project. Linos Cotsapas suggested that they list some of their accomplishments on these signs.

Early in the morning of 1 March, Elie and Linos dropped me off at the airport in Bahrain, and I boarded a Gulf Air jet for Frankfurt, Germany, where I met with JM and we headed on to Lagos, Nigeria.

Out of the frying pan into the fire.

22 AFTERWORD: THE BOTTOM OF THE 9th

At night when the streets of your cities
are silent and you think them deserted,
they will throng with the returning hosts
that once filled them and still love
this beautiful land.
The Whitemen will never be alone.
Let him be just and deal kindly
with my people,
for the dead are not powerless.
Dead --- I say? There is no death.
Only a change of worlds.

CHIEF SEATTLE

Old men in the bad old days used to
renounce, retire, take to religion, spend
their time reading, thinking - thinking!

ALDOUS HUXLEY in BRAVE NEW WORLD

It ain't over till its over

YOGI BERRA

Everybody has to believe in something, I believe I'll go fishing!

HENRY DAVID THOREAU

(April 2011 – At our mountain home in western North Carolina)

It is now early spring 2011, and the long-awaited restoration project in Saudi Arabia has been underway for over a year and a half. Our field crew of six permanent employees (plus several part-time consultants and employees of RPI as needed) has an office in Dammam on the shoreline of the Gulf in the Eastern Province. Although I paid that office a visit in the fall of 2009, I don't plan to participate in the field work that much, although, thanks to the on-site manager of the office and Chief Operating Officer of our International Projects, Linos Cotsapas, we are kept in close touch with the project, so we can fill in some scientific details if needed. JM participates in the project on a much more active basis than I do. I should also mention that our work on the *Deepwater Horizon* oil spill is still going at a significant pace, so the domestic activities of RPI are also pretty busy at this time.

Since my "semi-retirement" after completing the work discussed in the last chapter in this book, NIGHTMARE IN JUBAIL CITY, I have been working on a series of books that JM and I intend to publish covering several individual coastal states. We published the first one, *A Coast For All Seasons: A Naturalist's Guide to the South Carolina Coast* in 2008 (Hayes and Michel, 2008) and, as mentioned earlier, we published *A Coast To Explore: Coastal Geology and Ecology of Central California* in 2010 (Hayes and Michel, 2010). I have finished my part of the next one, *A Tide-swept Coast of Sand and Marsh: Coastal Geology and Ecology of Georgia*, but JM has been so busy with the *Deepwater Horizon* oil spill and other company projects that she has not been able to complete her part as yet. As soon as this particular book, *Coastal Heroes,* is finished, which should be soon, I plan to get back into writing our next book on the coast of southern Alaska. I have to admit that, to some extent, the idea for doing this series of books on state coastlines (upon the suggestion and urging of JM) follows in a similar way a series of books on roadside geology of the different states started by another graduate of the University of Texas Geology Department, David Alt, who finished up just before I got there. The book by Alt and Hyndman (1989) on the *Roadside Geology of Idaho* is an example of the books in that series.

Writing these books on the coastlines of the different states should keep me busy for a while. However, I can't help but say that as I approach the last out in the bottom of the ninth, I do match to some extent a comment in Aldous Huxley's book *Brave New World* that in the "bad old days" old men normally did a lot of reading and thinking as well as "take to religion" when they retired (see introductory quote). That is, I certainly do more "reading and thinking" than I did in the hubbub of my earlier years. I will discuss the "take to religion" part shortly.

Speaking of reading, while recently going through Samuel P. Huntington's book, *The Clash of Civilizations and the New World Order*, it became clear to me that not only had I crossed a lot of borders between countries, but I had also crossed borders between numerous "civilizations." Civilizations have been defined in slightly different ways by numerous scholars over the years, but Huntington asserted that, "Throughout history, civilizations have provided the broadest identifications for people." He showed on a map the distribution of the nine post-1990 civilizations that he concluded now exist in the world. These nine entities include Western, Latin American, African, Islamic, Sinic, Hindu, Orthodox, Buddhist, and Japanese civilizations. Either JM or I have worked within the confines of all of these civilizations except Hindu. In fact, we have had permanent business offices in four of them–Western, Latin American, Islamic, and African–which has added several diverse cultural experiences to our resumes.

Huntington further asserted that, "Of all the objective elements that define civilizations, however, the most important usually is religion." He also quoted some estimates on the proportion of the world's population adhering to major religious traditions in the year 2000, which showed about 30% being Christian and about 20% being Muslim, but he predicted that these percentages would be more-or-less equal by the year 2025. The number of nonreligious people was ranked at 17%, but that number becomes rather miniscule if you take out the Chinese Communists.

Another important point he made was that "The late twentieth century has seen a global resurgence of religions around the world. That resurgence has involved the increase in religious consciousness and the rise of fundamentalists' movements." This appears to become more evident every day. I spend a lot more time worrying about that (i.e., a takeover of secular societies by the Islamists or the Christianists) than I probably should. At least, until I actually do something about it.

These revelations gave me pause to think again about the role of religion in my life, which, as already noted, I more-or-less gave up a long time ago. To anyone who has read some of the classics, as well as the latest books and treatises, on human philosophy and religion, something I have done a little of lately, it would seem that one of the primary reasons so many humans devote their lives to religion is the unknown aspect of the **afterlife**, for some reason assuming that there is one. I put in the introductory quote by Chief Seattle as an example of this belief. Being a "born again" secular humanist, I can find no reason to believe this myself.

Anyway, with regard to that subject, the two options for the afterlife promised by the religion I grew up in (Southern Baptist) didn't seem all that enticing, considering that they were either: 1) Spending eternity swimming around in a burning lake of fire, or 2) Running up and down the heavenly city's streets paved with gold singing praises to our Lord and Master on into eternity. Living in a small cabin in the Corner of Gloryland, a plea in a song by Old Man Woods that I heard once in a revival meeting when I was in about the 11th grade, was another option we were presented with. When I was in my teens, even though I did make a try at getting "born again," as was discussed in the FOREWORD, I always thought that hitting baseballs into Mr. Edwards' pea patch and catching smallmouth bass with Hoojer Smith and Marvin Jones seemed a lot more exciting and enjoyable than any of those three alternatives. Or, to quote Yogi Berra again, *And then you die, and that doesn't sound like much fun.*

You might know that not long after I wrote the preceding paragraphs, on 21 May 2011, the projected day for the rapture, I was scanning that week's Book Review in the New York Times and came across another possibility. This was the review by Alison Gopnik of a book called *Soul Dust: The Magic Of Consciousness* by a psychology professor named Nicholas Humphrey. In this book, Humphrey offers the theory that consciousness is a show we stage for ourselves.

In the review, Gopnik asked the following question– "What is the evolutionary function of the experience of the ineffable and irreducible?" Presumably in an attempt to answer this question, Gopnik stated further that "Humphrey points to a feature of consciousness that has been surprisingly neglected. The bottom line about how consciousness changes the human outlook–as deep an existential truth as anyone could ask for–is this: We do not want to be zombies, we like being present, we like having it be like something to be me."

Continuing, Gopnik noted that "Humphrey ingeniously works out the many consequences of this apparently simple fact. He pointed out, for example, that we humans will work as hard to get a newer or more vivid or more intense experience as we will to get a meal or a mate. Almost as soon as we could use tools to make hearths and spears, we also used them to construct consciousness–expanding art installations in painted caves like Altamira. We fear death so profoundly, not because it means the end of our body, but because it means the end of our consciousness–better to be a spirit in heaven than a zombie on earth." I must say that this makes a certain amount of sense to me, so *Soul Dust* is next on my reading list.

Speaking of souls, sometimes at a dinner during one of my field seminars, especially if the conversation was boring me, I would sometimes ask, "Does a horse or a dog have a soul?" And if someone informed me that the Bible says no, God created man in his own likeness, or something like that, which I had heard on more than one occasion along with a number of word-for-word quotes from the Bible, I tried to find a quick escape so I could sit at another table. What I think is that, if there actually is such a thing as a soul, which I doubt, but am not too sure of, these animals do have one the same as you and me. Remember Ellik's story about his encounter with the female dolphin that I discussed in Chapter 20 – THE ENVIRONMENTAL SENSITIVITY INDEX (ESI)?

Okay, proof that a horse does indeed have a mythical soul. On 17 August 1975, our small field team investigating the *Metula* oil spill was riding a van across some gravel beach ridges on Punta Catalina, a truly spectacular cuspate foreland that guards the south entrance to the Strait of Magellan, like a mirror image of Cabo Dungeness on the north side (discussed in The Story in Chapter 8 – OIL SPILLS; see Figure 48B and Plate 5). The sun was setting down the strait to the west, backlighting the ridges as we drove across them and filling the western sky with a number of spectacular colors. Our Chilean associate, Rosee, told me that in a land so inhospitable, referring to the dry steppe of Tierra del Fuego, "We must have a sky like this."

As we approached the high scarp on the south side of the foreland, a single wild horse stood at the edge of the scarp staring intently to the west at the brilliant sunset. The horse never moved for the ten minutes or so it was still in our view. I thought, "That horse is enjoying the sunset as much, if not more so, than I am." I didn't think much about souls then, but, yes, if there is such a thing as a soul, that horse had one.

Once during a visit with my eye doctor, he was enthralled to learn that I was a geologist. One of the few animists I have ever met, he said he spent much of his down time in meditation and communication with his rock collection. Even I have some doubt that rocks have a soul.

I guess this all means that I'm leaning at least a little bit, and I do mean little, toward spirituality of some kind, which gives me a reason to think once again about the primal question, "Why were we created only to suffer and die?" This query was posed repetitiously by Kilgore Trout (Kurt Vonnegut, Jr.) in *Venus on the Half Shell.* A. Huxley wrote in *Brave New World* that the meaning of life is something similar to professor Humphrey's concept given in *Soul Dust.*

Well, so maybe there is an answer and maybe there isn't, but most of the time this science bum's philosophy has mirrored what I saw on a bumper sticker while driving to work in Columbia a few days ago – "The meaning of life is how you live it!"

[NOTE: Bringing up the subject of spirituality, perilous ground to explore if there ever was one, brought to mind a discussion of this subject by Daniel C. Dennett in his 2006 book *Breaking The Spell*, a book no self-respecting secular humanist would want to miss, by the way. He pointed out that in his research for this book, he found that one opinion universally expressed by humans was that "man has a deep need for spirituality," a need that is traditionally fulfilled by organized religion, New Age cults, pursuits in art and music, and BASEBALL!! Actually, I added baseball, which fulfills my spiritual needs four months out of the year when the Gamecocks play–very fulfilling in the last two years as they won back-to-back national championships! But what about the other eight months of the year? Well, Dennett proposed that the "secret of spiritually" has "nothing at all to do with believing in an immortal soul, or in anything supernatural." His answer was approaching the world's complexities, both its glories and its horrors, with an attitude of humble curiosity, acknowledging that however deeply you have seen, you have only just scratched the surface," and in doing so, "you will find worlds within worlds and beauties you could not heretofore imagine."]

On a night back in early August 2007, JM and I were spending ten days at our mountain place in North Carolina. We were at that 4,400-feet-high retreat to escape the +100 degree temperatures in Columbia and to do a little trout fishing. Just as we were getting ready to retire, we turned off all the lights and stepped out on the upstairs deck to admire an amazingly clear sky for that time of year. The light haze of the Milky Way angled obliquely across the sky and several constellations stood out, including the Little Dipper, with the North Star steadfastly in place over the crest of the roof. JM was muttering "billions and billions" to herself, referring to the comments Carl Sagan used to make about the numbers of stars even in our own galaxy, not to mention elsewhere in the universe. As we watched, the mystery of it all seemed to come flooding in, and I thought, "On a night like this, it is inconceivable to be dead certain and arrogant about anything."

For the record, I wrote this before I read Dennett's book. If you have made it this far, thanks for your interest.

He has outsoared the shadow of our night.
Envy and calumny and hate and pain,
and that unrest which men miscall delight,
can touch him not and torture not again.
From the contagion of the world's slow stain
he is secure; and now can never mourn
a heart grown cold, a head grown grey in vain—
Nor, when the spirit's self has ceased to burn,
with sparkless ashes load an unlamented urn.

PERCY BYSSHE SHELLEY – IN ELEGY
WRITTEN FOR JOHN KEATS IN 1821

NOTES:

1. I kept these books for many years, but some of them were lost eventually. This one was probably *The Physiographic Provinces of North America* (1940) by W. W. Atwood and C.L. Hubbs. Another one, by Sir Charles Lyell first published in 1833, was titled *Principles of Geology*.

2. I'm sorry I couldn't think of a more sophisticated way of saying this. I believe the downfall of this kind of religion for a scientist like me is learning about the discovery of the origin and makeup of the universe. Pretty hard to imagine the creator of the universe, with those millions of planets with life on them to worry about, choosing this rather insignificant one to send his son "down to" be sacrificed on the alter for "our sins." After writing this statement, I later ran across the following quote by Bertrand Russell (Russell, 1957), which posed the question in another, and no doubt much more profound, way. "There is to me something a little odd about the ethical evaluations of those who think that an omnipotent, omniscient, and benevolent Deity, after preparing the ground by many millions of years of lifeless nebulae, would consider Himself adequately rewarded by the final emergence of Hitler and Stalin and the H-bomb." However, I still get my kicks when I see another one of those bumper stickers–IN THE EVENT OF THE RAPTURE, THIS CAR WILL BE EMPTY. Always makes me pay a little more attention to my driving. Then, of course, there was Charles Darwin. No wonder the "creationists" are so paranoid about him.

3. Hayes, Miles O., 1999, *Black Tides*: Univ. Texas Press, Austin, Texas, 287 pp.

4. The following is a direct quote from Western Carolina University (2005):

   "A nearshore sonar survey of the area from Myrtle Beach to Folly Island (Gayes, 1991) showed extensive shore-perpendicular scour channels on the inner shelf. These scour channels indicated strong offshore bottom currents caused by the storm-surge ebb. Nearshore bars or shoals extending 50 to 75 meters further offshore in the gaps between multistory hotels than in the areas in front of large development, suggesting that sand was carried offshore by storm-surge ebb currents."

5. In 1961, Gerry Friedman published a paper in which he said sedimentary environments can be distinguished on the basis of statistical grain size parameters. For example, he concluded that beach sand tends to be negatively skewed and dune sand positively skewed. As shown in Figure 6, my results for Padre Island showed that skewness was more related to the mixing of grain size modes from different sources than to the environment of deposition. I gave a paper on this at a national meeting in St. Louis a short time after these results were first published (Hayes, 1964b). The great marine geologist, Francis P. Shepard, the so-called "Father of Marine Geology," was in the audience. At that time, Shepard was in a battle in the literature with Friedman, inasmuch as he had concluded that one could not distinguish environments on this basis. Anyway, as soon as my talk ended, Shepard jumped up, heaped praise on my

findings, and then lit into Friedman big time for several minutes, while I stood there with my proverbial thumb up my ass. Remember, I was just out of grad school at the time. Now I wouldn't want to imply by this story that my paper was a major contributor to this argument, because, in fact, it wasn't. However, this was my one and only time to ever meet the great Shepard, whose name adorns the award I won in 1997 for presumably knowing something about marine geology (called the Francis P. Shepard Medal in Marine Geology awarded by the Society of Sedimentary Geology).

6. My concept of the Seven Ages of Geotechnical Man was also presented in my earlier book, *Black Tides* (Hayes, 1999). The Ages are given in the following table. This concept may or may not apply to women; my observational sample is too small to be sure one way or the other.

Table 3. Seven Ages of Geotechnical Man.

Ages	Years of Age	Description
Innocence	0 – 12/15	Learning is rapid; ends when person is converted to a cause (e.g., religion, science, etc.); the First Conversion.
Wonderment	12/15 – late 20s or when he gets his Ph.D.	Greatest intellectual growth in a geotechnical sense; ends at Second Conversion when he knows he is the best and can conquer the world (i.e., gets his Ph.D.!).
Arrogance	Late 20s – late 30s Now called Dr. So-and-so by some	Feeling of superiority manifested by an overbearing manner or presumptuous claims; ends at Third Conversion, when he realizes that life has more to offer (e.g., mid-life crisis).
Paranoia	Late 30s – mid to late 40s	Realizes he is no longer master of all; ends when new achievable goals are established.
Realignment	Mid to late 40s – early/late 60s	With new dreams and goals established, he should have numerous intellectual pursuits with only learning and enjoyment in mind.
Space Walking	Early/late 60s – mid 70s	Hopefully still with numerous intellectual pursuits.
Senility	Mid 70s - The End	As the title says.

7. Murray Felsher was something of the Geology Department poet, writing songs we sang at the annual roast of the faculty that took place during a

departmental seminar toward the end of every spring semester. During my last semester there, he wrote a poem about me, which follows:

A SHORT HISTORY

by M. FELSHER, a short historian

He resides in a hole in the basement,
where he's spent about 2,000 days
amidst columns of maps, charts and textbooks.
Oh, how pleasant to know Mr. Hayes.

'Twas the latter end of the past decade
that he first took on Folkian ways,
and he vowed to be finished in two years.
Oh, how pleasant to know Mr. Hayes.

Then the barriers parted before him.
The languages went in a blaze,
then the courses, and finally prelims.
Oh, how pleasant to know Mr. Hayes.

Now he's finally ending his thesis,
and he smiles while the DRL pays
for his drafting and all of his typing.
Oh, how pleasant to know Mr. Hayes.

But the start of the coming semester
will test his Ashvillian ways,
as he tries to pronounce "Massachusetts."
Oh, how pleasant to know Mr. Hayes.

8. For example, Reguero, Marenssi, and Santillana (2002), in studying the Paleogene of southern Patagonia and the Antarctica Peninsula, concluded that the "late Paleocene Cross Valley flora (Seymour Island) contains ferns and other elements suggesting a much warmer climate than at this latitude today." The volumes of ice on the Earth started increasing dramatically during the Eocene. Even if the climate was warmer when the ice was gone, the growing season was obviously restricted somewhat, because of the long periods of darkness in the winter months. Some paleobotanists are working on this issue, but further discussion of their work is not warranted here.

9. See also Note 6.

10. This short discourse on my personal life may seem a bit abrupt, but some things are better left unsaid.

11. Orrin Pilkey and Edna Fraser published a book in 2003 titled *A Celebration of the World's Barrier Islands*. This is a thorough presentation of the subject that is very well written and illustrated. In the acknowledgments of this book, Pilkey referred to me as the "King of Barrier Islands." Whatever that means. He also does that whenever I am in the audience for one of his talks. I guess

that is an acknowledgment of all the work my grad students have done on the barrier islands of the South Carolina coast and elsewhere.

12. The oil-and-gas business in western Canada and the High Arctic has played a significant role in my professional life. On 15 October 1980, we formed a company called RPI/Canada located in Calgary, which had its profitable days at first, but that ended for us with the crash of oil prices in 1985/86. We also had many encounters with the tar sands of the Ft. McMurray area of northern Alberta, which is discussed in more detail in Chapter 14 - THE ROCK RECORD. Currently, the United States gets more oil from Canada than any other country. Canada has vast oil reserves, second only to Saudi Arabia (Canadian Oil Sands, 2011).

13. SCAT stands for Shoreline Cleanup Assessment Technique. The typical SCAT team leader is commonly a coastal geomorphologist, and the team usually contains at least one state agency representative with local knowledge, possibly a member of a local clean-up contractor's team, local landowners, etc. Their job is to map the distribution of oil along the impacted shoreline, typically subdividing the survey area into easily defined and recognizable geomorphic units to focus their recommendations on. Any special problems to be encountered within these individual units should be pointed out and, where appropriate, cleanup recommendations should be made. The largest of this type of survey, historically, was supervised by Ed Owens at the *Exxon Valdez* spill along the Alaskan coast in 1989, 1990, and 1991. Ed once told me that he had 40 geologists working on those teams. This effort has probably now been exceeded by the response to the *Deepwater Horizon* spill in the Gulf of Mexico, in which both Ed and JM are now heavily involved (as noted earlier). JM and I have taught numerous training courses on this subject to participants in the U.S. Coast Guard, state environmental agencies (e.g., in California and Texas), and industry.

14. I was born on 20 October 1934 in the small community of Oakley in the mountains of North Carolina. As fate would have it, two fellow geologists, Bill Kanes and Orrin Pilkey, who have had a significant impact upon my life and career, were born within a couple of days, plus or minus, of that same date. I met Bill Kanes, who was born in New York City and spent some years working for big oil, on a field trip to Jekyll Island, Georgia while I was still a grad student at UT. We dug a couple of trenches on the beach and discussed coastal geomorphology. I will never forget that. Bill played a prominent role in getting me hired to be a professor at USC-East, where we were colleagues for several years. This truly brilliant geologist, with his photographic memory and giant ego, remains one of my best friends. Orrin Pilkey, who was born in Chicago and ended up a long-time professor of marine geology at Duke University, has already been mentioned several times in this book. At this time, March 2011, we communicate on a regular basis by e-mail, and still get together now and then. He likes my pictures, using them regularly in some of his numerous books, and I like his passion and fanaticism for saving the shoreline, so we get along great. The three of us got together at one of Pilkey's talks at USC-East a few months ago and had a spirited argument about who was aging better, now that we are all close to the same number of days past 76 years old (plus or minus a couple of days).

15. This quote is from a book called *Cold Mountain: 100 Poems by the T'ang Poet Han-Shan*, a Chinese Buddhist poet-recluse who lived 900 years ago. This book was loaned to us by George the Mountain Climber one night in the summer of 1975 as our field team visited with him in our hut along the side of the airport runway in Yakutat, Alaska. According to Burton Watson, the translator of *Cold Mountain*, the washing of ears refers to a much older story about the sage recluse, Hsii Yu, who, when asked by Emperor Yao to take over the throne, hastily went and washed his ears to cleanse them of such a vile suggestion.

16. At this point in my life, I can't believe I was eating any of that stuff. JM and I had been together only a few short years when we gave up eating meat altogether. We are not pure vegetarians, eating fish about once a week, but it has been about a couple of decades since I ate my last piece of dead cow.

17. Despite his abundant skills as a fisherman, my Uncle Luster, who taught me how to fly fish, had two major flaws as piscatorial guru, an awesome temper and almost no patience. In fact, once in a while he would hook up his favorite fly on a tree limb and lose it. Standing there shaking in anger, he typically pronounced, "Damn it all to hell, I am going to throw this rod and all this other fishing gear I just bought out into the river and never go fishing again as long as I live, so help me God!"

    At which time, I would run over to him, grab hold of his excellent bamboo fly rod and exclaim, "Don't throw it Uncle Luster, I would really love to have this rod."

    And then, after calming down a bit, Luster would be back fishing just as avidly as before.

18. This remark was made by the famous tectonics geologist Robert S. Dietz in his acceptance speech for the award of the PENROSE MEDAL by the Geological Society of America in 1988. He was referring to his original mentor and supervisor of his Ph.D. dissertation, Francis P. Shepard, the "Father of Marine Geology," who is discussed briefly elsewhere in this text (e.g., footnote 5).

19. This past Christmas (2010), JM and I took a 15-day vacation in Australia. During that time, I took an interest in the fact that on 19 April 1770, Captain Cook and the crew on board his ship, the *Endeavour*, discovered and claimed that part of Australia (southeast shoreline) for Great Britain. During that particular sail around the world, during 1768-1771, Captain Cook also stopped off at Tahiti to observe the transit of Venus across the face of the Sun (Blainey, 2008). This observation, made on 3 June 1769, was done to increase the accuracy of determining longitude at sea. This and other skills of celestial navigation allowed Cook to map the eastern shore of South Island, New Zealand, I believe during that same cruise. Even to this day, that map is considered to be an accurate one. Now that is celestial navigation. Okay, so ours was a rather puny attempt at it, but we did, at least, get back to McClellanville in the dark that night.

20. I have already said quite a bit about Orrin Pilkey in footnote 14. In addition to his already published over 20 books on the subject of the human developments along shorelines, he has just recently completed two more books, *The Rising Sea*, with Rob Young, and *The World's Beaches*, with three co-authors, which

is still in press, but will be published in 2011. With regard to the shoreline, his agenda has almost always been "Let It Be," which, by the way, was one of the main themes in our tone poem, *Suzanne's Lament*.

21. Although many signs of the coming decline of the oil-and-gas industry in the U.S. were evident in the mid-1980s, we were too under-capitalized to diversify into the international oil exploration business in time to compensate for the disintegration of the domestic industry. When the price of a barrel of oil dropped to under ten dollars in early 1986, an almost two-hundred-percent drop in a few short months, oil centers like Denver and Houston went into a major catharsis the likes of which had not been seen there since the Great Depression. Just like most of the other companies in the so-called service industry for the oil-exploration business, we struggled on through 1986 while staring at the prospect of bankruptcy every day. Our company survived this first shock and the energy program was collapsed back to the Boulder office. This meant that yours truly and Dan Domeracki, who was the Chief Operations Officer of the newly formed holding company, had to travel around the country to the individual offices and close them down, laying off the entire staff in some cases. This meant the demise of RPI/ Texas in Austin, RPI/ Pacific in Redwood City, California, and RPI/Canada in Calgary. The Boulder office (RPI/Colorado) survived for several more years, but it was eventually shut down in the early 1990s. We restructured the environmental unit into Research Planning, Inc. (RPI) in July 1990. Therefore, the environmental program survived the crisis, about a ten-year struggle if you go back to 1986 to start counting, and it is doing okay at this time (2011). But you never know for sure, 'biness is 'biness.

22. I started to use the quote "Religion is the opium of the poor" from a Hemingway book (*The Gambler, The Nun, and The Radio*). However, I later learned that this quote was not totally original with him, inasmuch as, apparently, Karl Marx broached the idea first in 1843, when referring to not just the poor, but to the people in general, or, as some translate it, the masses – hence *"Die Religion ... ist das Opium des Volkes."*

    The very morning I started to put this quote in the first draft of this chapter (31 January 2011), I searched for the quote on Google and found an article by Olaewe Ewegbemi titled *Religion–The Opium of the Poor in Nigeria*. This article was based on a visit he made to Nigeria in 2003, in which he found the country to be "saturated with religion." However, the comments I have written in this chapter are based on my journals of the 1980s and were not inspired in any way by this article. Nonetheless, it is interesting that Ewegbemi made the same connection.

23. As a generalization, in those days anyway, European firms had an advantage over U.S. firms when it came to getting jobs in Africa. For one thing, the European firms could take risks that the U.S. firms couldn't, because the European firms were backed financially by their country's government if the Africans for some reason did not pay their bills. Not so for American companies. U.S. firms were better off, of course, if the funds for the project came for sources such as USAID. Projects funded by the European Communities, the United Nations, or the government of the country in question were a lot more risky.

24. The tidal range is about the same, but the waves are considerably larger in places. The mix of freshwater and tidal (saltwater) streams was another similarity. The tidal inlets were amazingly similar. Walter J. and I felt right at home there as far as the coastal geomorphology was concerned.

25. While we were there, it didn't quite sink in for me just how crucial this place was to the history of Nigeria. This is truly a fascinating story. I am sorry I didn't spend more time there. I started to say that I would like to go back there some time, but that probably would not be such a great idea, considering the political unrest and other problems they have in Nigeria at the moment.

26. During this episode, for the first time in my life, I think, I was observing the true meaning of the word subservient.

27. I am telling the following story from memory. I had been in Caryville, Tennessee before. When I was an undergraduate at Berea College, I did not have a car, and neither did any other student at Berea College, as far as I knew. Accordingly, when I wanted to go home to North Carolina, I hitchhiked my way there. U.S. Highway 25, formerly known as the Dixie Highway, connected Lexington, Kentucky with Knoxville, Tennessee, passing through Berea. I think either my sophomore or junior year, I was hitchhiking my way home for Christmas when I was picked up somewhere south of Berea by a black man, pulling a car hitched to the one he was driving. He was on his way from Detroit to deliver the backup car to an automobile sales company in Knoxville, Tennessee. This is how they did it in those days. I don't know how legal this practice was. I did know that driving truckloads of coal across the state border into North Carolina at night was illegal. Doing so during daylight hours was illegal as well, of course, but apparently, it was easier to get away with at night. I had ridden across the border in the dark with a couple of those guys before, but that's another story.

    One of the first things the driver said to me after he picked me up was, "Hey man, would you mind getting me some food? I can't get anybody to serve me."

    "Sure. Stop as soon as you see a good place."

    We stopped, and I bought us some food and something to drink, probably a couple of hamburgers. When I got back to the car, he pulled out a twenty-dollar bill to pay for the food.

    "No. This is on me. After all, you are giving me a free ride all the way to Knoxville."

    As we approached the small town of Caryville in East Tennessee, a cop following us turned on his siren and signaled us to stop. As soon as my new friend heard the siren, he stuffed the twenty into his shoe. I will call him James. I don't remember his real name. The cop, a local, not the state highway patrol, came alongside and ordered us to follow him, which James did, pulling up in front of the County Jail.

    I was left on my own as the cops took James inside for "interrogation." I wandered into the quaint Jail, being summoned to the bars by a couple of inmates asking for cigarettes, which I didn't have, of course. It was a spooky place.

Finally, James and the cops came back outside and they ordered him to unhitch the backup car, telling him that sort of thing was not allowed in Tennessee. Then the cops walked away.

After they left, James took out his wallet and showed me that they had taken all of his cash. I never found out the fate of the twenty in his shoe. But, there was still the small problem of getting the backup car to Knoxville, about 30 miles further down U.S. 25.

"I can drive the other car," I told him. "You lead the way."

About an hour later, we pulled into the automobile dealership, the owner of which was very glad to see us.

As James and I parted company, he said, "I never expected."

Then I gave him the bills I had in my wallet, probably about forty dollars.

Back on U.S. 25, I hitchhiked my way to Asheville without any problems, where I spent the remaining change in my pocket for a bus ticket out to Oakley and to home. My mother puzzled about why I showed up broke, but I never told her why. In fact, this is one story that I have told only three or four times. However, I did tell it to JM again that night in Caryville.

28. On a Wednesday evening in January 1980, JM and I walked from the Gregg St. house near the USC-East campus a few blocks to the local art theater, the Nickelodeon, and watched a movie called *The King of Marvin Gardens*, a masterpiece on the great American dream. In that story, the hero decided to go into business for himself and create a development on an island in Hawaii. Unfortunately, in a jealous rage, his mistress shot the hero, played by Bruce Dern. She looked amazingly like Leita Jean, and they lived exactly the way she and I did between 1970-1974.

As we walked back home, I wondered, "Which mistress is going to shoot me, Ms. Future Shock?"

Two days earlier, we had the first of our one-day marketing meetings at RPI. After it was over, I wrote in my journal: "I think it was successful. I guess the major goal was $2M in sales in 1980, and expansion to 30 people with an office in Colorado. It was a tight meeting. Not too much yelling and screaming."

That night I sat at home in front of the fire and read back issues of the Wall Street Journal, while JM calculated our monetary net worth. It had jumped $100K in the past year.

The following Sunday night, I was in an airplane in the Chicago airport writing the following in my journal:

"Sometimes I feel like I'm living some uncontrolled fantasy, a nightmare or a space dream - $270K real estate deals, anticipated profits of $200-400K in one year, completely uproot and move to the plastic mountains (i.e., Boulder, Colorado), follow the money! I've never felt so confused about how to manage my affairs. Maybe it's because I never had any affairs to manage. I can picture losing it all and being out in the cold on my own again, back to 1974-75."

29. A document published by Syncrude states: "The bitumen in the oil sand mined at the Mildred Lake site averages 10.5 percent by weight, and is made up of 50 to 60 percent oil, 30 to 35 percent resins, and 15 to 25 percent asphaltenes. In its raw state, bitumen is a black, asphalt-like oil–as thick as molasses. It has to be upgraded to be transported by pipeline. The upgraded product is a light, sweet (i.e., sulphur free) crude."

30. Of course, I am referring here to Frank Herbert's great science fiction novel, *Dune*, which was first published in 1965.

31. Syncrude 1994 Annual Report.

32. The name, Dr. Bobby, is not original with me, and by using it, I mean no disrespect to Dr. Peele, whom I greatly admire. Actually, it was his relative, Walter J. Sexton, who got me started saying it when he asked me one day, "How are things going with Dr. Bobby?" as we were riding along on the boat.

33. Our friend and fellow coastal geologist, Shea Penland, was the head man in the Pontchartrain Institute of Environmental Studies (PIES) on the campus of the University of New Orleans (UNO). We had several discussions with him regarding our starting a company in New Orleans to work with him, and his Institute, on issues such as restoring the barrier islands of Louisiana and other environmental projects. Thus, in the fall of 2004, JM and I moved to New Orleans and established a small office of our new company, RPI/Louisiana. We still retained the bulk of the staff in the main RPI office in Columbia, South Carolina. Our new office in New Orleans was located on the UNO campus, on the 5th floor of the same building in which PIES was located, on the shore of Lake Pontchartrain. We were pretty successful at first. For example, Dallon Weathers and I made what we called Reach Sensitivity Index (RSI) maps of several of the streams draining into Lake Pontchartrain. We were also beginning to make plans for restoration projects on the barrier islands. And then, on Monday, 29 August 2005, Hurricane *Katrina* paid a visit.

    Our exquisite historic house in Uptown suffered very little damage from the hurricane; it was located above sea level on the old natural levee of the Mississippi River. However, we could not say the same for our office, with the first floor of the building it was located in being flooded, resulting in a severe case of mold and other problems, making our 5th floor facilities uninhabitable. We moved the office to the outskirts of Baton Rouge, about a 1.5-hour drive from our house at that time, because of the severe traffic problems resulting from the storm. Also, most of the available government funds were directed to rebuilding the levees, a process in which we had no expertise. In addition to these problems, Shea took sick, eventually dying some time thereafter, much to the sorrow of his many friends and acquaintances.

    We moved back to Columbia in the spring of 2006, sadly leaving behind our historic home, which sold within a few days after we put it on the market. You may recall that there was a pretty severe shortage of housing in NOLA at that time.

    It was a nice try, as they say, that didn't cost us much. In fact, JM and other members of the RPI staff in Columbia responded to a large number of oil spills caused by Hurricane *Katrina*. And, oh yes, we were vacationing at our mountain home in North Carolina (at 4,400 feet elevation) when *Katrina* crossed the Louisiana shore.

34. As I discussed in footnote 16, JM and I have not eaten meat of any kind for at least a couple of decades, although we do eat fish about once a week, in addition to the usual veggies. Looks like I was backsliding a bit there in San Diego.

35. Ann Hayward Walker had been a NOAA Scientific Support Coordinator from 1980 to 1990, which means that our group had worked with her on a number of spills. Great to work with. The title of her talk at this conference was "*Subsistence Fisheries and the Exxon Valdez: Human Health Concerns*."

36. In this joke, I was going to allude to "The Great Rock Washer Debate" that took place the preceding summer in Anchorage. During the "debate," both the state and Exxon people were driving me crazy, referring to the GRAVEL (pebbles, cobbles, and boulders) the state people were talking about washing as ROCKS! Notice that they said the debate was about "washing rocks." I kept telling anyone who would listen that a rock is fastened to the earth, and that gravel is broken up pieces of those rocks. They were proposing to wash gravel, not rocks! Anyway, at this point, I can't remember exactly how the joke went, but it was certainly a sore point with me, and I was more than a little upset that the projectionist had screwed up my opportunity to make that point.

37. This quote is in reference to an incident that I am not particularly proud of, by the way, but it went like this. The date was 25 September 1994, and I was in the beginning stages of doing the fieldwork for an ESI map of the Jordanian shoreline of the Gulf of Aqaba. This process began with a meeting in the offices of the Port Authority in the town of Aqaba. During this first meeting, one of the Jordanian engineers began to explain how oil exports from the Port had flourished for a time, beginning during the Iraq/Iran conflict as the Iraqis trucked oil across the hills in Jordan to Aqaba. And then, of course, all of that stopped when the Americans bombed the Iraqi trucks into oblivion during the first Gulf War.

    After sort of absolving myself of any blame for that particular bombing exercise, I said, "Hey, let me tell you a story."

    Then I told them about the time when I was visiting North Korea on a UN mission in February 1993. Once during that tour, our in-country guide, whom I called Deion in my notes, kind of blamed me for the fact that the American GI's had leveled Pyongyang, the city that we were meeting in, during the Korean War. I said that I had told Deion not to blame me, because "I was only fifteen years old at the time." I was hoping the assembled Jordanians would find that story somewhat amusing, and a few of the engineers did laugh a little.

    However, upon the completion of that story is when Saleh, the quiet one, who was sitting close by on my left, said softly, "But, you are the American."

REFERENCES CITED:

Anan, F.S., Farrell, S.C., Hayes, M.O., and Greer, S.A., 1969. Glossary of primary sedimentary structures of the intertidal zone. In: Hayes, M.O. (ed.), Coastal Environments – NE Massachusetts and New Hampshire: Field Trip Guidebook for Eastern Section of SEPM, May 9-11, 1969, pp. 455-457.

Alt, D. and Hyndman, D.W., 1989. Roadside Geology of Idaho. Mountain Press Pub. Co., Missoula, MT, 394 pp.

Barry, B., Obuobie, E., Adreini, M., Andah, W., and Pluquet, M., 2005. The Volta River Basin. IWMI International Water Management Institute. Available at: http://www.iwmi.cgiar.org/assessment/files_new/research_projects/River_Basin_Development_and_Management/VoltaRiverBasin_Boubacar.pdf

Barwis, J.H. and Hayes, M.O., 1985. Antidunes on modern and ancient washover fans. Journal of Sedimentary Research, 55:907-916.

Bascom, W., 1954. Characteristics of natural beaches. In: Proc. 4th Conference on Coastal Engineering, American Society of Civil Engineers, New York, pp. 163-180.

Blainey, G., 2008. Sea of Dangers: Captain Cook and His Rivals. Penguin Books, 420 pp.

Bluck, B.J., 1967. Sedimentation of beach gravels: Examples from South Wales. Journal of Sedimentary Petrology, 37:128-156.

Boothroyd, J.C., 1969. Hydraulic conditions controlling the formation of estuarine bedforms. In: Hayes, M.O. (ed.), Coastal Environments – NE Massachusetts and New Hampshire: Field Trip Guidebook for Eastern Section of SEPM, May 9-11, 1969, pp. 417-427.

Boothroyd, J.C., 1976. Braided streams and alluvial fans. In: Hayes, M.O. and Kana, T.W. (eds.). Terrigenous Clastic Depositional Environments. Technical Report 11-CRD, Geology Department, University of South Carolina, Columbia, SC, pp.1-17-1-26.

Boothroyd, J.C. and Ashley, G.M., 1975. Process, bar morphology and sedimentary structures on braided outwash fans, northeastern Gulf of Alaska. Society of Economic Paleontologists and Mineralogists Spec. Publ. No. 23, pp. 193-222.

Boothroyd, J.C. and Hubbard, D.K., 1975. Genesis of bedforms in mesotidal estuaries. In: Cronin, L. E. (ed.), Estuarine Research, Acad. Press, New York, pp. 217-234.

Canadian Oil Sands, 2011. Developing America's natural resources. Available at: http://energytomorrow.org/issues/energy-security/throughout-the-world/canadian.

Carter, R.W.G., 1988. Coastal Environments. Academic Press, London, 617 pp.

DaBoll, J.M., 1969. Holocene sediments of the Parker River Estuary, Massachusetts. In: Hayes, M.O. (ed.), Coastal Environments – NE Massachusetts and New Hampshire: Field Trip Guidebook for Eastern Section of SEPM, May 9-11, 1969, pp. 337-355.

Davies, J.L., 1973. Geographical Variation in Coastal Development. Hafner Publ. Co., New York, 204 pp.

Delft Hydraulics, 1961. Coastal Erosion in the Bight of Benin – A Critical Review. Available at: http://www.widelft.nl/rnd/publ/docs/Pub455.pdf.

Dennett, Daniel C., 2006. Breaking the Spell: Religion as a Natural Phenomenon. Penguin Books, London, 448 pp.

Duc, A.W., 1981. Back-barrier stratigraphy of Kiawah Island, South Carolina. Ph.D. dissertation, Dept. of Geology, University of South Carolina, 253 pp.

Emery, K.O., 1965. A simple method for measuring beach profiles. Limnology and Oceanography, 6:1:90-93.

e Notes, 2011. World of Earth Science – Porosity and Permeability. Available at: http://www.enotes.com/earth-science/porosity-permeability.

Ewegbemi, O., 2011. Religion – The Opium of the Poor in Nigeria. Available at: http://www.dawodu.com/ewegbemi2.htm.

Finley, R.J., 1975. Hydrodynamics and tidal deltas of North Inlet, South Carolina. In: Cronin, L.E. (ed.), Estuarine Research 3. Academic Press, New York, pp. 227-292.

Finkelstein, K., 1979. Morphologic variations and sediment transport in crenulate bay beaches, Kodiak Island, Alaska. M.S. thesis, Dept. of Geology, University of South Carolina, 96 pp.

Finley, R.J., 1978. Ebb-tidal delta morphology and sediment supply in relation to seasonal wave energy flux, North Inlet, South Carolina. Journal of Sedimentary Petrology, 48:227-238.

FitzGerald, D.M., 1977. Hydraulics, morphology, and sediment transport at Price Inlet, South Carolina. Ph.D. dissertation, Dept. of Geology, University of South Carolina, 84 pp.

FitzGerald, D.M. and Hayes, M.O., 1980. Tidal inlet effects on barrier island management. Coastal Zone '80, American Society of Civil Engineers, New York, pp. 2355-2379.

FitzGerald, D.M., Nummedal, D., and Kana, T.W., 1976. Sand circulation pattern at Price Inlet, South Carolina. 15th Coastal Eng. Conf., American Society of Civil Engineers, New York, pp. 1973-1994.

FitzGerald, D.M., Hubbard, D.K., and Nummedal, D., 1978. Shoreline changes associated with tidal inlets along the South Carolina coast. Coastal Zone '78, American Society of Civil Engineers, New York, pp. 1973-1994.

Friedman, G.M., 1961. Distinction of dune, beach, and river sands from their textural characteristics. Journal of Sedimentary Petrology, 31:514-429.

Galvin, C.J. and Hayes, M.O., 1969. Winter beach profiles and wave climate. Abstract Program of the Annual Meeting, Geological Society of America, pp. 74-75.

Gayes, P.T., 1991. Post-Hurricane *Hugo* nearshore sidescan sonar survey: Myrtle Beach to Folly Beach, South Carolina. In: Finkl, C.W. and Pilkey, O.H. (eds.),

Impacts of Hurricane *Hugo*: September 10-22, 1989. Journal of Coastal Research, Special Issue 8:95-111.

Goldsmith, D. and Owen, T., 2002. The Search for Life in the Universe. Third Edition, Univ. Science Books, Sausalito, CA, 573 pp.

Gundlach, E.R. and Hayes, M.O., 1978. Classification of coastal environments in terms of potential vulnerability to oil spill damage. Marine Technology Society Journal, 12:18-27.

Gundlach, E.R., Domeracki, D.D., and Thebeau, L.C., 1982. Persistence of *Metula* oil in the Strait of Magellan six and one-half years after the incident. Oil and Petrochemical Pollution, 1:37-48.

Gundlach, E.R., Ruby, C. H., Hayes, M.O., and Blount, A.E., 1978. The *Urquiola* oil spill, La Coruña, Spain: Impact and reaction on beaches and rocky coasts. Environmental Geology, 2:131-143.

Gustavson, T.C. and Boothroyd, J. C., 1987. A depositional model for outwash, sediment sources, and hydrologic characteristics, Malaspina Glacier, Alaska: A modern analog of the southeastern margin of the Laurentide Ice Sheet. GSA Bulletin, 99:187-200.

Hand, B.M., Wessel, J.M., and Hayes, M.O., 1969. Antidunes in the Mt. Toby Conglomerate (Triassic), Massachusetts. Journal of Sedimentary Research, 39:1310-1316.

Hayes, M.O., 1963. Petrology of the Krebs subgroup (Pennsylvanian, Desmoinesian) of western Missouri. AAPG Bulletin, 47:1537-1551.

Hayes, M.O., 1964a. Summary of geological effects of Hurricanes *Carla*, 1961, and *Cindy*, 1963, on the south Texas coast: In: Field Trip Guidebook Depositional Environments, South-Central Texas Coast, Gulf Coast Assoc. Geol. Societies, Ann. Mtg., Corpus Christi, TX, pp. 121-127.

Hayes, M.O., 1964b. Grain size modes in Padre Island sands: In: Field Trip Guidebook Depositional Environments, South-Central Texas Coast, Gulf Coast Assoc. Geol. Societies, Ann. Mtg., Corpus Christi, TX, pp. 127-136.

Hayes, M.O. 1964c. Lognormal distribution of inner continental shelf widths and slopes. Deep-Sea Research, 11:53-78.

Hayes, M.O., 1965. Sedimentation on a semiarid, wave-dominated coast (south Texas); With emphasis on hurricane effects. Ph.D. dissertation, Dept. of Geology, University of Texas, 350 pp.

Hayes, M.O., 1967a. Hurricanes as Geological Agents: Case Studies of Hurricanes *Carla* (1961) and *Cindy* (1963). University of Texas, Austin, Bureau of Economic Geology Report 61, 56 pp.

Hayes, M.O., 1967b. Hurricanes as geological agents; South Texas coast. AAPG Bulletin, 51:937-941.

Hayes, M.O., 1967c. Relationship between coastal climate and bottom sediment type on the inner continental shelf. Marine Geology, 5:111-132.

Hayes, M.O., 1969. (ed.) Coastal Environments: NE Massachusetts and New Hampshire. Guidebook, Field Trip for Eastern Section of SEPM, 9-11 May 1969, 462 pp.

Hayes, M.O., 1972. Forms of sediment accumulation in the beach zone. In: Meyer, R.E. (ed.), Waves on Beaches, Academic Press, New York, pp. 297-356.

Hayes, M.O., 1975. Morphology of sand accumulation in estuaries; An introduction to the symposium. In: Cronin, L.E. (ed.), Estuarine Research, Vol. II, Academic Press, New York, pp. 3-22.

Hayes, M.O., 1976. Part 1. Lecture Notes. In: Hayes, M.O. and Kana, T.W., (eds.), 1976. Terrigenous Clastic Depositional Environments. Tech. Report 11-CRD, Dept. of Geology, University of South Carolina, pp.1-132.

Hayes, M.O., 1977. Development of Kiawah Island, South Carolina. Coastal Sediments '77, American Society of Civil Engineers, New York, pp. 828-847.

Hayes, M.O., 1979. Barrier island morphology as a function of tidal and wave regime. In: Leatherman, S. (ed.), Barrier Islands, from the Gulf of St. Lawrence to the Gulf of Mexico. Academic Press, New York, pp. 1-27.

Hayes, M.O., 1980a. General morphology and sediment patterns in tidal inlets. Sedimentary Geology, 26:139-156.

Hayes, M.O., 1980b. Oil spill vulnerability, coastal morphology, and sedimentation of the Outer Kenai Peninsula and Montague Island, Alaska. National Oceanic and Atmospheric Administration, Outer Continental Shelf Environmental Assessment Program, Seattle, WA, 51 pp.

Hayes, M.O., 1985a. Beach Erosion. In: Clark, J.R. (ed.). Coasts: Coastal Resources Management: Development Case Studies. Coastal Publication No. 3, Renewable Resources Information Series. National Park Service, Washington, DC, pp. 67-202.

Hayes, M.O., 1985b. Atlantic USA – South. In: Bird, E.C.F. and Schwartz, M.L. (eds.), The World's Coastline. Van Nostrand Reinhold, New York, pp. 207-211.

Hayes, M.O., 1991. Geomorphology and sedimentation patterns in tidal inlets: A Review. Coastal Sediments '91, American Society of Civil Engineers, New York, pp. 1343-1355.

Hayes, M.O., 1994. Georgia Bight. Chapter 7 In: R.A. Davis, Jr. (ed.), Geology of the Holocene Barrier Island System. Springer-Verlag, Berlin, pp. 233-304.

Hayes, M.O., 1999. Black Tides. University of Texas Press, Austin, TX, 287 pp.

Hayes, M.O., 2005a. Fan Deltas as Exploration Targets. Prepared by RPI Louisiana, Inc., New Orleans, LA for Devon Energy Corporation, 97 pp.

Hayes, M.O., 2005b. Tide-dominated Coasts. In: Swartz, M. (ed). Encyclopedia of Coastal Science. Elsevier, pp. 982-984.

Hayes, M.O., 2005c. Wave-dominated Coasts. In: Swartz, M. (ed). Encyclopedia of Coastal Science. Elsevier, pp. 1053-1056.

Hayes, M.O., 2005d. Barrier Islands. In: Swartz, M. (ed). Encyclopedia of Coastal Science. Elsevier, pp. 117-119.

Hayes, M.O., 2009. Barrier islands. In: Gillespie, R. and Clague, D. (eds.) Encyclopedia of Islands. University California Press, Berkeley, CA, pp. 82-88.

Hayes, M.O. and Baird, W.F., 1993. Shoreline erosional/depositional patterns in Oman. Coastal Engineering Considerations. In: Coastal Zone Management. American Society of Civil Engineers, New York, pp. 144-158.

Hayes, M.O. and Boothroyd, J.C., 1969. Storms as modifying agents in the coastal environment. In: Hayes, M.O. (ed.), Coastal Environments: NE Massachusetts and New Hampshire: Guidebook, Field Trip for Eastern Section of SEPM, May 9-11, 1969, pp. 245-265.

Hayes, M.O. and Gundlach, E.R., 1975. Coastal Geomorphology and Sedimentation of the *Metula* Oil Spill Site in the Strait of Magellan. Dept. Geol., University of South Carolina. Prepared for National Science Foundation, Washington, DC, 103 pp.

Hayes, M.O. and Kana, T.W., 1976. Terrigenous Clastic Depositional Environments. Tech. Report 11-CRD, Dept. of Geology, University of South Carolina, Columbia, SC, 315 pp.

Hayes, M.O. and Michel, J., 1982. Shoreline sedimentation within a forearc embayment, Lower Cook Inlet, Alaska. Journal of Sedimentary Petrology, 52:251-263.

Hayes, M.O. and Michel, J., 1989. Modern Clastic Depositional Systems of South-Central Alaska. Field Trip Guidebook T101, 28th International Geological Congress, American Geophysical Union, Washington, DC, 42 pp.

Hayes, M.O. and Michel, J., 1992. Chapter 6. Field methods for oil spill response. In: An Introduction to Coastal Habitats and Biological Resources for Oil Spill Response, Report No. HMRAD 92-4, NOAA, Seattle, WA, pp. 6-1–6-10.

Hayes, M.O. and Michel, J., 1997. Evaluation of the Condition of Prince William Sound Shorelines following the *Exxon Valdez* Oil Spill and Subsequent Shoreline Treatment: 1997 Geomorphological Monitoring Survey. NOAA, Hazardous Materials Response and Assessment Division, Seattle, WA, 109 pp.

Hayes, M.O. and Michel, J., 1999. Factors determining the long-term persistence of *Exxon Valdez* oil in gravel beaches. Marine Pollution Bulletin, 38:92-101.

Hayes, M.O. and Michel, J., 2008. A Coast for All Seasons: A Naturalist's Guide to the Coast of South Carolina. Pandion Books, Columbia, SC, 285 pp.

Hayes, M.O. and Michel, J., 2010. A Coast to Explore: Coastal Geology and Ecology of Central California. Pandion Books, Columbia, SC, 338 pp.

Hayes, M.O. and Montello, T.M., 1993a. Coastal inlet protection strategies for oil-spill response. Vol. I. San Diego, Orange, Los Angeles, and Santa Barbara Channel Planning Areas. Marine Spill Response Corporation and California Department of Fish and Game Office of Oil Spill Prevention and Response. Prepared by Research Planning, Inc., Columbia, SC.

Hayes, M.O. and Montello, T.M., 1993b. Coastal Inlet Protection Strategies for Oil-Spill Response. Vol. II. Central Coast and San Francisco Bay/Delta Planning Areas. Marine Spill Response Corporation and California Department of Fish and Game Office of Oil Spill Prevention and Response. Prepared by Research Planning, Inc., Columbia, SC.

Hayes, M.O. and Ruby, C.H., 1994. Chapter 10: Pacific Coast of Alaska. In: Davis, R.A., Jr. (ed.), Geology of the Holocene Barrier Island System, Springer Verlag, New York, pp. 395-433.

Hayes, M.O. and Sexton, W.J., 1989. Field Trip Guidebook T371: Modern Clastic Depositional Environments, South Carolina. 29th Int. Geol. Cong., American Geophysical Union, Washington, DC, 85 pp.

Hayes, M.O., Anan, F.S., and Bozeman, R.N., 1969. Sediment dispersal trends in the littoral zone: A problem in paleogeographic reconstruction. In: Hayes, M.O. (ed.), Coastal Environments: NE Massachusetts and New Hampshire: Guidebook, Field Trip for Eastern Section of SEPM, May 9-11, 1969, pp. 290-315.

Hayes, M.O., Gundlach, E.R., and D'Ozouville, L., 1979. Role of dynamic coastal processes in the impact and dispersal of the *Amoco Cadiz* oil spill (March 1978), Brittany, France. Proc. 1979 Oil Spill Conference, American Petroleum Institute, Washington DC, pp. 193-198.

Hayes, M.O., Gundlach, E.R. and Getter, C.D., 1980. Sensitivity ranking of energy port shorelines. Specialty Conference Proc. American Society of Civil Engineers, New York, pp. 697-709.

Hayes, M.O., Kana, T.W., and Barwis, J.H., 1980. Soft designs for coastal protection at Seabrook Island, SC. Proc. 17th Conference Coastal Engineering, American Society of Civil Engineers, New York, pp. 897-912.

Hayes, M.O., Michel, J., and Betenbaugh, D.V., 2009. The intermittently exposed, coarse-grained gravel beaches of Prince William Sound, Alaska: Comparison with open-ocean gravel beaches. Journal of Coastal Research, 26:4-30.

Hayes, M.O., Michel, J., and Brown, F.J., 1977. Vulnerability of coastal environments of Lower Cook Inlet, Alaska to oil spill impact. Proc. 4th International Conference on Port and Ocean Engineering Under Arctic Conditions, Memorial University of Newfoundland, Canada, 12 pp.

Hayes, M.O., Michel, J., and Fichaut, B., 1990. Oiled gravel beaches: A special problem. Proc. Conference on Oil Spills: Management and Legislative Implications, Newport, RI, pp. 444-457.

Hayes, M.O., Michel, J., and Kana, T.W., 1986, Coastal processes and geomorphology along the Kuwait City waterfront. In: Halwagy, R. Clayton, D., and Behbehani, M. (eds.), Marine Environmental and Pollution, Kuwait University and Kuwait Foundation for Advancement of Science, pp. 25-42.

Hayes, M.O., Michel, J. and Murday, M., 1985. Environmental Baseline Studies for the Establishment of Control Criteria and Standards Against Petroleum-Related Pollution in Nigeria. Nigerian National Petroleum Corporation, Prepared by Research Planning Institute, Inc., Columbia, SC, 1013 pp.

Hayes, M.O., Michel, J., and Noe, D., 1991. Factors controlling initial deposition and long-term fate of spilled oil on gravel beaches. Proc. 1991 International Oil Spill Conference, American Petroleum Institute, Washington, DC, pp. 453-460.

Hayes, M.O., Montello, T.M., and Nairn, R., 1997. Role of geomorphology/Holocene stratigraphy in design engineering for beach erosion in Ghana. Abstract, NE Section Geological Society of America.

Hayes, M.O., Sexton, W.J., and Sipple, K.N., 1994. Fluid-bearing capacity of strandline deposits-implication for hydrocarbon exploration (Abstract). AAPG Bulletin, 68:485.

Hayes, M.O., Ruby, C.H., Stephen, M.F., and Wilson, S.J., 1976a. Geomorphology of the southern coast of Alaska. 15th Conference on Coastal Engineering, American Society of Civil Engineers, New York, pp. 1992-2008.

Hayes, M.O., FitzGerald, D.M., Hulmes, L.J., and Wilson, S.J., 1976b. Geomorphology of Kiawah Island, South Carolina. In: Hayes, M.O. and Kana, T. W. (eds.), Terrigenous Clastic Depositional Environments. Tech Rep 11-CRD, Dept. of Geology, University of South Carolina, Columbia, SC, pp. II-80-II-100.

Hayes, M.O., Montello, T.M., Schrader, R.J., and Levine, E.A., 1999. Tidal inlets – A major hurdle to effectively protecting sensitive coastal resources. Proc. 1999 International Oil Spill Conference, American Petroleum Institute, Washington, DC, pp. 1239-1243.

Hayes, M.O., Owens, E.H., Hubbard, D.K., and Abele, R.W., Jr., 1973. Investigation of form and processes in the coastal zone: In: Coates, D.R. (ed.), Coastal Geomorphology, Proc. 3rd Ann. Geomorphology Symposia Series, Binghamton, NY, pp. 11-41.

Hayes, M.O., Wilson, S.J., FitzGerald, D.M., Hulmes, L.J., and Hubbard, D.K., 1975. Coastal processes and geomorphology. In: Environmental Inventory of Kiawah Island. Prepared by Environmental Research Center, Inc., Columbia, SC, 165 pp.

Hayes, M.O., Michel, J., Montello, T.M., Aurand, D.V., Thayer, G.W., Al-Mansi, A., and Al-Momen, A.H., 1993a. Mechanisms for the enhancement of the preservation of subsurface oil at the Gulf War spill sites. Proc. 1993 International Oil Spill Conference, American Petroleum Institute, Washington, DC, pp. 443-451.

Hayes, M.O., Michel, J., Montello, T.M. Aurand, D.V., Al-Mansi, A.M., Al-Momen, A.H., Sauer, T.C., and Thayer, G.W., 1993b, Distribution and weathering of shoreline oil one year after the Gulf War oil spill. Marine Pollution Bulletin, 27:135-142.

Hayes, M.O., Michel, J., Montello, T.M., Aurand, D.V., Sauer, T.C., Al-Mansi, A.M., and Al-Momen, A.H., 1995. Distribution and weathering of oil from the Iraq-Kuwait conflict oil spill within intertidal habitats - two years later. Proc. 1995 International Oil Spill Conference, American Petroleum Institute, Washington, DC, pp. 443-451.

Henry, R.L., 1971. Morphology of wave-formed spits, Cape Suckling to Icy Bay, southeastern coastline of Alaska. Senior Honors Thesis, Dept. of Geology, University of Massachusetts, 69 pp.

Herbert, F., 1965. Dune. Clinton Book Company, Philadelphia, PA, 507 pp.

Hoyt, J.H. and Weimer, R.J., 1963. Comparison of modern and ancient beaches, central Georgia Coast. AAPG Bulletin, 47:529-531.

Hubbard, D.K., Oertel, G., and Nummedal, D., 1979. The role of waves and tidal currents in the development of tidal-inlet sedimentary structures and sand body geometry; examples from North Carolina, South Carolina, and Georgia. Journal of Sedimentary Petrology, 49:1072-1092.

Humphrey, N., 2011. Soul Dust: The Magic of Consciousness. Princeton Univ. Press, Princeton, NJ, 243 pp.

Huntington, S.P., 2003. The Clash of Civilizations and the Remaking of the World Order. Simon & Schuster Paperbacks, New York, 367 pp.

Imperato, D.P., Sexton, W.J., and Hayes, M.O., 1988. Stratigraphy and sediment characteristics of a mesotidal ebb-tidal delta, North Edisto Inlet, South Carolina. Journal of Sedimentary Petrology, 58:950-958.

Inman, D.L. and Nordstrom, C.E., 1971. On the tectonic and morphologic classification of coasts. Journal of Geology, 79:1-21.

James Dobbin Associates, Inc., 1992. Preliminary Draft Report – Draft Regulations for the Prevention of Coastal Erosion in the Sultanate of Oman. For the Ministry of Regional Municipalities and Environment, Sultanate of Oman, Written in cooperation with Baird and Associates, Research Planning, Inc., Walter Clark, Wolfgang Burhene, and Rodney Salm.

Kana, T.W., 1979. Suspended sediments in breaking waves. Ph.D. dissertation, Dept. of Geology, University of South Carolina, 153 pp.

Kana, T.W., 1989. Erosion and beach restoration at Seabrook Island, South Carolina. Shore and Beach, 57:3-18.

Kana, T.W., 2011. Coastal Erosion and Solutions: A Primer. 2nd Edition, Coastal Science and Engineering, Inc., Columbia, SC, 37 pp.

Kana, T.W. and Knoth, J., 1977. Longshore sediment transport rates in South Carolina. In: Nummedal, D. (ed.), Beaches and Barriers of the Central South Carolina coast: Field Trip Guidebook. 5th Symposium, Coastal Sediments '77, American Society of Civil Engineers, New York, pp. 39-44.

Kana, T.W. and Mason, J.E., 1988. Evolution of an ebb-tidal delta after inlet relocation. In: Aubrey, D.G. and Weishar, L. (eds.), Hydrodynamics and Sediment Dynamics of Tidal Inlets. Springer-Verlag, New York, pp. 382-411.

Kana, T.W., and McKee, P., 2003. Relocation of Captain Sam's Inlet – 20 years later: Coastal Sediments '03, American Society of Civil Engineers, New York, 12 pp.

Komar, P.D., 1976. Beach Processes and Sedimentation. Prentice-Hall, Inc., Englewood Cliffs, NJ, 429 pp.

Lennon, G., 2000. Folly Beach, South Carolina: Tomorrow's coastal problems today. In: A Compendium of Field Trips of South Carolina Geology with emphasis on the Charleston, South Carolina area. SC Dept. of Natural Resources, Geol. Survey, Columbia, SC, pp. 39-46.

McLaren, P., 1977. The Coasts of Eastern Melville and Western Byam Martin Islands: Coastal processes and related geology of a High Arctic environment. Ph.D. dissertation, Dept. of Geology, University of South Carolina, 303 pp.

Matz, D.B. and Hayes, M.O., 1966, Sedimentary petrology of Beacon sediments. Antarctic Journal of the United States, 1:134-135.

Meyer, R.E. (Ed.), 1972. Waves on Beaches. Academic Press, New York, 462 pp.

Michel, J. and Hayes, M.O., 1999. Weathering patterns of oil residues eight years after the *Exxon Valdez* oil spill. Marine Pollution Bulletin, 38:855-863.

Michel, J., Hayes, M.O., and Brown, J.F., 1978. Application of an oil spill vulnerability index to the shoreline of Lower Cook Inlet, Alaska. Environmental Geology, 2:107-117.

Michel, J., Hayes, M.O., Getter, C.D. and Cotsapas, L., 2005. The Gulf War oil spill twelve years later: Consequences of eco-terrorism. Proc. 2005 International Oil Spill Conference, American Petroleum Institute, Washington, DC, (CD-ROM), pp. 1-5.

Michel, J., Hayes, M.O., Keenan, R.S., Sauer, T.C., Jensen, J.R., and Narumalani, S., 1993a. Contamination of nearshore subtidal sediments of Saudi Arabia from the Gulf War oil spill, Marine Pollution Bulletin, 27:109-116.

Michel, J., Hayes, M.O., Keenan, R.S., Jensen, J.R. and Narumalani, S., 1993b. Oil in nearshore subtidal sediments of Saudi Arabia from the Gulf War spill. Proc. 1993 International Oil Spill Conference, American Petroleum Institute, Washington, DC, pp. 383-388.

Monroe, J.S. and Wicander, R., 1998. Physical Geology. (3rd edition). Wadsworth Publishing Company, Belmont, CA, 646 pp.

Moslow, T.F., 1980. Stratigraphy of mesotidal barrier islands. Ph.D. dissertation, Dept. of Geology, University of South Carolina, 186 pp.

Nairn, R. and Hayes, M.O., 1997. Large scale coastal evolutions in the vicinity of Keta Lagoon, Ghana Abstract. Proc. International Conference on Coastal Research through Large Scale Experiments, University of Plymouth.

New World Encyclopedia, 2010. Rifting and breakup of Pangaea. Available at: http://www.newworldencyclopedia.org/entry/Pangaea.

NEDECO, 1978. Etude detailes de l'erosion littorale: phause 2. Rapport Final, submitted to Government of Togo, French West Africa.

NOAA/RPI, 1983. Environmental Sensitivity Index (ESI) Maps of the Shoreline of the Shelikof Strait Region, Alaska. Research Planning, Inc. (RPI) under contract to the National Oceanic and Atmospheric Administration, NOS Data Explorer, 40 maps.

NOAA/RPI, 1993-2001. Environmental Sensitivity Index (ESI) Maps of the Shoreline of Southeast Alaska (four volumes). Research Planning, Inc. (RPI) under contract to the National Oceanic and Atmospheric Administration, NOS Data Explorer, 237 maps.

NOAA/RPI, 1994. Environmental Sensitivity Index (ESI) Maps of the Shoreline of Northern California. Research Planning, Inc. (RPI) under contract to the National Oceanic and Atmospheric Administration, NOS Data Explorer, 39 maps.

NOAA/RPI, 1995. Environmental Sensitivity Index (ESI) Maps of the Shoreline of Southern California. Research Planning, Inc. (RPI) under contract to the National Oceanic and Atmospheric Administration, NOS Data Explorer, 51 maps.

NOAA/RPI, 1999. Environmental Sensitivity Index (ESI) Maps of the Shoreline of Massachusetts. Research Planning, Inc. (RPI) under contract to the National Oceanic and Atmospheric Administration, NOS Data Explorer, 55 maps.

NOAA/RPI, 2000. Environmental Sensitivity Index (ESI) Maps of the Shoreline of Prince William Sound, Alaska. Research Planning, Inc. (RPI) under contract to the National Oceanic and Atmospheric Administration, NOS Data Explorer, 46 maps.

NOAA/RPI, 2004. Environmental Sensitivity Index (ESI) Maps of the Shoreline of New Hampshire. Research Planning, Inc. (RPI) under contract to the National Oceanic and Atmospheric Administration, NOS Data Explorer, 7 maps.

NOAA/RPI, 2005. Environmental Sensitivity Index (ESI) Maps of the Shoreline of the North Slope of Alaska. Research Planning, Inc. (RPI) under contract to the National Oceanic and Atmospheric Administration, NOS Data Explorer, 24 maps.

NOAA/RPI, 2006a. Environmental Sensitivity Index (ESI) Maps of the Shoreline of Central California. Research Planning, Inc. (RPI) under contract to the National Oceanic and Atmospheric Administration, NOS Data Explorer, 41 maps.

NOAA/RPI, 2006b. Environmental Sensitivity Index (ESI) Maps of the Shorelines of Puget Sound and the Strait of Juan de Fuca. Research Planning, Inc. (RPI) under contract to the National Oceanic and Atmospheric Administration, NOS Data Explorer, 111 maps.

Oak, H.L., 1984. The boulder beach: A fundamentally distinct sedimentary assemblage. Annals of the Association of American Geographers, 74:71-82.

Orford, J.D., Forbes, D.L., and Jennings, S.C., 2002. Organizational controls, typologies and timescales of paraglacial gravel-dominated coastal systems. Geomorphology, 48:51-85.

Owens, E.H., 1975. The geodynamics of two beach units in the Magdalen Islands, Quebec, within the framework of coastal environments of the Southern Gulf of St. Lawrence. Ph.D. dissertation, Dept. of Geology, University of South Carolina, 650 pp.

Owens, E.H., 1977. Temporal variation in beach and nearshore dynamics. Journal of Sedimentary Research, 47:191-198.

Pararas-Carayannis, G., 2010. The Tsunami Page. Available at: http://www.drgeorgepc.com/Tsunami1964GreatGulf.html.

Plafker, G., 1969. Tectonics of the 27 March 1964 Alaska Earthquake. U.S. Geological Survey Prof. Paper 543-1, 74 pp.

Plafker, G., 1971. Possible future petroleum resources of Pacific Margin Tertiary Basin, Alaska. AAPG Bulletin, 15:120-135.

Pilkey, O.H. and Fraser, M.E., 2003. A Celebration of the World's Barrier Islands. Columbia University Press, New York, 309 pp.

Pilkey, O.H. and Young, R., 2009. The Rising Sea. Island Press/Shearwater Books, 203 pp.

Pilkey, O.H., Neal, W.J., Kelley, J.T., and Cooper, J.A.G., 2011. The World's Beaches. University California Press, Berkeley, CA, 283 pp.

Pinet, P.R., Matz, D. B., and Hayes, M.O., 1967. Petrology of the upper division of the Beacon Sandstone. Antarctic Journal of the United States, 2:108-109.

Pinet, P. R., Matz, D.B., and Hayes, M.O., 1971. An upper Paleozoic tillite in the Dry Valleys, South Victoria Land, Antarctica. Journal of Sedimentary Research, 41:835-838.

Prior, D.B., Wiseman, W.J., Jr., and Gilbert, R., 1981. Submarine slope processes on a fan delta, Howe Sound, British Columbia. Geo-Marine Letters, 1:85-90.

Prior, D.B., Wiseman, W.J., Jr., and Bryant, W.R., 1981. Submarine chutes on the slopes of fjord deltas. Nature, 290:326-328.

Prior, D.B., Bornhold, B.D., Coleman, J.M. and Bryant, W.R., 1982. Morphology of a submarine slide, Kitmat Arm, British Columbia. Geology, 10:588-592.

Reel, C.L., 1982. The origin and geologic significance of bubble sand, Bull Island, South Carolina. M.S. thesis, Dept. of Geology, University of South Carolina, 134 pp.

Reguero, M.A., Marenssi, S.A., and Santillana, S.N., 2002. Antarctic Peninsula and Southern South America (Patagonia) terrestrial Paleogene faunas and environments: Biogeographic relationships. Paleogeography, Paleoclimatology, Paleoecology, 179/(3-4):189-210.

Reimnitz, E., 1966. Late Quaternary history and sedimentation of the Copper River Delta and vicinity, Alaska. Ph.D. dissertation, University California, San Diego, CA, 160 pp.

Research Planning, Inc., 2003. Oiled Shoreline Survey In Support of the Marine and Coastal Damage Assessment: Final Report. Submitted to Presidency of Meteorology and Environment, Jeddah, Saudi Arabia, 387 pp.

Rosen, P.S., 1979. Boulder barricades in Central Labrador. Journal of Sedimentary Petrology, 49:1113-1124.

Ruark, R., 1950. Poor No Moore. Henry Holt and Company, New York, 706 pp.

Ruby, C.H., 1977. Coastal geomorphology, sedimentation, and oil spill vulnerability – northern Gulf of Alaska. M.S. thesis, Dept. of Geology, University of South Carolina, 223 pp.

Ruby, C.H., 1981. Clastic facies and stratigraphy of a rapidly retreating cuspate foreland, Cape Romain, South Carolina. Ph.D. dissertation, Dept. of Geology, University of of South Carolina, 207 pp.

Ruby, C.H. and Hayes, M.O., 1978. Oil spill vulnerability index, Copper River Delta, Alaska. Proc. Coastal Zone '78, American Society of Civil Engineers, New York, pp. 2204-2220.

Russel, B., 1957. Why I am not a Christian. A Touchstone Book, Simon & Schuster, Inc., New York, 266 pp.

Scott, A.J., Hayes, M.O., Andrews, P.B., Siler, W.L., and Behrens, E.W., 1964. Field Trip Guidebook: Depositional Environments, South-Central Texas Coast. Gulf Coast Assoc. Geol Soc., Annual Meeting, Corpus Christi, TX, 170 pp.

Sexton, W.J., 1987. Morphology and sediment character of mesotidal shoreline depositional environments. Ph.D. dissertation, Dept. of Geology, University of South Carolina, 197 pp.

Sexton, W.J. and Hayes, M.O., 1996. Holocene deposits of reservoir-quality sand along the Central South Carolina coastline. AAPG Bull., 80/6:831-855.

Shepard, F.P., 1950. Beach cycles in Southern California. Tech. Memo 20, Beach Erosion Board. U.S. Army Corps. of Engineers, Washington, DC, 26 pp.

Simons, D.B. and Richardson, E.V., 1962. Resistance to flow in alluvial channels. Transactions, American Society of Civil Engineers, New York, 127:927-953.

Stephen, M.F., Brown, P.J., FitzGerald, D.M., Hubbard, D.K., and Hayes, M.O., 1975. Beach Erosion Inventory of Charleston County, South Carolina: A Preliminary Report. South Carolina Sea Grant Tech. Rep. 4, South Carolina Sea Grant Program, Charleston, SC, 79 pp.

Timson, B.S., 1969. Post-storm profile and particle-size changes of a New Hampshire gravel beach: a preliminary report. In: Hayes, M.O. (ed.), Coastal Environments – NE Massachusetts and New Hampshire: Field Trip guidebook for Eastern Section of SEPM, May 9-11, 1969, pp. 391-402.

Toffler, A., 1970. Future Shock. Random House, London, 541 pp.

Tye, R.S., 1984. Geomorphic evolution and stratigraphy of Price and Capers Inlets, South Carolina. Sedimentology, 31:655-674.

U.N. Technical Group, 1979. Report of the workshop on cases and possible solutions to coastal erosion in Benin and Togo. UNEP, Lomé, 29 January to 9 February 1979.

Wentworth, C.K., 1922. A scale of grade and class terms for clastic sediments. Journal of Geology, 30:377-439.

Western Carolina University, 2005. Coastal Hazards Information Clearinghouse. Chapter 3: Storm Processes at Coastal Hazards. Available at: http://www.wcu.edu/coastalhazards/Libros/libroschapter3.htm.

Zabawa, C.F., 1976. Estuarine sediments and sedimentation processes in Winyah Bay South Carolina. In: Hayes, M.O. and Kana, T.W. (eds.), Terrigenous Clastic Depositional Environments. Tech Rep 11-CRD, Dept. of Geology, University of South Carolina, pp. II-64-II-78.

Zabawa, C.F., 1978. Flocculation in the turbidity maximum of Northern Chesapeake Bay. Ph.D. dissertation, Dept. of Geology, University South Carolina, 123 pp.

APPENDIX A

M.S. Theses and Ph.D. Dissertations
University of Massachusetts and University of South Carolina

University of Massachusetts – Ph.D. Dissertations (1965-1973):

Anan, Fayez Shaban, 1971. Provenance and statistical parameters of sediments of the Merrimack Embayment, Gulf of Maine. 377 pp.

Farrell, Stewart C., 1972. Present coastal processes, recorded changes, and the post-Pleistocene geologic record of Saco Bay, Maine. 296 pp.

Goldsmith, Victor, 1972. Coastal processes of a barrier island complex and adjacent ocean floor: Monomoy Island - Nauset Spit, Cape Cod, Massachusetts. 469 pp.

McCormick, Charles Larry, 1968. Holocene stratigraphy of the marshes at Plum Island, Massachusetts. 104 pp.

University of Massachusetts – Masters Theses (1965-1973):

Abele, Ralph Warren, 1973. Short-term changes in beach morphology and concurrent dynamic processes, summer and winter periods, 1971-1972, Plum Island, Massachusetts. 166 pp.

Farrell, Stewart C., 1970. Sediment distribution and hydrodynamics Saco River and Scarboro Estuaries, Maine. 128 pp.

Hartwell, Allan D., 1970. Hydrography and Holocene sedimentation of the Merrimack River Estuary, Massachusetts. 170 pp.

Hine, Albert C., 1972. Sand deposition in the Chatham Harbor Estuary and on the neighboring beaches, Cape Cod, Massachusetts. 187 pp.

Hobbs, Carl H., 1972. Sedimentary environments and coastal dynamics of a segment of the shoreline of Cape Cod Bay, Massachusetts. 154 pp.

Kaczorowski, Raymond T., 1972. Offset tidal inlets, Long Island, New York. 150 pp.

Matz, David Boyer, 1968. Stratigraphy and petrology of the Beacon Sandstone, southern Victoria Land, Antarctica. 110 pp.

Nilsson, Harold Daniel, 1973. Coastal and submarine morpology of eastern Cape Cod Bay. 178 pp.

Pinet, Paul Raymond, 1970. Petrology of the upper division of the Beacon Sandstone, Dry Valley Region, South Victoria Land, Antarctica. 113 pp.

Rhodes, Eugene G., 1971. Three-dimensional analysis and interpretation of Pleistocene/Holocene deposits by the seismic reflection-refraction method, Merrimack Embayment, Gulf of Maine. 134 pp.

Rosen, Peter S., 1972. Evolution and processes of Coatue Beach, Nantucket Island, Massachusetts: A cuspate spit shoreline. 203 pp.

Sancton, Alan J., 1970. The East Berlin formation of Massachusetts: A Triassic alluvial fan complex. 132 pp.

Wessel, James McCandless, 1969. A paleocurrent analysis and petrographic study of late Triassic Turners Falls Sandstone and Mount Toby Conglomerate in north central Massachusetts. 157 pp.

University of South Carolina – Ph.D. Dissertations (1972-1999):

Al-Sarawi, Mohammad, 1980. Morphology of Jal-Az-Zor escarpment and Holocene-Pleistocene sedimentation along the northern Kuwait Bay. 170 pp.

Barwis, John H., 1979. The sedimentologic and stratigraphic characteristics of beach ridge and tidal channel deposits in mesotidal barrier systems. 134 pp.

Boothroyd, Jon Cushing, 1974. Outwash fan sedimentation, Northeast Gulf of Alaska. 215 pp.

Finley, Robert James, 1975. Morphologic development and dynamic processes at a barrier island inlet, North Inlet, South Carolina. 272 pp.

FitzGerald, Duncan Martin, 1977. Hydraulics, morphology and sediment transport at Price Inlet, South Carolina. 84 pp.

Gundlach, Erich R., 1979. Oil spill impact on temperate shoreline environments based on study of the *Urquiola* (May 1976) and *Amoco Cadiz* (March 1978) oil spills. 242 pp.

Hine, Albert Case, 1975. Shallow carbonate bank margin structure and depositional processes: northwestern Little Bahama Bank, Bahamas. 124 pp.

Hubbard, Dennis K., 1977. Variations in tidal inlet processes and morphology in the Georgia Embayment. 79 pp.

Kaczorowski, Raymond T., 1977. The Carolina Bays and their relationship to modern oriented lakes. 124 pp.

Kana, Timothy W., 1979. Suspended sediment in breaking waves. 153 pp.

McLaren, Patrick, 1977. The coasts of eastern Melville and western Byam Martin Islands: Coastal processes and related geology of a High Arctic environment. 304 pp.

Moslow, Thomas F., 1980. Stratigraphy of mesotidal barrier islands. 187 pp.

Murday, Maylo, 1986. Beach erosion in West Africa. 101 pp.

Owens, Edward Henry, 1975. The geodynamics of two beach units in the Magdalen Islands, Quebec, within the framework of coastal environments of the southern Gulf of St. Lawrence. 650 pp.

Stephen, Michael F., 1981. Effects of seawall construction on beach and inlet morphology and dynamics at Caxambas Pass, Florida. 196 pp.

Ward, Larry Guy, 1978. Physical and sedimentological processes in a salt marsh tidal channel: Kiawah Island, South Carolina. 180 pp.

Zabawa, Christopher Francis, 1978. Flocculation in the turbidity maximum of Northern Chesapeake Bay. 123 pp.

University of SC – Masters Theses (1972-1999):

Blount, Anna E., 1978. Two years after the *Metula* oil spill: Strait of Magellan, Chile: Oil interaction with coastal environments. 217 pp.

Brown, P.J., 1975. Coastal morphology of South Carolina. 122 pp.

Domeracki, Daniel D., 1982. Stratigraphy and evolution of the Pawleys Island area, South Carolina. 101 pp.

Eckard, Timothy L., 1986. Stratigraphy and depositional history of the Santee River Delta, South Carolina. 103 pp.

Fico, C., 1973. Influence of wave refraction on coastal geomorphology - Bull Island to Isle of Palms, South Carolina. 143 pp.

Finkelstein, Kenneth, 1979. Morphologic variations and sediment transport in crenulate bay beaches, Kodiak Island, Alaska. 96 pp.

Fischer, Ian A., 1977. Tidal flat sedimentation in a macrotidal embayment, Bahia de Lomas, Strait of Magellan, Chile. 128 pp.

Freeman, Thomas H., 1982. Bedform distributions, bedform dynamics, and preserved bedding types on a mesotidal sand ridge, Port Royal Sound, South Carolina. 35 pp.

Griffin, Martha M., 1981. Feldspar distributions in river, delta, and barrier sands of central South Carolina. 122 pp.

Hodge, John Adams, 1981. Erosion of the North Santee River Delta and development of a flood-tidal delta complex. 151 pp.

Hubbard, Dennis K., 1974. Tidal inlet morphology and hydrodynamics of Merrimack Inlet, Massachusetts. 144 pp.

Humphries, Stanley M., 1977. Seasonal variation in morphology at North Inlet, S.C. 97 pp.

Imperato, Douglas P., 1984. Sandy depositional environments of the North Edisto Tidal Basin. 134 pp

Johnson, Helen Mary, 1983. Channel-fill stratigraphy at the fluvial-tidal interface. 131 pp.

Kana, Timothy W., 1976. Coastal processes and sediment transport at Price Inlet and North Inlet, South Carolina. 144 pp.

Knoth, Jeffrey, 1978. Analysis of longshore sediment transport on Bull Island, South Carolina, using a fluorescent tracer. 41 pp.

Levey, Raymond Allen, 1977. Characteristics of coarse-grained point bars, Upper Congaree River, South Carolina. 61 pp.

McCants, C.Y., 1982. Evolution and stratigraphy of a sandy tidal flat complex within a mesotidal estuary. 108 pp.

McCreesh, Catherine A., 1982. A beach process-response study at Hunting Island, South Carolina. 72 pp.

Munroe, Hugh Douglas, 1978. Historical development and present day variations in the morphology of Neguac Island. 109 pp.

Nelligan, David F., 1982. Ebb-tidal delta stratigraphy, Breach Inlet, S.C. 221 pp.

Reel, Christopher L., 1982. The origin and geologic significance of bubble sand, Bull Island, South Carolina. 134 pp.

Reinhart, Peter Joseph, 1980. Morphology, sedimentology and stratigraphy of intertidal shoals within a tidal embayment, St. Helena Sound, South Carolina. 135 pp.

Ruby, Christopher H., 1977. Coastal morphology, sedimentation and oil spill vulnerability, northern Gulf of Alaska. 223 pp.

Sexton, Walter J., 1981. Natural bar-bypassing of sand at Captain Sam's Inlet, South Carolina. 136 pp.

Shipp, R. Craig, 1980. Morphology and sedimentology of a single-barred nearshore system, Eastern Long Island, New York. 169 pp.

Svetlichny, Michael, 1982. Short term impact of beach scraping and backshore restoration, Myrtle Beach, South Carolina. 107 pp.

Tye, Robert S., 1981. Geomorphic evolution and stratigraphic framework of Price and Capers inlets, South Carolina. 144 pp.

Vogel, Michael J., 1984. Tidal dominance and the effects of inlet stabilization and dredging at Moriches Inlet, Long Island, New York. 116 pp.

Wach, Grant D., 1984. The depositional history of the McMurray Formation: section 3, 4, 9, and 10, TWP. 93, R. 173 pp.

Ward, Larry Guy, 1974. The morphology and hydrology of the Skeidará and Skaftafellsá River distributaries, Skeidarársandur, Iceland. 91 pp.

ABOUT THE AUTHOR

Miles O. Hayes

Dr. Miles O. Hayes is a coastal geomorphologist with over 50 years of research experience. He has authored over 250 articles and reports and five books on numerous topics relating to tidal hydraulics, river morphology and processes, beach erosion, barrier island morphology, oil pollution, and petroleum exploration. Hayes' teaching experience includes both undergraduate and graduate courses while a Professor at the Universities of Massachusetts and South Carolina. Sixty five graduate students received their degrees under his supervision, most of whom are now leaders in their respective academic, government, and industry positions. He is considered to be the "Father of Coastal Geology." A review of his 1999 book – *Black Tides*, said "A skilled raconteur, Hayes tells engrossing stories of responding to most of the recent, headline-grabbing oil spills, including the Gulf War spills, the *Exxon Valdez*, the *Amoco Cadiz* spill in France, and the *Ixtoc I* blowout in Mexico. Interspersed among them are personal events and adventures, such as his survival of a plane crash while mapping a remote part of Alaska." He is Chairman of the Board of Research Planning, Inc. (RPI), a science technology company located in Columbia, SC.

He is pictured here with his wife and business partner Jacqui Michel (JM). Together they have begun a series of books on the coastal geology and ecology of the coastal states of the U.S. The first two books in the series are:

1) Hayes, M.O. and J. Michel. 2008. *A Coast for all Seasons: A Naturalist's Guide to the Coast of South Carolina*. Pandion Books, Columbia, SC. 235 pp.
2) Hayes, M.O. and J. Michel. 2010. *A Coast to Explore: Coastal Geology and Ecology of Central California*. Pandion Books, Columbia, SC. 338 pp.